T0361427

Optical Communications in the 5G Era

Optical Communications in the 5G Era

XIANG LIU
Formerly with Futurewei Technologies, Bridgewater, NJ, United States

Academic Press is an imprint of Elsevier
125 London Wall, London EC2Y 5AS, United Kingdom
525 B Street, Suite 1650, San Diego, CA 92101, United States
50 Hampshire Street, 5th Floor, Cambridge, MA 02139, United States
The Boulevard, Langford Lane, Kidlington, Oxford OX5 1GB, United Kingdom

Notices

Knowledge and best practice in this field are constantly changing. As new research and experience broaden our understanding, changes in research methods, professional practices, or medical treatment may become necessary.

Practitioners and researchers must always rely on their own experience and knowledge in evaluating and using any information, methods, compounds, or experiments described herein. In using such information or methods they should be mindful of their own safety and the safety of others, including parties for whom they have a professional responsibility.

To the fullest extent of the law, neither the Publisher nor the authors, contributors, or editors, assume any liability for any injury and/or damage to persons or property as a matter of products liability, negligence or otherwise, or from any use or operation of any methods, products, instructions, or ideas contained in the material herein.

British Library Cataloguing-in-Publication Data
A catalogue record for this book is available from the British Library

Library of Congress Cataloging-in-Publication Data
A catalog record for this book is available from the Library of Congress

ISBN: 978-0-12-821627-9

For Information on all Academic Press publications
visit our website at https://www.elsevier.com/books-and-journals

Publisher: Mara Conner
Acquisitions Editor: Tim Pitts
Editorial Project Manager: Ruby Smith
Production Project Manager: Nirmala Arumugam
Cover Designer: Greg Harris

Typeset by MPS Limited, Chennai, India

Dedications

To the people around the world who have risked their own lives in saving and supporting others during the COVID-19 pandemic.

For my wife Hongou, our children (Daniel, Frank, and Jack), and my parents (Zhongren and Shuirong) with love.

Contents

About the author

Dr. Xiang Liu is the Vice President for Optical Transport and Access at Futurewei Technologies. He had been with Bell Labs working on high-speed optical transmission technologies for 14 years. Dr. Liu earned a Ph.D. degree in Applied Physics from Cornell University in 2000. He has authored more than 350 publications and holds more than 100 US patents. He has served as a General Co-Chair of 2018 Optical Fiber Communication Conference (OFC) and a Co-Editor of IEEE Communications Magazine's Optical Communications and Networks Series. He is currently serving as a Deputy Editor of Optics Express and an Advisory Board Member of the Next-Generation Optical Transport Forum. He was elected Fellow of the Optical Society of America (OSA) in 2011 for *contributions to fundamental research in optical fiber communications that have been incorporated in commercial systems, including high-speed phase-shift keyed transmission and nonlinearity mitigation*, and Fellow of the Institute of Electrical and Electronics Engineers (IEEE) in 2017 for *contributions to broadband optical fiber communication systems and networks*. Since 2017, he has been teaching an OFC short course on Optical Communication Technologies for 5G. He has also served as a co-organizer of OSA's first 5G Summit held in 2019 and second 5G Summit held in 2021.

Preface

The fifth-generation technology standard for cellular networks (5G) is extensively regarded as the key driver for modern telecommunications in the decade of the 2020s. To better support 5G, optical communication networks are being redesigned and refined. This book aims to provide an up-to-date overview of the important optical communication technologies in the 5G era. A number of attempts have been made to capture the recent technological advances in various optical network segments, such as front-haul, access, metro, data center, backbone, and undersea network segments. The emerging fifth-generation fixed network (F5G) initiative is also discussed.

This book is intended for a diverse readership including but not limited to researchers, engineers, professors, graduate students, and undergraduate students, to gain knowledge of and insights into modern optical communication technologies that are crucial to 5G and F5G, which are impacting our society in a profound way. This book also aims to help the readers to broaden their knowledge on the emerging applications of optical networks in 5G mobile networks, deepen their understanding of the state-of-the-art optical communication technologies, and explore new R&D opportunities in the field of converged fixed-mobile networks.

With the exciting advances in this dynamic field of optical communications in the 5G era, It is my hope that the readers will find this book helpful, enlightening, and enjoyable.

Acknowledgments

I wish to thank many current and past colleagues at Futurewei Technologies for their collaboration. Among them are Naresh Chand, Ning Cheng, Frank Effenberger, Aihua Guo, Guangzhi Li, Linlin Li, Yuanqiu Luo, Sharief Megeed, Andy Shen, and Huaiyu Zeng. I also wish to acknowledge past collaboration with a number of Huawei colleagues working in both optical and wireless research departments. Among them are Tianhai Chang, Ning Deng, Guangxiang Fang, Shengping Li, Huafeng Lin, Dekun Liu, Jiang Qi, Liang Song, Xiao Sun, Minghui Tao, Yin Wang, Xuming Xu, Zhuhong Zhang, Jun Zhou, Lei Zhou, and Min Zhou. Many of the past collaborations have resulted in publications that are cited in this book.

In addition, I am grateful to many of my former colleagues at Bell Laboratories, New Jersey, for fruitful collaboration and valuable discussion. I am particularly grateful for the support from Drs. S. Chandrasekhar, Andrew Chraplyvy, Randy Giles, Wayne Knox, Robert Tkach, and Peter Winzer.

Special thanks are due to Dr. Tingye Li and Prof. Alan Willner, who extended their kind support to me during the early days of my career and invited me to contribute book chapters in three recent editions of the classic *Optical Fiber Telecommunications* book series.

The progress in the field of optical communications for 5G delineated in this book implicitly represents the works of many researchers and engineers around the world. Many of their works are cited in the book as references. Particularly, I would like to acknowledge valuable discussion with experts from China Mobile, China Telecom, China Unicom, ETRI, Finisar (now II–VI), Lumentum, Nokia, NTT, and Orange Labs.

Finally, I wish to thank Tim Pitts, Ruby Smith, Nirmala Arumugam, and the editorial team of Elsevier for their professional support throughout the publication process.

Acknowledgements

CHAPTER 1

Introduction

1.1 The 5G era

1.1.1 Evolution of mobile networks

In today's information age, information and communication technology (ICT) has been impacting our global society in countless ways. Mobile and optical communication networks are at the foundation of the ICT. Mobile communication networks have experienced dramatic advances over the last 40 years. Fig. 1.1 shows the evolution of mobile networks. The first-generation (1G), second-generation (2G), third-generation (3G), and fourth-generation (4G) mobile networks were primarily deployed in the 1980s, 1990s, 2000s, and 2010s, respectively [1–4]. They have made a remarkable impact on our society by providing important means of communication such as mobile voice communication, text messaging, mobile internet access, social media applications, and video streaming. The fifth generation of mobile network (5G) started to be deployed in some countries in 2019 and a widespread availability of 5G is expected in the decade of the 2020s.

During the 1G era, mobile networks were mainly used for providing voice services with mobility. There were no international standards, which in turn limited the widespread usage of 1G technologies. The underlying technology for 1G is based on frequency-division multiple access (FDMA) and analog signal processing. The typical downstream data rate per user is limited to 2 kb/s [4].

Realizing the need for a global standard on mobile networks, the European telecommunications standards institute (ETSI) developed the global system for mobile communications (GSM) standard, which defined the protocols for 2G cellular networks used by mobile devices such as mobile phones and tablets. This enabled standardized technologies and interfaces that allowed international roaming and interoperability among different vendors. The underlying technology for 2G is based on time-division multiple access (TDMA) and digital signal processing. The typical downstream data rate per user was limited to 64 kb/s [4].

Subsequently, 3G was developed by the 3rd Generation Partnership Project (3GPP), which comprised a number of standards organizations that develop protocols for mobile telecommunications. 3GPP has seven national or regional telecommunications standards organizations as primary members, which are enumerated as follows.

- Association of Radio Industries and Businesses, Japan
- Alliance for Telecommunications Industry Solutions, the United States

Optical Communications in the 5G Era
DOI: https://doi.org/10.1016/B978-0-12-821627-9.00012-7

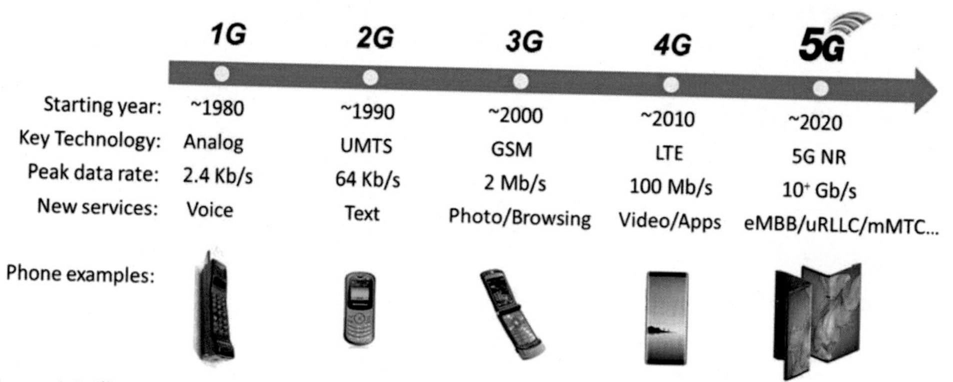

Figure 1.1 Illustration of the evolution of mobile networks from 1G to 5G.

- China Communications Standards Association, China,
- European Telecommunications Standards Institute, Europe
- Telecommunications Standards Development Society, India
- Telecommunications Technology Association, South Korea
- Telecommunication Technology Committee, Japan

Data services and internet access were introduced with 3G technology. The underlying technology for 3G is primarily based on wideband code division multiple access (CDMA) radio access technology to offer greater spectral efficiency and bandwidth to mobile users. The typical downstream data rate per user was 2 Mb/s [4].

For 4G, the International Telecommunications Union-Radio communications sector (ITU-R) specified a set of requirements, namely, the International Mobile Telecommunications Advanced (IMT-Advanced) specifications. These guidelines define peak speed requirements for 4G service at 100 Mb/s for high mobility communication (such as for trains and cars) and 1 Gb/s for low mobility communication (such as for pedestrians and stationary users) [4]. 3GPP issued multiple technical specification releases for 4G, ranging from Release 8 (R8) to Release 14 (R14). The underlying technology for 4G is orthogonal frequency-division multiple access (OFDMA) for downlink, which offers high spectral efficiency even in the presence of multipath interference. Single-carrier orthogonal frequency-division multiplex access (SC-OFDMA) is used for uplinking to achieve both high spectral efficiency and power efficiency. The peak data rate is further improved by using a wider spectrum bandwidth and antenna arrays that offer multiple-input multiple-output (MIMO) communications.

For 5G, ITU-R specified a set of requirements, namely, the IMT-2020, for 5G networks, devices, and services in Refs. [1–4]. 3GPP also issued multiple technical specification releases starting from Release 15 (R15) for 5G, which represent a significant evolution from 4G and aim to meet the unprecedented communication demands

Table 1.1 Key aspects of five generations of mobile networks.

Generation	1G	2G	3G	4G	5G
Reference downlink data rate	2 kb/s	64 kb/s	2 Mb/s	100 Mb/s	10 Gb/s
Reference specifications	(No global standards)	GSM	3GPP R4–R7	IMT-2010, 3GPP R8–R14	IMT-2020, 3GPP R15–R17
Reference technologies	FDMA	TDMA	WCDMA	OFDMA, SC-OFDMA, MIMO	OFDMA, DFT-s-OFDM, m-MIMO

DFT-s-OFDM, Discrete Fourier transform spread OFDM; *FDMA*, frequency-division multiple access; *GSM*, global system for mobile; *MIMO*, multiple-input multiple-output; *OFDMA*, orthogonal frequency-division multiple access; *SC-OFDMA*, single-carrier orthogonal frequency-division multiplex access; *TDMA*, time-division multiple access; *WCDMA*, wideband code-division multiple access.

for a fully connected intelligent world. In 5G, the peak data rate is further improved to 20 Gb/s by using even wider spectrum bandwidth and massive MIMO (m-MIMO) communications. For downlink transmission, the modulation and multiplexing are still based on OFDMA. For uplink transmission, the modulation and multiplexing are based on OFDMA or discrete Fourier transform spread OFDM (DFT-s-OFDM), which is a slightly enhanced version of SC-OFDMA. In addition to offering higher data rates, 5G offers lower network latency for time-sensitive applications and more connections for applications such as the Internet of Things (IoT).

Table 1.1 summarizes the key aspects of the five generations of mobile networks. It is worth noting that each generation lasted about 10 years, and the mobile data rate scaled up roughly 30 times per generation, which means about 1.5 dB increase per year, or doubling per two years. The mobile data rate's doubling per two years tracks well with the Moore's Law, which states that "the number of transistors in a dense integrated circuit (IC) doubles about every two years" [5]. This is reasonable considering that the mobile data are eventually processed by ICs.

1.1.2 5G Scope and applications

5G represents a significant evolution from 4G by meeting the unprecedented communication demands for a fully connected intelligent world. As illustrated in Fig. 1.2, the former has been designed to support a diverse set of applications.

The applications supported by 5G include but are not limited to:

- Smartphones with 5G connectivity
- Wearables that monitor sport activities and health conditions
- Augmented reality (AR) and virtual reality (VR)
- Cloud office for work-from-home and remote team collaborations
- Online education allowing teachers to teach students remotely
- Remote medical diagnostics and treatments
- Smart manufacturing, which improves productivity and safety
- Smart home with autonomous optimization of home appliances and other features

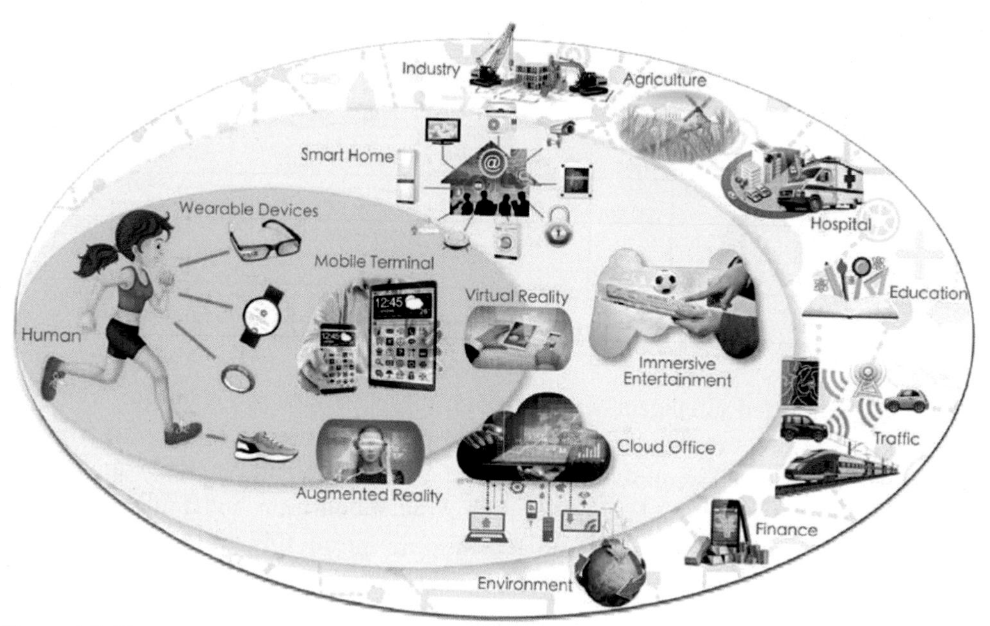

Figure 1.2 Illustration of a diverse set of applications supported by 5G. *Courtesy of Dr. Chih-Lin I from her keynote presentation at the 1st IEEE 5G Summit, Princeton, New Jersey, United States, May 26, 2015.*

- Smart city with autonomous driving, transportation, etc.
- Billions of connections for IoT.

The main 5G application categories are:

- Enhanced mobile broadband (eMBB), supporting applications including ultra-high definition video, 3D video, work and play in the cloud, AR and VR
- Ultra-reliable and low-latency communications (uRLLC), supporting applications such as self-driving cars and drones, industry automation, remote medical procedures, and other mission-critical applications
- Massive machine-type communications (mMTC), supporting applications such as smart home, smart building, smart city, and massive IoT.

Fig. 1.3 is a well-known triangle diagram that illustrates the three main categories of 5G applications. For the eMBB category, the reference peak data rate can reach beyond 10 Gb/s, providing over 100 times faster data speeds than 4G. New applications include fixed wireless internet access for homes and buildings, outdoor broadcasting of large events without the need for broadcast vans, and greater connectivity for people on the move.

For the uRLLC category, the end-to-end network latency can be controlled to be within 1 ms, which is about 30 times less than that in 4G, while the network reliability is also improved from 99.99% to 99.999%. This enables mission-critical applications

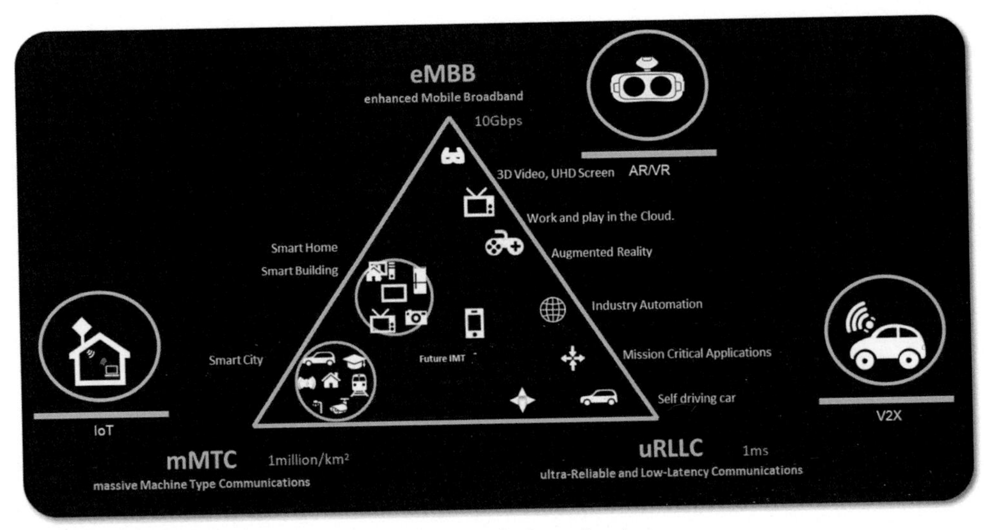

Figure 1.3 Illustration of three main categories of 5G applications.

such as vehicle to anything (V2X) communications, real-time control of industrial robots, and remote medical care, procedures, and treatments.

For the mMTC category, over 1 million devices can be connected within 1 km^2, which represents a 100-fold increase over 4G. This enables massive IoT connections that connect billions of devices without human intervention at an unprecedented scale, assisting the intelligent operation of future factories, farms, mines, ports, hospitals, schools, homes, and cities.

The above three categories of 5G applications collectively support a greatly expanded space of service types and are forward-compatible with many new services that are expected to emerge in the 5G era, thereby revolutionizing the way how ICT is serving our global society in the 2020s.

1.1.3 5G Standards developments

Over the last few years, standardization of 5G technologies has been intensively carried out by 3GPP and ITU-R. The global standardization efforts have been essential for defining the new use cases and specifying the new technologies to support these new use cases. Fig. 1.4 shows the major standards releases leading to the global launch of 5G in 2020.

ITU-R's IMT-2020 has put forth the key requirements for 5G new radio (NR), which focus on fulfilling three key performance indicators:

- >10 Gb/s peak data rates for eMBB
- <1 ms latency for uRLLC
- >1 Million connections per square kilometer for mMTC

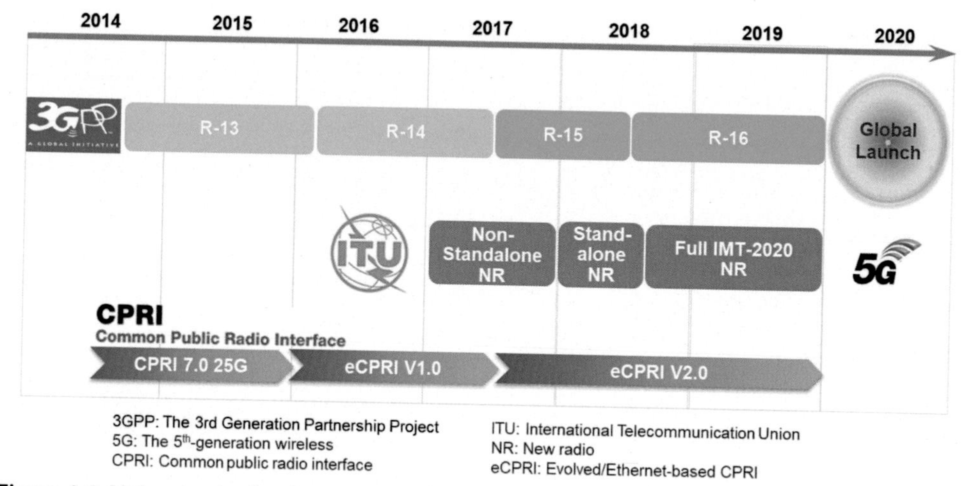

Figure 1.4 Major standards releases leading to the global launch of 5G in 2020.

To truly implement the full version of NR, a massive amount of new hardware needs to be deployed. To continue using existing 4G hardware, a phased approach has been proposed. In the initial phase, a nonstandalone (NSA) architecture based on the 4G core network can be used. In the later phase, a standalone (SA) architecture based on the new 5G core network can be adopted.

3GPP Release 15 (R15) focused on the eMBB category of applications and was completed in 2018 [1]. R15 is of tremendous importance as it introduced NR technical specifications for the first time. It focuses on eMBB use cases with very high data rate and also provides some features for uRLLC applications.

3GPP Release 16 (R16) enhanced R15 specifications and focused on the uRLLC category. R16 was substantially completed in 2019 [1]. It covers new verticals and deployment scenarios such as time-sensitive networking, intelligent transportation systems, V2X communications, accurate user equipment positioning, integrated access and backhaul, and industrial IoT.

3GPP Release 17 (R17) is currently under development and is anticipated to be completed in late 2021. It will provide further enhancements to R15 and R16 for all the three categories of eMBB, uRLLC, and mMTC. Its target is to support the expected growth in mobile data traffic and to enhance NR for use cases such as autonomous driving, manufacturing, public safety, smart home, and smart city, in accordance with new requirements emerging from large-scale 5G deployment.

In addition, the Common Public Radio Interface (CPRI) industry cooperation group defines publicly available specification for the key internal interface of radio base stations between the radio equipment control and the radio equipment to facilitate the deployment of m-MIMO and cloud radio access networks (C-RAN) [6]. Most of the CPRI interfaces are based on optical fiber connections. With the increase of the interface data rates in the 5G era, evolved CPRI (eCPRI) is being developed to improve the interface bandwidth efficiency.

Recently, the Open Radio Access Network (O-RAN) Alliance was formed in 2018 with the aim to transform RAN to become open, intelligent, virtualized, and fully interoperable [7]. It was founded by network operators to clearly define the requirements and help build a supply chain ecosystem to realize its objectives with two core principles, namely, openness and intelligence.

For the principle of openness, open interfaces are defined to enable network operators to introduce their own services and customize their networks to suit their own unique needs. Open interfaces also enable multivendor deployments and a competitive supply chain. In addition, open source software and hardware reference designs are introduced to encourage cooperative innovation. For the principle of intelligence, the O-RAN Alliance aims to leverage emerging learning-based technologies to automate and optimize the operation of networks and reduce the operational expenditure (OPEX).

With network operators and suppliers from different parts of the world working together for a common vision, it is expected that the global standardization and specification efforts will continue to help drive the evolution of 5G to better serve our society in the forthcoming years.

1.2 5G Deployment status and societal impacts

1.2.1 Initial 5G deployment status

During the 2018 Winter Olympics held in Pyeongchang, South Korea, in February 2018, 5G wireless technologies were showcased to general public as part of a collaboration between domestic wireless sponsor Korea Telecom and worldwide sponsor Intel [8]. 5G networks exhibited impressive features such as live high-dynamic-range 8K video streaming, camera feeds from bobsleds, and multicamera views from cross-country and figure skating events. Since then, the deployment of 5G networks has been accelerated in countries such as China, Japan, and the United States.

On June 6, 2019, China's Ministry of Industry and Information Technology (MIIT) officially issued licenses for the launch of commercial 5G networks in China. The 5G licenses were granted to China Mobile, China Telecom, China Unicom, and China Broadcasting Network (CBN). The initial wireless spectra allocated are [9]:

- 2.515–2.675 GHz (or 160 MHz in bandwidth) for China Mobile
- 3.4–3.5 GHz (or 100 MHz in bandwidth) for China Telecom
- 3.5–3.6 GHz (or 100 MHz in bandwidth) for China Unicom

Later, China Mobile was given another 100 MHz of bandwidth in the 4.9-GHz band, while China Telecom, China Unicom, and CBN were given the 3.3–3.4 GHz band for indoor 5G usage [10].

The year 2019 witnessed the initial deployment of 5G base stations worldwide. By the end of 2019, China alone had deployed about 130,000 base stations [11]. The deployment markedly accelerated in 2020. According to China's MIIT, over 500,000 5G base stations are to be put into service during 2020. To support the global 5G deployment, system vendors such as Ericsson, Huawei, and Nokia have been ramping up their production. For example, the 5G base stations produced by Huawei in year 2020 were estimated to reach 1.5 million units [12].

In addition to 5G base stations, 5G-ready mobile phones have started to reach general public in 2020. By May 2020, China alone had shipped over 60 million 5G mobile phones [13]. The shipments of 5G phones exceeded 180 million by the end of 2020. Furthermore, according to a recent report by the GSMA [13], it is forecasted that 28% of China's mobile connections will be running on 5G networks by 2025, accounting for about one-third of all 5G global connections. This provides a glimpse of the exciting 5G era in this decade.

There has been a long-standing issue of *digital divide*, where different groups of people have different levels of access to the advancements in ICT [14]. In the deployment of 5G, it may be economically challenging to deploy 5G networks in rural areas where the return of investment tends to be less favorable than that in densely populated urban areas. Thus efforts need to be taken to prevent the increase of the digital divide during 5G deployment.

The ITU strongly advocates to bridge the digital divide. In a 2018 ITU report on 5G, it was suggested that policymakers could use a range of legal and regulatory actions to facilitate 5G network deployment such as

- Supporting the use of affordable wireless coverage (e.g., through sub-GHz bands) to reduce the digital divide, and
- Commercial incentives such as grants or public–private partnerships to stimulate investment in 5G networks.

According to a recent report [15], China has effectively implemented universal telecommunications service as a feature of the development of human rights in the country. In 2019 there were over 6 million 4G base stations worldwide, more than half of which were in China. As a result, more than 800 million Chinese farmers have been enjoying high-quality and low-priced telecommunications services. With the ongoing network upgrade from 4G to 5G, more significant impact to rural areas can be expected. For example, widespread 5G connectivity throughout a country can enable new applications such as ubiquitous coverage for connected and autonomous vehicles on national highways.

As a remarkable example of initial 5G deployment, China Mobile and Huawei successfully deployed 5G base stations at a record high elevation of 6500 m on Mount Everest in April 2020 [16], as shown in Fig. 1.5.

These 5G base stations on Mount Everest ensure reliable telecommunication for the activities of mountain climbing, high-definition live streaming, scientific research, environmental monitoring, as well as the 2020 effort of Mount Everest

Figure 1.5 Picture of 5G base station deployment at a record high elevation of 6500 m on Mount Everest (after [16]).

remeasurement. The 5G download speed exceeded 1.66 Gb/s, while the upload speed topped 215 Mb/s. The key deployment highlights include:

- 5G active antenna unit with compact size, light weight, and low power consumption
- m-MIMO with three-dimensional beamforming
- Simultaneous operation in both SA and NSA modes to leverage existing 4G networks
- Live high-definition video streaming via an intelligent software-defined video surveillance system
- 10-Gb/s passive optical network (XG-PON) for connecting multiple access nodes
- 200-Gb/s high-speed long-haul transmission for connecting 5G base stations to national backbone networks

A press release regarding this milestone achievement concluded as follows [16]:

> Huawei strongly believes that technology means to make the world better. The beauty of Mount Everest can be displayed via 5G high-definition video and VR experience, which also provides further insights for mountaineers, scientists and other specialists into the nature. The ground-breaking establishment on Mount Everest once again proves that 5G technology connect mankind and the Earth harmoniously.

1.2.2 5G Values being delivered to our society

Even during the very early stage of 5G deployment, the promised values of 5G have started to be delivered to our global society. The COVID-19 pandemic is a public health crisis that has impacted a number of countries. In fighting the COVID-19 pandemic, a profound impact has been made by 5G networks, as well as the supporting optical communication networks.

The foremost task during the pandemic is to save lives. At the epicenter of Wuhan City, Huoshenshan hospital was built within a few days in February 2020 to accommodate a rapid rise in infected patients. A 5G network was established to provide high-bandwidth and ultra-reliable connectivity for the hospital to communicate with other hospitals in Beijing and other major cities to conduct mission-critical tasks such as remote diagnosis. The entire process of network planning, 5G base station deployment, fiber network installation, and network commissioning was completed in just three days [17]. Since its completion, the 5G network enabled the hospital to carry out important services such as high-speed medical data communication, remote diagnosis, and remote patient monitoring.

Contact tracing is an effective approach to contain the spread of the coronavirus. Mobile applications and big data analytics have been utilized, together with 5G networks, to realize contact tracing in countries such as China and South Korea. In addition, self-reporting of health status, acquiring information about the surrounding conditions, and receiving medical advice were all supported by a reliable communication network infrastructure. More fundamentally, network connectivity ensured food supplies to reach the homes of hundreds of millions of house-bound people in China during the pandemic.

During the pandemic, stay-at-home directives have caused unprecedented surges in network traffic volume. Thanks to a joint effort from governments, network operators, companies and general public, communication networks have sustained the surges and ensured the crucially needed connectivity. As an example, GSMA issued "Eleven Regulatory Recommendations to Sustain Connectivity During the COVID-19 Crisis" [18], which covers the following key aspects:

- Network resilience
- Flexibility for personnel and prioritization of crisis-related activities
- Responsible approaches to digital connectivity

On the aspect of network resilience, network operators are encouraged to proactively increase network capacity in order to accommodate the significant increase in traffic demand as a result of more people working from home, interacting online, and accessing digital services. On the aspect of flexibility for personnel and prioritization of crisis-related activities, telecommunications workers are regarded as essential workers and are given flexibility to realize social distancing and prioritize crisis-related activities in accordance with the circumstances. On the aspect of responsible approaches to digital connectivity, network operators are committed to pragmatic, sustainable, and responsible approaches to meet the vital connectivity needs, especially for those who are vulnerable during this crisis.

In this epidemic, the social value of the ICT applications has surfaced in an unprecedented manner. Various applications developed based on 5G, cloud services, artificial intelligence (AI), and big data analytics, have played a great role in many areas, as illustrated in Fig. 1.6. Some key applications include:

- Critical hospital communications for tasks including remote diagnosis
- Contact tracing and health condition monitoring and reporting
- Remote working and video conferencing
- Online education for students of all ages
- Online shopping and entertainment

As an example, Huawei's cloud service platform has provided free services such as electronic whiteboard for remote collaboration and video conferencing for companies with fewer than 1000 employees to run online meetings with up to 100 participants [19]. In addition, the cloud service platform has provided online teaching and learning services for primary and secondary schools, training centers, and colleges, meeting the online education needs of millions of students. Aided by 5G connectivity, the Huawei cloud platform has served over 10,000 hospitals, healthcare centers, schools, and colleges during the epidemic [19].

In the United States, the Federal Communications Commission announced the *Keep Americans Connected Initiative* on March 13, 2020, to address the challenges that many Americans faced during the COVID-19 pandemic [20]. This included a number of measures to meet the increased demand for broadband connectivity, such as

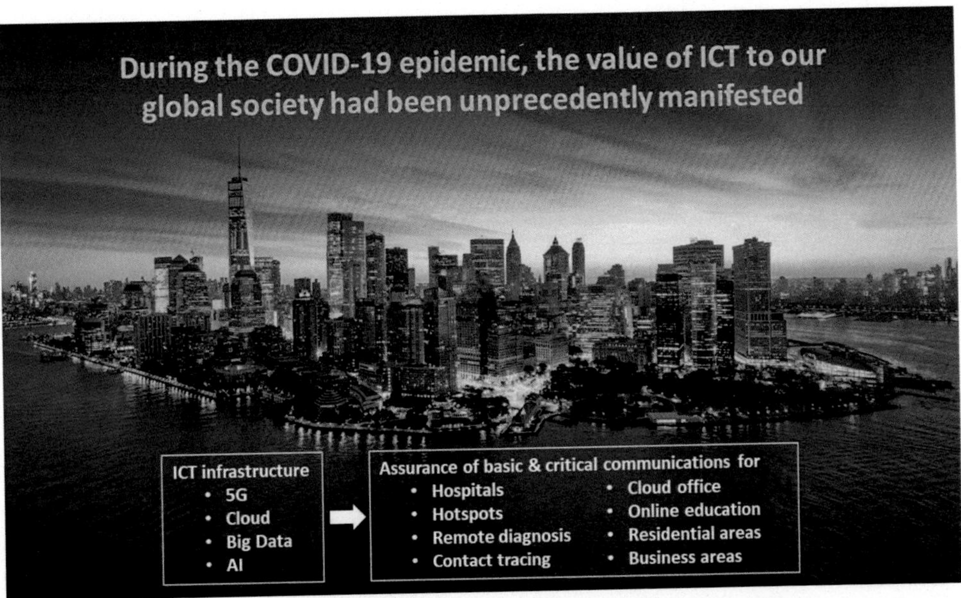

Figure 1.6 Illustration of important ICT applications during the COVID-19 pandemic.

- Establishing the COVID-19 Telehealth Program to help health care providers furnish connected care services to patients at their homes or mobile locations
- Promoting remote learning with the Department of Education
- Addressing the digital divide with the Institute of Museum and Library Services
- Granting service providers additional spectrum to support increased broadband usage

Remarkably, more than 800 US companies and associations participated in this initiative. Many US telecommunications companies have gone above and beyond the pledge by expanding low-income broadband programs and relaxing data usage limits in appropriate circumstances such as telehealth and remote learning [21].

Indeed, the profound value of ICT has been highlighted during the COVID-19 crisis. As a key element of ICT, 5G not only supports the many use cases originally conceived but also enables new applications to protect our citizens and save lives that are critical in crises such as the COVID-19 pandemic. The strong cooperation among policymakers, regulatory authorities, and the ICT community has been essential in helping the global society navigate through this public health crisis.

1.2.3 Societal impact of 5G in the 2020s

Many new applications of 5G have emerged during the global fight against COVID-19, as illustrated in Fig. 1.7. These applications include contactless experience for consumers, intelligent operation in factories, farms, mining sites, and shipping ports, in

Figure 1.7 Illustration of the profound impact and value of 5G to our society.

addition to previously mentioned applications such as remote medical diagnosis, online education, remote working, and online shopping.

The South Korean government has promoted the development of a 5G-based *Untact* ecosystem for contactless consumer experience and industry automation [22]. As a result, even under very difficult circumstances in the first quarter of 2020, the gross domestic product (GDP) of South Korea increased by 1.3% in comparison with its 2019 counterpart.

On intelligent operation in factories, industrial robots, self-driving vehicles, AI, machine vision, and other related technologies are being utilized. As an example, Commercial Aircraft Corporation of China used 5G, AI, and machine vision to realize human-free inspection of numerous layers of the carbon fiber materials used in the manufacture of an aircraft fuselage, thus reducing the inspection time by 40-folds and the waste of materials by 10-folds in comparison to traditional human inspection [19]. To better support machine vision, software-defined cameras (SDCs) capable of dynamically adjust imaging settings to produce high-quality images in real time are key enablers. New benchmarks for intelligent manufacturing to further reduce human labor and improve production efficiency are constantly emerging.

On intelligent operation in farms, unmanned autonomous harvesting is one of the exciting applications. As an example, China Mobile in partnership with Huawei deployed 5G technologies for the Hefeng Unmanned Farm in Linzi District of Zibo City, Shandong Province, China [19]. The farm covers an area of about 33 ha and is a national exemplary farm to showcase how modern, high-efficiency farming can be integrated with the new generation of ICT. In the past, farmers would work long hours each day during wheat harvest time to conduct labor-intensive operations. Now, hectares of wheat can be harvested by remotely controlling 5G-connected autonomous harvesters. In the future, various 5G-connected unmanned agricultural machines are anticipated to be used for plowing, planting, and other essential operations in the farming process. More and more new technologies such as image recognition, big data analytics, and AI-assisted smart decision-making are expected to be applied to improve land yield, resource usage, and productivity, thus realizing the full potential of intelligent farming [23].

On intelligent operation of mining sites, Shanxi Xinyuan Coal Mine in China has built a 5G network in a 534-m underground mine [19]. The first purpose of the underground 5G network is to connect hundreds of high-definition SDCs for real-time video surveillance, which was not possible in the past owing to network bandwidth limitations. With simultaneous transmission of all the high-definition surveillance videos, the entire underground mine system can be monitored with great details in real time. This has considerably helped improve the safety of the miners. The second purpose of the 5G network is to ensure a highly reliable communication between the miners and the workers above the group. The third purpose of the 5G network is to enable unmanned mining operation and self-driving transportation vehicles in the future, which will further improve both the safety of the miners and the mining efficiency. There are over 5000 coal mine sites in China alone, so there is assuredly a tremendous societal benefit of 5G in future intelligent mining.

On intelligent operation of shipping ports, 5G networks are being utilized to improve operation efficiency and reduce human intervening. As an example, Ningbo Zhoushan Port is one of the world's largest ports and had a preliminarily transformation to a 5G Intelligent Port in early 2020 [19]. Remote operation of mobile harbor cranes has been enabled by 5G connectivity, allowing 90% of the operation to be completed without human intervention. This enables one operator to simultaneously operate multiple mobile harbor cranes, thus greatly improving the operation efficiency and lowering the OPEX. In addition, 5G has enabled unmanned self-driving container trucks, intelligent cargo handling, unmanned drone inspection, etc. There are over 4000 ports in the world, so there is also a substantial societal benefit of 5G in future intelligent ports.

It is worthwhile to estimate the overall economic impact of 5G. According to China's MIIT, China had 3.72 million 4G base stations in 2018 [24]. It was estimated that over 7 million 5G base stations would be deployed in China [24]. This indicates that on average, one 5G base station is needed for every 200 citizens. To achieve the same coverage as the 4G network, the investment to build 5G networks in China would be 4 trillion yuan (or about $0.57T). According to the China Academy of Information and Communications Technology, the commercial use of 5G is expected to directly bring 10.6 trillion yuan (or about $1.5T) in economic output between 2020 and 2025 [24]. In addition, these 5G networks are expected to bring indirect values to relevant sectors. Furthermore, these deployed 5G networks would contribute even more greatly to the GDP between 2025 and 2030. Therefore the return of investment for 5G deployment in China is well justified. The same can also be true for many other countries in the world.

It is also worth noting that while 5G networks are being globally deployed, it is beneficial to unleash their capabilities to bring tangible values to general public, so that people can enjoy the so-called 5G technology dividend along the way [19]. The year 2020 already witnessed the initial deployment of 5G as well as the new values it brings. It can be expected that more profound values will be offered to our global society in the current decade.

1.3 Evolution of optical communications

1.3.1 Key milestones in optical communications

Optical communication networks are supporting a wide range of communication services including residential services, enterprise services, and mobile services that include the emerging 5G services. Optical communications have benefited from many ground-breaking scientific advances in the field of optics, photonics, electronics, and digital signal processing (DSP) [25–27]. The following list of milestones highlights some of the major breakthroughs and events [25,28].

1880: Alexander Graham Bell invented Photophone.

1948: Claude Shannon formulated the Shannon limit of a communication channel [29].

1957: Townes and Schawlow outlined the principles of laser operation [30].

1966: Kao concluded that the fundamental limit on glass transparency is below 20 dB/km, which would be practical for communications. Hockham calculated that clad fibers should not radiate much light. They published a paper proposing fiber-optic communications [31].

1970: First continuous-wave room-temperature semiconductor lasers were made in early May by Zhores Alferov's group at the Ioffe Physical Institute in Leningrad (now St. Petersburg) and on June 1 by Mort Panish and Izuo Hayashi at Bell Labs.

1987: Payne's group reported making the first erbium-doped optical fiber amplifier (EDFA) at the University of Southampton. Desurvire and Giles developed a model to predict the behavior of erbium optical amplifier at Bell Labs [32,33].

1988: Mollenauer of Bell Labs demonstrated soliton transmission through 4000 km of single-mode fiber.

1993: Chraplyvy et al. at Bell Labs transmitted at 10 Gb/s on each of eight wavelengths through 280 km of dispersion-managed fiber [34].

1996: Commercial wavelength-division multiplexing (WDM) systems were introduced with pioneering contribution from Dr. Tingye Li [35].

2002: Differential phase-shift keying (DPSK) was first demonstrated for 40 Gb/s long-haul transmission by Bell Labs [36,37].

2002: Nonlinearity compensation in fiber transmission was introduced for phase-modulated signals by Bell Labs [38].

2003: Gigabit-capable Passive Optical Networks (G-PON) was standardized by the Telecommunication Standardization Sector of the ITU (ITU-T) [39].

2004: The concept of DSP-based coherent optical detection was introduced by Michael Taylor of University College London [40].

2007: DSP-based coherent optical detection was experimentally demonstrated at 40 Gb/s and 100 Gb/s by Nortel, Bell Labs, and Coreoptics/Eindhoven University of Technology/Siemens.

2009: Superchannel concept was introduced and experimentally demonstrated at 1.2 Tb/s by Chandrasekhar and Liu et al. at Bell Labs [41].

2010: The Shannon limit for nonlinear fiber-optical transmission was studied by Essiambre et al. at Bell Labs [42].

2010: 10-Gb-capable XG-PON was standardized by ITU-T [43].

2011: Spatial multiplexing for optical transport capacity scaling was studied by Winzer et al. at Bell Labs [44].

2011: 100 Gb/s coherent transceivers were commercialized by Alcatel-Lucent with contributions from Bell Labs.

2012: Flexible-grid WDM was standardized by ITU-T [45].

2012: Constellation shaping was experimentally demonstrated for high-speed optical transmission by Bell Labs [46].

2014: 200 Gb/s coherent transceivers were commercialized by Huawei, Acacia, Ciena, etc.

2016: Low-loss low-nonlinearity optical fibers were specified by ITU-T [47].

2017: Silicon Photonics was introduced to digital coherent transceivers by Acacia, Huawei, etc.

2018: Low-loss M × N CDC-WSS (colorless, directionless, and contentionless-wavelength-selective switch) was developed by Huawei and Lumentum [48,49].

2019: Super-C-band transmission with 6-THz optical bandwidth was demonstrated by Huawei [50].

2020: The fifth-generation fixed network (F5G) started to be defined by the F5G Industry Specification Group in the ETSI [51].

Over the past 30 years, optical fiber transmission systems have advanced rapidly. As evident from Table 1.2, the single-channel rate of fiber-optic transmission has increased 160 times from 2.5 Gb/s in around 1989 to 400 Gb/s in 2019. The main technologies include high-speed electro-optical modulation, high-speed optical detection, hard-decision forward-error correction (HD-FEC), DPSK and quadrature phase-shift keying (DQPSK) [52,53], DSP for coherent optical detection (oDSP) [54–58], soft-decision FEC (SD-FEC), polarization-division multiplexing (PDM), high-order quadrature-amplitude modulation (QAM), constellation shaping (CS) such as probabilistic CS (PCS), advanced oDSP such as faster-than-Nyquist (FTN) detection, 100-Gbaud-class high-speed opto-electronic (OE) devices. With the introduction of the superchannel technology, the aggregated channel data rate can readily exceed 1 Tb/s [41,59,60].

In addition to the increase in data rate per channel, the number of channels per fiber is also increased through WDM or dense WDM (DWDM) to further improve the overall transmission capacity. Besides the demand for high capacity, optical transport networks desire long unregenerated optical transmission distance to effectively support metropolitan, regional, and national network applications. The advent of fiber-optic amplifiers, such as broadband EDFAs and Raman amplifiers, has made cost-effective simultaneous optical amplification of multiple WDM channels a reality. Moreover, flexible wavelength management, enabled by elements such as reconfigurable optical add/drop multiplexers (ROADM), is often utilized to make optical networks transparent, flexible, and scalable. With these advances, fiber-optic transmission

Table 1.2 Key technical advances in fiber transmission over the last 30 years.

Year	1980	1990	2000	2010	2020
Channel rate	2.5 Gb/s	10 Gb/s	40 Gb/s	100 Gb/s	200/400 Gb/s
Modulation format (typical)	OOK (NRZ)	OOK (RZ)	DPSK DQPSK	PDM-QPSK	PDM-nQAM, CS, PAM4, DMT
System features (newly added)	Single-span, Single-channel	Multispan with EDFAs, WDM	DWDM, Raman, ROADMs	1:N WSS, CDC-ROADMs	Flexible-grid WDM, M:N WSS, Super-C EDFA
System capacity (typical)	2.5 Gb/s (single channel)	400 Gb/s (40 WDM channel)	1.6 Tb/s (40 WDM channel)	8 Tb/s (80 DWDM channel)	16–32 Tb/s (Fixed-grid or flex-grid channels)
System reach (typical)	100 km (single span)	1000 km	1000 km @40G 3000 km @10G	4000 km @100G	2000 (1000) km @200 (400) G
Enabling technologies	Directly modulated laser	High-speed modulation, HD-FEC	DPSK	Coherent detection with DSP, SD-FEC	100-Gbaud-class OE device and advanced oDSP

CS, Constellation shaping; *DPSK*, differential phase-shift keying; *DQPSK*, differential quadrature phase-shift keying; *DWDM*, dense wavelength-division multiplexing; *EDFA*, erbium-doped fiber amplifier; *HD-FEC*, hard-decision forward-error correction; *DSP*, digital signal processing; *PDM*, polarization-division multiplexing; *QPSK*; quadrature phase-shift keying; *ROADMs*, reconfigurable optical add/drop multiplexers; *SD-FEC*, soft-decision FEC; *OOK*, on-off keying; *NRZ*, non-return-to-zero; *nQAM*, n-point quadrature amplitude modulation; *PAM4*, 4-level pulse amplitude modulation; *DMT*, discrete multi-tone; *WSS*, wavelength-selective switch; *CDC*, colorless, directionless and contentionless.

system has also evolved from early single-span single-wavelength transmission system to today's optically amplified multispan WDM system with reconfigurable optical add/drop sites and flexible channel grids. The single-fiber transmission capacity has increased significantly from 2.5 Gb/s in 1989 to 32 Tb/s in 2019 that is, 12,800 times. The average single-fiber transmission capacity increase over the last 30 years is thus at a remarkable rate of over 30% per year.

It is worth noting that the growth rate of single-fiber transmission capacity is slightly below that of the mobile data rate, which tracks closely with the Moore's Law, as discussed at the end of Section 1.1.1. This indicates that more fiber links are needed to support mobile communications and other communications such as data center communications. This is indeed true as the number of activated fiber links has been increasing over the last 30 years. Based on a recent report, the amount of optical fiber installed per year globally has increased with a compound annual growth rate (CAGR) of 14% for 20 years, from 1997 to 2017 [61]. In 1997 the total installed fiber was 37 million fiber-km, while it reached 492 million fiber-km in 2017. Going forward, the CAGR of the global fiber-optic cable market is expected to grow at a CAGR of 23.2% and reach $87.58 billion by 2023 [62]. This further indicates the increased need of optical fiber networks in supporting the communication demands of this decade.

1.3.2 5G-Oriented optical networks

1.3.2.1 5G-Oriented optical network overview

The 5G wireless networks bring to modern optical networks new requirements such as high bandwidth, low latency, accurate synchronization, and the ability to perform network slicing. The requirement for high bandwidth is driven by emerging wireless applications such as m-MIMO, while the requirements for low latency and accurate synchronization are mainly driven by applications such as C-RAN and coordinated multipoint (CoMP). The requirement for network slicing is aimed at optimizing the resource utilization for any given applications. All these requirements are to be addressed in the so-called 5G-oriented optical networks, as shown in Fig. 1.8. It illustrates a typical end-to-end optical communication network consisting of access, aggregation, and core optical network segments.

For 5G, there is a well-known triangle diagram to illustrate its three main features, eMBB, uRLLC, and mMTC, as shown in Fig. 1.3. Similarly, we can use a triangle diagram to illustrate three main features of 5G-oriented optical transport networks:

- enhanced fiber transmission capacity (eFTC),
- intelligent network operation and maintenance (iO&M), and
- massive optical cross-connections (mOXC),

as shown in Fig. 1.9.

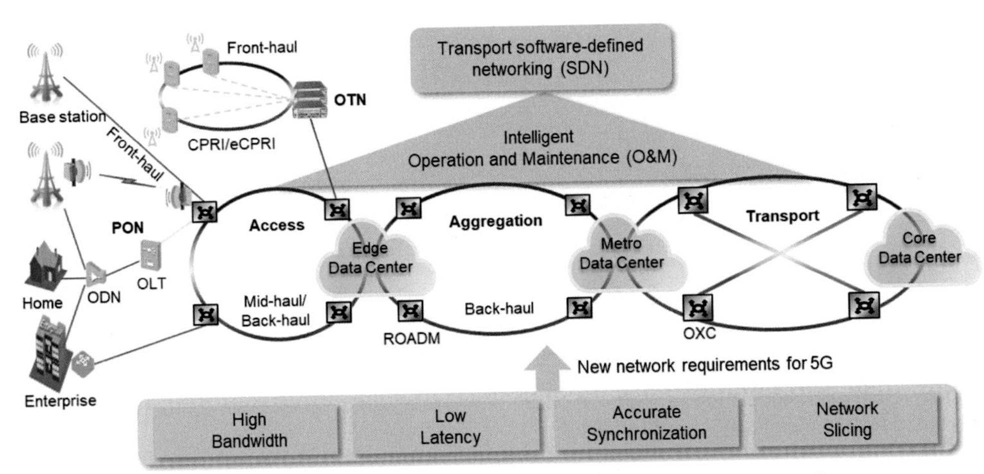

Figure 1.8 Illustration of a 5G-oriented end-to-end optical network consisting of access, aggregation, and core network segments to support diverse applications. *CPRI*, Common public radio interface; *eCPRI*, evolved CPRI; *ODN*, optical distribution network; *OLT*, optical line terminal; *OTN*, optical transport network; *OXC*, optical cross-connect; *ROADM*, reconfigurable optical add/drop multiplexer.

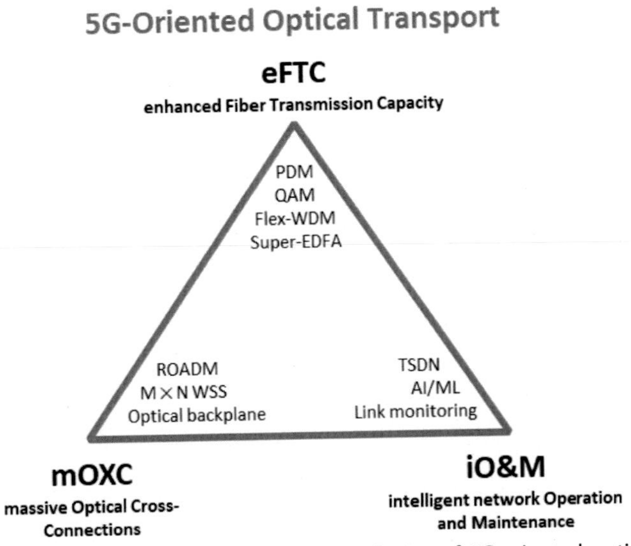

Figure 1.9 Illustration of the main features and technologies of 5G-oriented optical transport networks. *AI*, artificial intelligence; *EDFA*, Erbium-doped fiber amplifier; *ML*, machine learning; *PDM*, polarization-division multiplexing; *QAM*, quadrature-amplitude modulation; *ROADM*, reconfigurable optical add/drop multiplexer; *TSDN*, transport software-defined networking; *WSS*, wavelength-selective switch.

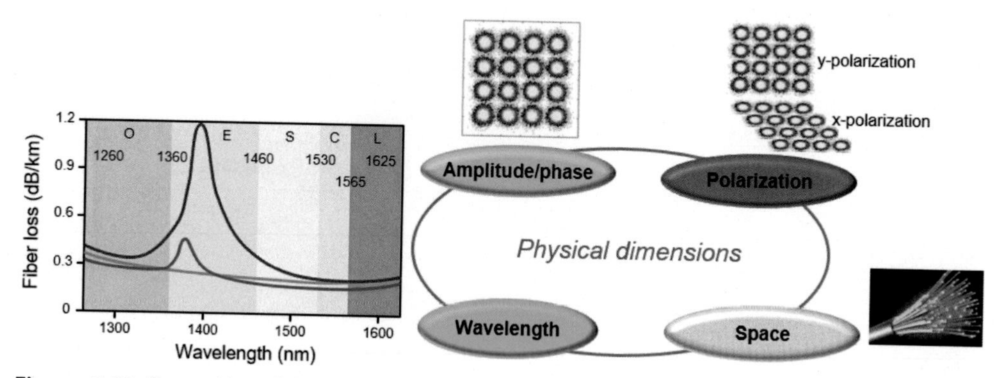

Figure 1.10 Illustration of four main physical dimensions that have been utilized to achieve enhanced fiber transmission capacity (eFTC).

1.3.2.2 Enhanced fiber transmission capacity

To meet the demands of eMBB in 5G, optical networks need to provide more transmission capacity, and do so cost-effectively. This calls for eFTC. The key is to increase per-fiber capacity, which can be realized by fully utilizing the following four physical dimensions of a light wave traveling along a fiber transmission link, as illustrated in Fig. 1.10,

- Amplitude and phase of a given optical carrier, for example, via quadrature-amplitude modulation and coherent detection
- Polarization, via PDM with dual-polarization modulation and detection
- Wavelength, via WDM over a broad optical amplification window
- Space, for example, by using multiple single-mode fibers in a same fiber cable

Fig. 1.11 illustrates the recent advances in per-fiber capacity from 8 Tb/s typically in 2013 to 32 Tb/s in 2019 representing a remarkable increase of about 1 dB per year. In 2013 digital coherent detection with SD-FEC enabled 100-Gb/s PDM-QPSK wavelength channels to be transmitted over long distances, for example, over 4000 km. With the typical C-band transmission with EDFAs having an amplification bandwidth of 4 THz, 80 50-GHz-spaced wavelength channels could be transmitted by a single optical fiber link, leading to a per-fiber capacity of 8 Tb/s. The spectral efficiency (SE) of the WDM system was thus typically 2 b/s/Hz. The OE devices needed for modulation at the transmitter and detection at the receiver were typically specified at 32 Gbaud, as it allowed PDM-QPSK to provide a raw data rate of 128 Gb/s, sufficient to guarantee a net data rate of 100 Gb/s after removing overheads used for FEC and other purposes. The oDSP was typically based on 40-nm CMOS technology.

In 2016 higher-level modulation such as PDM-16QAM was commercially deployed to double the per-channel data rate to 200 Gb/s, the WDM system SE to 4 b/s/Hz, and the per-fiber capacity to 16 Tb/s. To achieve longer transmission distance for 200-Gb/s channels, 64-Gbaud PDM-QPSK was commercialized, thanks to the availability of OE devices with doubled speed and the deployment of flexible-grid WDM in which the channel spacing can be flexibly adjusted at the increment of 12.5 GHz. A typical channel

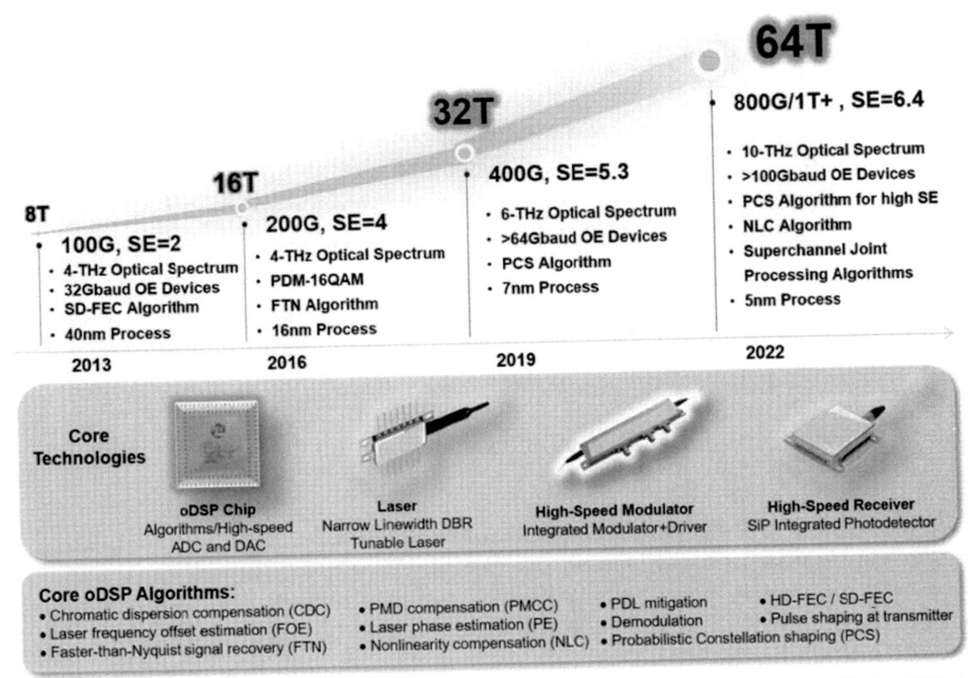

Figure 1.11 Illustration of recent advances in per-fiber capacity from 8 Tb/s typically in 2013 to 32 Tb/s in 2019.

spacing for 200-Gb/s PDM-QPSK channels is 75 GHz, which results in a WDM SE of 2.67 b/s/Hz. To allow the optical signal to be tolerant to bandwidth limitations due to cascaded optical filtering, FTN algorithm was developed. The oDSP for 200-Gb/s coherent receivers was typically based on 16-nm CMOS technology.

In 2019 even higher-level QAM modulations, such as PDM-64QAM, were introduced to further increase the transmission system SE. CS, particularly PCS, was introduced to further improve coherent optical transceiver performance to better approach the Shannon capacity limit at high SE [63,64]. Field trails on the use of real-time PCS-enabled coherent transceivers had also been demonstrated, achieving a twofold increase in reach when PCS is activated [65]. 400-Gb/s wavelength channels were transmitted over a 75-GHz-spaced flexible-grid WDM system, leading to a further increased WDM SE of 5.3 b/s/Hz. The oDSP for 400-Gb/s coherent receivers was typically based on 7-nm CMOS technology. With an innovative design, the spectral bandwidth of the C-band of EDFA was broadened from its typical value of 4 THz by 50% to 6 THz, forming the so-called Super-C-band [50]. In the Super-C-band, 80 75-GHz-spaced 400-Gb/s channels can be supported, thus achieving a further doubled per-fiber capacity of 32 Tb/s.

Going forward, we expect further advances in broadening the amplification bandwidth of EDFAs, increasing the OE device speeds, and adopting more advanced oDSP algorithms, as shown in Fig. 1.11. With the use of both the extended C-band and the extended L-band, the amplification bandwidth of EDFA can be increased to 10 THz and beyond. With advancement in OE device technologies, the modulation baud rate may exceed 100 Gbaud. More advanced oDSP algorithms such as PCS for even higher-level QAM, nonlinearity compensation, and joint processing of superchannel constituents may be implemented with further improved CMOS technologies. With the above-mentioned advances, further doubling of per-fiber capacity to 64 Tb/s can be achieved in commercial systems by 2022. Note again that the growth rate of single-fiber transmission capacity is slightly below that of the mobile data rate or the Moore's Law, so more optical fiber links are to be deployed to meet the communication capacity demands in the 5G era.

1.3.2.3 Intelligent network operation and maintenance

The second feature of 5G-oriented optical networks, iO&M, is important to ease and optimize network operation and maintenance. Fig. 1.12 illustrates an intelligently managed optical network with iO&M. In the network deployment phase, iO&M enables fast network planning, service initiation, and provisioning. In the operation phase, traffic forecast, troubleshooting, and failure warning can be supported. In the optimization phase, adaptive resource management and network slicing are used to partition the network in such a way that a given service is adequately supported by a

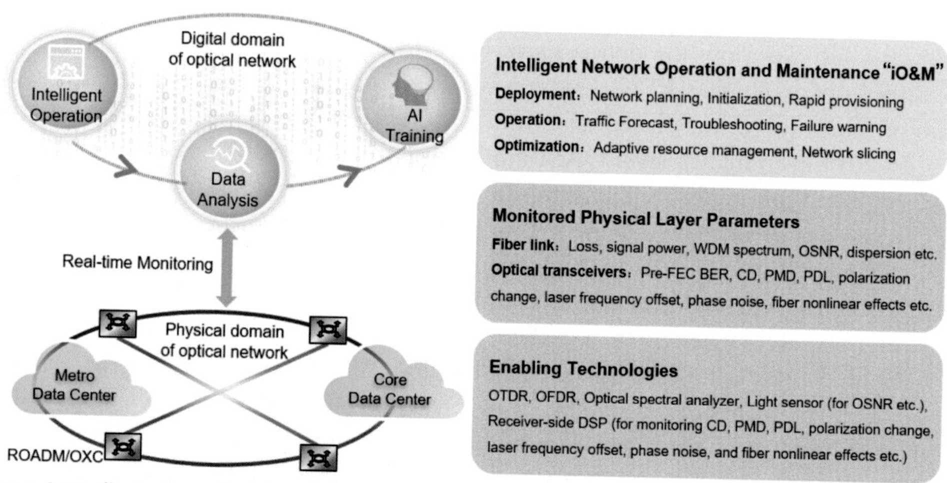

Figure 1.12 Illustration of an intelligently managed optical network with iO&M. *BER*, Bit-error rate; *CD*, chromatic dispersion; *FEC*, forward-error correction; *OFDR*, optical frequency-domain reflectometer; *OSNR*, optical signal-to-noise ratio; *OTDR*, optical time-domain reflectometer; *PDL*, polarization-dependent loss; *PMD*, polarization-mode dispersion.

network slice in a resource-efficient manner [25]. With the introduction of AI in the network cloud engine, it is feasible to predict faults in optical networks and better anticipate future network resource requirements, thus improving O&M efficiency and marking a critical step toward the so-called "zero-touch" optical networks. To realize the above-mentioned iO&M features, the network controller needs to know the key parameters of the underlying physical layer of the optical network, as shown in Fig. 1.12. These physical-layer parameters are obtained from "light sensors" based on optical time-domain reflectometer, optical frequency-domain reflectometer, etc.

1.3.2.4 Massive optical cross-connections

The third feature of 5G-oriented optical networks, mOXC, is important to realize transparent mesh optical connections for low-latency and energy-efficient transmission. 5G applications impose a stringent requirement on overall network latency. To effectively transport WDM channels with low latency, it is highly preferred to adopt optical wavelength switching as much as possible to achieve the direct wavelength pass-through at the optical layer (L0) over various nodes of the 5G-oriented optical networks. Thus massive optical cross-connections are needed to provide the full switching capability for all the WDM channels in all the fibers connected to an optical node. In addition, optical routing on the wavelength level avoids unnecessary optical–electrical–optical (O–E–O) conversions that are power consuming, thereby improving the optical network energy efficiency. The industry has evolved from the conventional ROADM era to the multidegree optical cross-connect (MD-OXC) era.

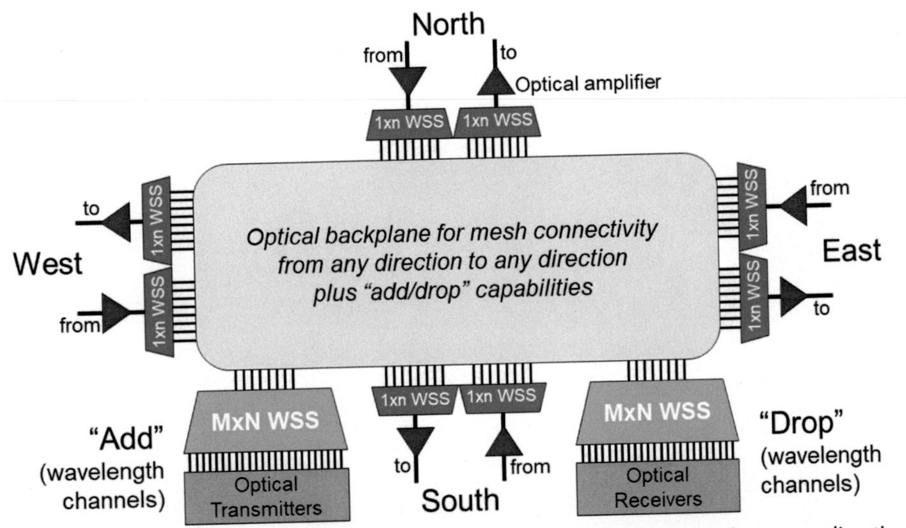

Figure 1.13 Illustration of an MD-OXC node supporting mesh connectivity from any direction to any direction, as well as adding/dropping any wavelength channel to/from any direction.

Fig. 1.13 illustrates a MD-OXC node supporting the mesh connectivity from any direction (e.g., north, south, east, or west) to any direction, as well as the adding/dropping of any wavelength channel to/from any direction. The WDM channels coming from a given direction first pass a 1 × n WSS so that the wavelength channels intended for different directions are switched to different output ports that are connected via a compact optical backplane to their corresponding directions. Similarly, the WDM channels to be outputted from a given direction use a 1 × n WSS to combine all the wavelength channels coming from different directions via the optical backplane. In effect, the optical backplane provides a full mesh connectivity from any direction to any direction. Moreover, the optical backplane also provides extra connections to add local wavelength channels to any direction and to drop wavelengths from any direction to this MD-OXC node. A novel contention-less N × M WSS was developed to realize the "add/drop" capabilities with low insertion loss, low power consumption, and small form factor [48,49].

1.3.3 The vision of fiber-to-everywhere in the 5G era

In the 5G era, optical fiber connectivity is expected to play an increasingly important role in supporting broadband access to 5G base stations, homes, offices, shopping centers, business buildings, factories, and smart cities. By the first half of 2019, 570 million fiber-to-the-home users have been registered worldwide, according to an Omdia report [66]. It is also estimated that 700 million households will have implemented optical access by 2023. Reaching deeper to final access points at locations such as rooms, office desks, and factory machines, optical fiber will realize its full potential to support a fully connected intelligent world.

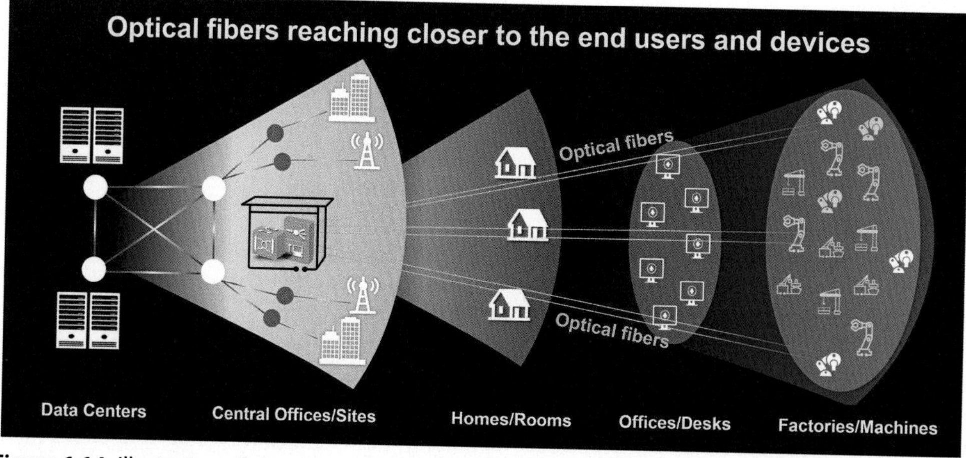

Figure 1.14 Illustration of the vision of fiber-to-everywhere in the 5G era.

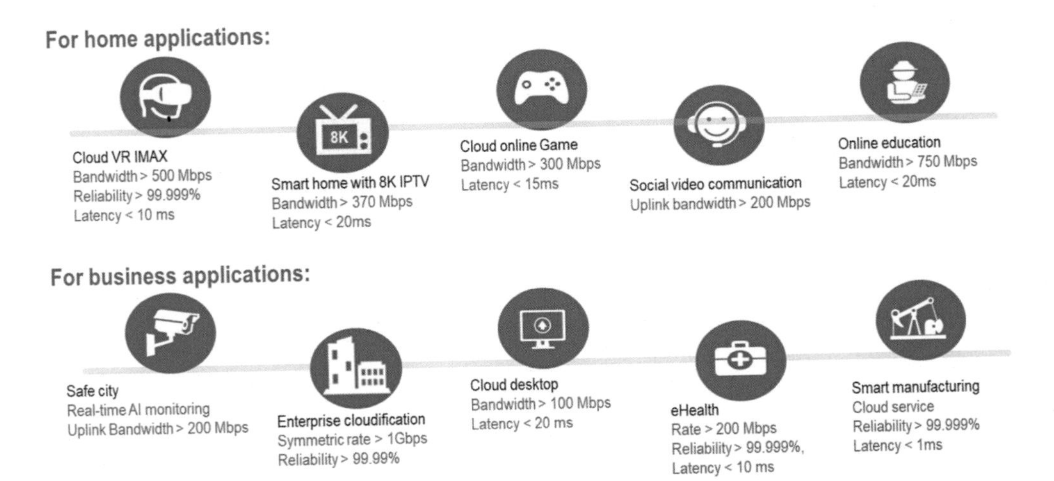

Figure 1.15 Illustration of major use cases of F5G.

Fig. 1.14. illustrates such a fiber-to-everywhere vision in the 5G era. Teaming up with advanced wireless technologies such as Wi-Fi 6, fiber-to-everywhere technologies can provide high-bandwidth, high-reliability, and low-latency connectivity to people and machines with certain flexibility and mobility. Similar to mobile networks, fixed networks had entered their fifth generation, F5G, around year 2020, characterized by the use of 10-Gb/s XG-PON for the optical access network segment and 5G-oriented optical transport for the optical core and metro network segments. The ETSI has started the definition and specification of F5G since 2020 [51]. With the fiber-to-everywhere vision, F5G aims to transform how people and machines communicate in the 5G era.

Fig. 1.15 illustrates major use cases of F5G. For home users, F5G aims to enhance the user experience of cloud VR, high-definition video such as 8K internet protocol television, cloud online game, social video communication, and online education. For business users, F5G aims to aid the realization of safe cities, enterprise cloudification, cloud desktop, telehealth, and smart manufacturing. Fiber-to-everywhere will help factories, mining sites, shipping ports, farms, and oilfields implement industrial automation. Automated machinery and robots with precise and coordinated controls will eliminate intensive manual labor and dramatically improve workplace safety and production efficiency. Compared to the previous generation of fixed networks, F5G aims to provide 10 times more bandwidth, 100 times more connections, and 10 times better in user experiences judged by performance indicators such as reliability and latency.

On the basis of the above, we can expect that wireless and optical communication technologies will continue to support and complement each other to help realize the vision of a fully connected, intelligent world for the benefit of our global society in the exciting era of 5G and F5G. In the following chapters of this book, an up-to-date review of optical communication technologies, systems and networks in the 5G era is provided. Also presented are in-depth descriptions of the key technological advances in various optical network segments, such as front-haul, access, aggegation, metro, data center, backbone, and undersea network segments. Future trends in optical communications are also discussed.

References

[1] The 3rd Generation Partnership project (3GPP), <http://www.3gpp.org/>.

[2] ITU-T Technical Report. Transport network support of IMT-2020/5G, <https://www.itu.int/pub/T-TUT-HOME-2018>; 2018.

[3] ITU towards IMT for 2020 and beyond, <https://www.itu.int/en/ITU-R/study-groups/rsg5/rwp5d/imt-2020/Pages/default.aspx>.

[4] See for example, <https://en.wikipedia.org/wiki/1G>, <https://en.wikipedia.org/wiki/2G, https://en.wikipedia.org/wiki/3G>, <https://en.wikipedia.org/wiki/4G>, and <https://en.wikipedia.org/wiki/5G>.

[5] See for example, <https://en.wikipedia.org/wiki/Moore's_law>.

[6] China Mobile Research Institute. C-RAN: the road towards green RAN, whitepaper v. 2.6; Sep. 2013.

[7] See for example, <https://www.o-ran.org/>.

[8] See for example, <https://www.olympic.org/news/fans-of-the-olympic-winter-games-2018-to-experience-world-s-first-broad-scale-5g-network>.

[9] See for example, <https://www.rcrwireless.com/20190606/5g/chinese-government-issues-commercial-5g-licenses>.

[10] See for example, <https://5gobservatory.eu/china-has-released-5g-licences-for-indoor-usage-and-pushes-for-network-sharing/>.

[11] See for example, <https://techblog.comsoc.org/2019/11/22/2019-world-5g-convention-in-beijing-china-has-built-113000-5g-base-stations-130000-by-the-end-of-2019/>.

[12] See for example, <http://www.fintechbd.com/huawei-projects-more-than-1-5-million-5g-base-stations-to-be-deployed-globally-by-2020/>.

[13] See for example, <https://www.rcrwireless.com/20200609/5g/china-end-2020-over-600000-5g-base-stations-report>.

[14] See for example, <https://en.wikipedia.org/wiki/Digital_divide>.

[15] See for example, <https://www.globaltimes.cn/content/1172513.shtml>.

[16] See for example, <https://www.huawei.com/en/press-events/news/2020/4/china-mobile-huawei-deliver-world-highest-5g>.

[17] See for example, <https://www.criticalcomms.com/news/coronavirus-huawei-installs-5g-network-in-wuhans-huoshenshan-hospital>.

[18] See for example, <https://www.gsma.com/newsroom/blog/eleven-regulatory-recommendations-to-sustain-connectivity-during-the-covid-19-crisis/>.

[19] See for example, <https://mp.weixin.qq.com/s/0n9X8viEC9DCv2T-WC4rvA>.

[20] See for example, <https://www.fcc.gov/keep-americans-connected>.

[21] See for example, <https://www.fcc.gov/companies-have-gone-above-and-beyond-call-keep-americans-connected-during-pandemic>.

[22] See for example, <https://www.bloomberg.com/news/articles/2020-06-10/south-korea-untact-plans-for-the-post-pandemic-economy>.

[23] See for example, <https://carrier.huawei.com/en/success-stories/Industries-5G/Agriculture>.

[24] See for example, Around 20,000 5G base stations constructed across the country. China Daily, <http://124.127.52.76/a/201908/09/WS5d4d0d24a310cf3e35564d5a.html>; August 9, 2019.

[25] Liu X. Evolution of fiber-optic transmission and networking toward the 5G era iScience 2019;22:489–506[Online]. Available from:. Available from: https://www.cell.com/iscience/pdf/S2589-0042(19)30476-6.pdf.

[26] Liu X, Deng N. Chapter 17: Emerging optical communication technologies for 5G. In: Willner A, editor, Optical fiber telecommunications VII; 2019.

[27] Liu X, Optical communication technologies for 5G wireless. In: Optical fiber communication conference (OFC), short course 444; 2019.
[28] J Hecht. City of light: the story of fiber optics. Sloan Technology; 2004.
[29] Shannon CE. A mathematical theory of communication. Bell Syst Tech J 1948;27(3):379−423.
[30] Schawlow A, Townes C. Infrared and optical masers. Phys Rev 1958;112(6):1940−9.
[31] Kao KC, Hockham GA. Dielectric-fibre surface waveguides for optical frequencies. Proc IEE 1966;113(7):1151−8.
[32] Mears RJ, Reekie L, Jauncey IM, Payne DN. Low-noise Erbium-doped fiber amplifier at 1.54μm. Electron Lett 1987;23:1026−8.
[33] Giles CR, Desurvire E. Modeling erbium-doped fiber amplifiers. J Lightwave Technol 1991;9 (2):271−83.
[34] Chraplyvy AR, Gnauck AH, Tkach RW, Derosier RM. 8 x 10 Gb/s transmission through 280 km of dispersion-managed fiber. IEEE Photon Technol Lett 1993;5(10):1233−5.
[35] See for example, <https://en.wikipedia.org/wiki/Tingye_Li>.
[36] Gnauck AH, et al. 2.5 Tb/s (64x42.7 Gb/s) transmission over 40x100 km NZDSF using RZ-DPSK format and all-Raman-amplified spans. In: Optical fiber communication conference (OFC), PDP FC2; 2002.
[37] Xu C, Liu X, Wei X. Differential phase-shift keying for high spectral efficiency optical transmissions. IEEE J Sel Top Quant Electron 2004;10(2):281−93.
[38] Liu X, Wei X, Slusher RE, McKinstrie CJ. Improving transmission performance in differential phase-shift-keyed systems by use of lumped nonlinear phase-shift compensation. Opt Lett 2002;27:1616−18.
[39] ITU-T G.984.1: gigabit-capable passive optical networks (GPON).
[40] Taylor MG. Coherent detection method using DSP for demodulation of signal and subsequent equalization of propagation impairments. IEEE Photon Technol Lett 2004;16(2):674−6.
[41] S Chandrasekhar, X Liu, B Zhu, DW Peckham. Transmission of a 1.2-Tb/s 24-carrier no-guard-interval coherent OFDM superchannel over 7200-km of ultra-large-area fiber. In: Proceedings of European conference on optical communication (ECOC), Sept., paper PD2.6; 2009.
[42] Essiambre R-J, Kramer G, Winzer PJ, Foschini GJ, Goebel B. Capacity limits of optical fiber networks. J Lightwave Technol 2010;28(4):662−701.
[43] ITU-T G.987.3: 10-Gigabit-capable passive optical networks (XG-PON).
[44] Winzer PJ. Energy-efficient optical transport capacity scaling through spatial multiplexing. IEEE Photon Technol Lett 2011;23(13):851−3.
[45] Recommendation ITU-T G.694.1, <https://www.itu.int/rec/T-REC-G.694.1/>; 2012.
[46] Lotz TH, Liu X, et al. Coded PDM-OFDM transmission with shaped 256-iterative-polar-modulation achieving 11.15-b/s/Hz intrachannel spectral efficiency and 800-km reach. J Lightwave Technol 2013;31:538−45.
[47] ITU-T Recommendation G.654.E.
[48] L Zong, H Zhao, Z Feng, Y. Yan. Low-cost, degree-expandable and contention-free ROADM architecture based on M × N WSS. In: Optical fiber communication conference (OFC), paper M3E.3; 2016.
[49] PD Colbourne, S McLaughlin, C Murley, S Gaudet, D Burke. Contentionless Twin 8 × 24 WSS with low insertion loss. In: 2018 Optical fiber communications conference and exposition (OFC), PDP Th4A.1; 2018.
[50] See, e.g., press release, Huawei's ON2.0 leads the commercial use of all-optical networks in partnership with operators worldwide, <tagitnews.com/en/article/36282>.
[51] See for example, <https://www.etsi.org/technologies/fifth-generation-fixed-network-f5g>.
[52] Ho K-P. Phase-modulated optical communication systems. New York: Springer; 2005.
[53] Winzer PJ, Essiambre R-J. Advanced optical modulation formats. Proc IEEE 2006;94(5):952−85.
[54] Noe R. PLL-free synchronous QPSK polarization multiplex/diversity receiver concept with digital I&Q baseband processing. IEEE Photon Technol Lett 2005;17(4):887−9.
[55] Ly-Gagnon D-S, Tsukamoto S, Katoh K, Kikuchi K. Coherent detection of optical quadrature phase shift keying signals with carrier phase estimation. J Lightwave Technol 2006;24(1):12−21.

[56] Leven A, Kaneda N, Klein A, Koc U-V, Chen Y-K. Real-time implementation of 4.4 Gbit/s QPSK intradyne receiver using field programmable gate array. Electron Lett 2006;42(24):1421–2.
[57] Savory S. Digital filters for coherent optical receivers. Opt Express 2008;16:804.
[58] Sun H, Wu K-T, Roberts K. Real-time measurements of a 40 Gb/s coherent system. Opt Express 2008;16(2):873–9.
[59] Liu X, Chandrasekhar S, Winzer PJ. Digital signal processing techniques enabling multi-Tb/s superchannel transmission: an overview of recent advances in DSP-enabled superchannels. IEEE Signal Process Mag 2014;31(2):16–24.
[60] Y.R. Zhou, et al. Field trial demonstration of real-time optical superchannel transport up to 5.6 Tb/s over 359 km and 2 Tb/s over a live 727 km flexible grid link using 64G Baud software configurable transponders. OFC'16, PDP C-1; 2016.
[61] See for example, <https://www.crugroup.com/knowledge-and-insights/insights/2019/what-is-behind-the-chinese-optical-cable-contraction-and-outlook-post-2019/>.
[62] See for example, <https://www.businesswire.com/news/home/20200406005320/en/Global-Fiber-Optic-Cable-Market-Report-2020>.
[63] Buchali F, Steiner F, Böcherer G, Schmalen L, Schulte P, Idler W. Rate adaptation and reach increase by probabilistically shaped 64-QAM: an experimental demonstration. J Lightwave Technol 2016;34(7):1599–609.
[64] Cho J, Chen X, Chandrasekhar S, Raybon G, Dar R, Schmalen L, et al. Trans-Atlantic field trial using high spectral efficiency probabilistically shaped 64-QAM and single-carrier real-time 250-Gb/s 16-QAM. J Lightwave Technol 2018;36(1):103–13.
[65] J Li, A Zhang, C Zhang, X Huo, Q Yang, J Wang, et al. Field trial of probabilistic-shaping-programmable real-time 200-Gb/s coherent transceivers in an intelligent core optical network. In: Asia communications and photonics conference (ACP), PDP Su2C.1; 2018.
[66] Omdia report, Global fiber development index analysis 2020, <https://omdia.tech.informa.com/OM014270/Global-Fiber-Development-Index-Analysis-2020>.

CHAPTER 2

5G Wireless technologies

2.1 Overview of 5G technical requirements

5G promises to offer substantial improvements over 4G in all the three major application scenarios, namely, enhanced mobile broadband (eMBB), ultra-reliable and low-latency communications (uRLLC), and massive machine type communications (mMTC). Table 2.1 shows the key technical requirements of 5G specified by International Mobile Telecommunications (IMT)-2020 [1,2].

The above specifications enable the following capabilities in the three major application scenarios.

- *eMBB*: Peak download speeds of 20 Gb/s and a reliable 100 Mb/s user experience data rate in urban areas which supports bandwidth-hungry services such as virtual reality/augmented reality
- *uRLLC*: 1 ms latency and high reliability and mobility to support services such as autonomous vehicles and mobile healthcare
- *mMTC*: One million internet of things (IoT) connections per square kilometer with long battery life and wide coverage including inside buildings

Compared with 4G, the 5G specifications above aim to bring the following improvements

- 100 times increase in data rate
- 1000 times increase in throughput per unit area
- 30 times reduction in latency
- 10 times improvement in reliability
- 150 km/hour higher speed in mobility support (e.g., to support mobile connections for high-speed trains traveling at up to 500 km/hour)
- 100 times increase in connections per unit area

On the architecture level, cloud radio access network (C-RAN) [3,4] with mobile edge computing (MEC) [5,6] is introduced in 5G to enable computing near the antenna sites to provide improved real-time control and network performance by techniques such as coordinated multipoint (CoMP), as illustrated in Fig. 2.1.

In addition, network slicing [7–9] is implemented in 5G to allow for logically isolated network partitions (LINPs) to optimize network resource utilization for any given set of applications. Fig. 2.2 illustrates 5G network slicing for eMBB, uRLLC, and mMTC applications. Software-defined network, network function virtualization,

Optical Communications in the 5G Era
DOI: https://doi.org/10.1016/B978-0-12-821627-9.00013-9

Table 2.1 Key technical requirements of 5G.

Capability	Description	Requirement	Usage scenario
Downlink peak data rate	Maximal data rate supported	20 Gbit/s	eMBB
Downlink user experienced data rate	Data rate in dense urban test environment 95% of time	100 Mbit/s	eMBB
Downlink average spectrum efficiency	Throughput per unit wireless bandwidth and per network cell	30 bit/s/Hz	eMBB
Uplink peak data rate	Maximal data rate supported	10 Gbit/s	eMBB
Uplink user experienced data rate	Data rate in dense urban test environment 95% of time	50 Mbit/s	eMBB
Uplink average spectrum efficiency	Throughput per unit wireless bandwidth and per network cell	15 bit/s/Hz	eMBB
Area traffic capacity	Total traffic across coverage area	10 Mbit/s/m^2	eMBB
Latency	Radio network contribution to packet travel time	4 ms 1 ms	eMBB uRLLC
Mobility	Maximum speed for handoff and quality of service (QoS) requirements	500 km/h	uRLLC
Reliability	The success probability of data transmission within a given time	99.999%	uRLLC
Connection density	Total number of devices per unit area	10^6/km^2	mMTC

eMBB, Enhanced mobile broadband; *mMTC*, massive machine type communications; *uRLLC*, ultra-reliable and low-latency communications.

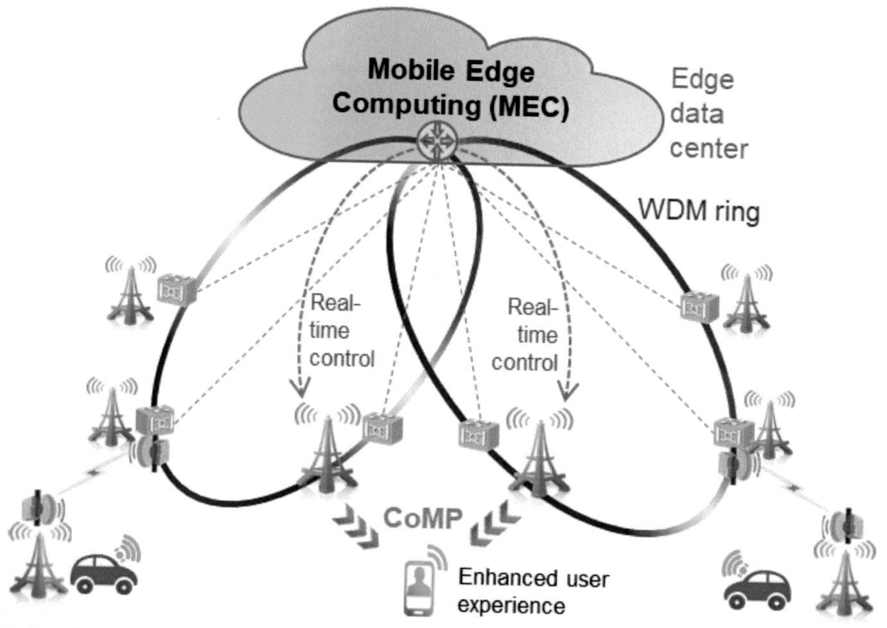

Figure 2.1 Illustration of coordinated multipoint (CoMP) for enhanced user experience via mobile edge computing (MEC) with real-time low-latency control.

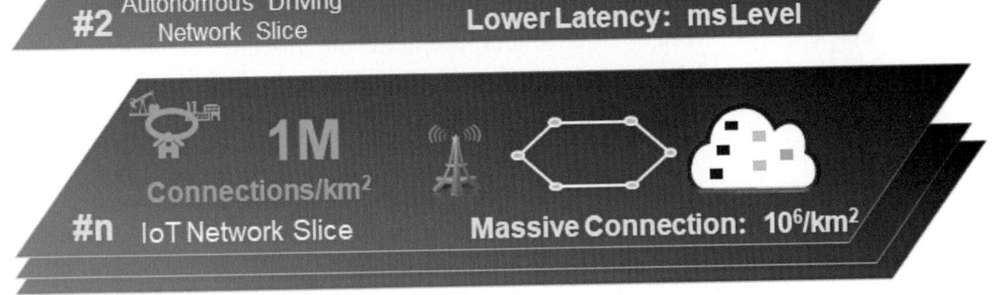

Figure 2.2 Illustration of 5G network slicing for eMBB, uRLLC, and mMTC applications.

as well as flexible orthogonal frequency-division multiple access frame designs, are key technologies that are being applied to support network slicing.

2.2 Overview of 5G wireless technologies

To meet the 5G specifications described in the previous section, multiple new technologies have been developed. Four major technological advances are enumerated as follows [10,11]:

- Harmonized spectrum allocation offering increased radio frequency (RF) bandwidth
- New radio (NR) supporting enhanced performance and flexible frame design
- Massive multiple-input and multiple-output (m-MIMO) supporting increased network capacity and coverage
- Cloud-RAN supporting MEC and CoMP

The NR air interface defined by the 3rd Generation Partnership Project (3GPP) for 5G is divided into two frequency ranges, the 1st frequency range (FR1), which is below 6 GHz, and the 2nd frequency range (FR2), which is above 24 GHz. To meet the key technical requirements described in the previous section, new frequency bands have been allocated to 5G by European Union, the Federal

Communications Commission, and other regulatory agencies [12]. Together with the previously allowed frequency bands in the 4G era, the spectrum available for 5G may cover the sub-6-GHz band and the millimeter wave bands ranging from 24 to ~100 GHz. The broadened spectrum enables 5G to support all the promised use cases as follows.

- Low-frequency bands below 1 GHz support wide coverage across urban, suburban, and rural areas, and help support mMTC services with low energy consumption.
- Mid-frequency bands between 2 and 6 GHz offer a good balance between coverage and capacity. The majority of commercial 5G networks rely on spectrum within the 3–5 GHz range to support eMBB and uRLLC services.
- High-frequency bands above 24 GHz allow each operator to use over 1 GHz of spectral bandwidth, thereby supporting ultra-high broadband speeds envisioned for fixed wireless access, which aims to complement the fiber-to-the-home type of services for homes and businesses. Currently, 26, 28, and 40 GHz have the most international support and momentum.

It is desirable to provide a harmonized mobile spectrum for global 5G deployment. Most regulators are aiming to make available around 100 MHz of contiguous spectrum per operator in the prime 5G mid-frequency bands and around 1 GHz of contiguous spectrum per operator in the high-frequency bands.

According to Shannon's theorem [13], the mobile communication capacity per unit area can be approximately expressed as [14]

$$C = nm\mathrm{B}\log\left(1 + \frac{S}{N + I}\right) \tag{2.1}$$

where C is the transmission capacity per unit area, n is the total number of cells per unit area, m is the bandwidth enhancement factor from MIMO, B is the RF spectrum bandwidth used, S is the signal power, N is the noise power, and I is the interference power. The second term in the parenthesis, that is, $S/(N+I)$ is the commonly known the signal-to-noise-plus-interference ratio (SNIR). Without any loss of generality, the above formula assumes that all the mobile cells are identical.

Fig. 2.3 illustrates the three main directions that enable the increase of mobile transmission capacity per unit area C,

- Increase in spectral bandwidth B via newly allocated frequency bands
- Increase in spectral efficiency (SE) via increased MIMO bandwidth enhancement factor (m) and SNIR
- Increase in the number of cells per unit area (n) via cell site densification.

To achieve an improvement of 1000 times in transmission capacity per unit area (C), one can increase n via densification of cell sites, m via multiuser m-MIMO, B via allocating more RF spectrum, and SNIR via beamforming and CoMP, etc. For

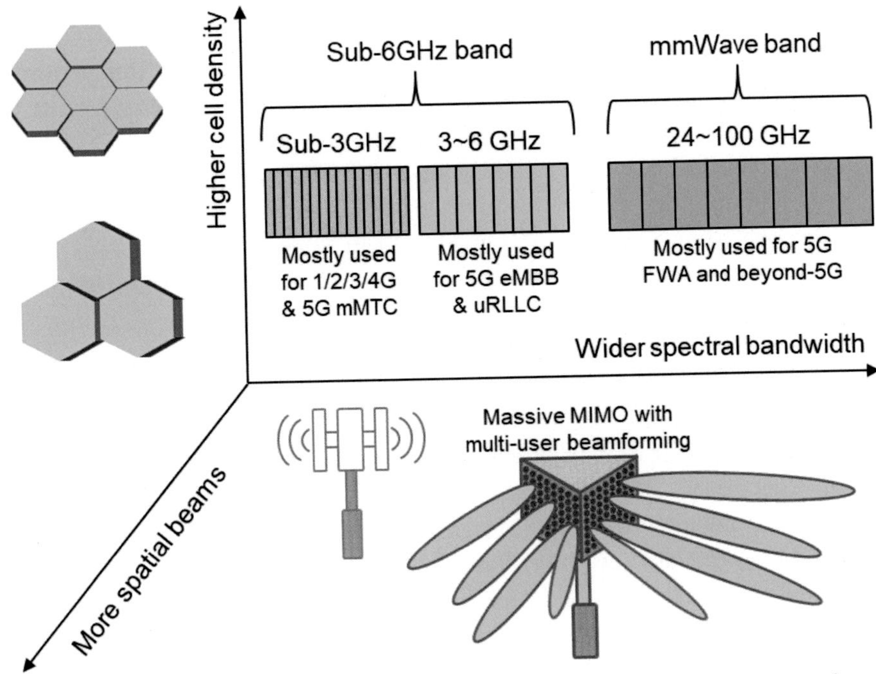

Figure 2.3 Illustration of three main directions for increasing mobile transmission capacity per unit area.

illustration, we have the following two implementation examples that offer a 1000-fold increase in C over a typical 4G scenario where $m=1$, $B=20$ MHz, SNIR supporting up to 16-QAM (quadrature amplitude modulation):

- 5G implementation example (1): 12.5-fold increase in n, $m=8$, $B=100$ MHz, and SNIR supporting up to 256-QAM;
- 5G implementation example (2): 3.125-fold increase in n, $m=16$, $B=200$ MHz, and SNIR supporting up to 256-QAM.

It is worth noting that increasing n means more cell sites per unit area and thus more installation and maintenance cost. While densification of cell sites is needed for 5G deployment, it is often more economic for 5G and 4G to share the same cell sites as much as possible. Thus m-MIMO with larger bandwidth enhancement factor and antennas with wider RF bandwidth are key enablers for achieving the capacity increase needed in the 5G era. In the following sections, we will describe in more depth the NR interface that supports enhanced performance and flexible frame design, m-MIMO supporting multiuser beamforming, C-RAN supporting mobile edge computing, and CoMP. Early 5G deployment advances will also be presented. Finally, an outlook for future 5G deployment will be provided.

2.3 5G New radio

The 5G NR air interface is much enhanced over 4G to support better performance and more flexible frame design [15]. Table 2.2 highlights the key enhancements.

In terms of the maximum RF bandwidth per component carrier (CC), 5G NR offers improvements of 5 and 20 times in FR1 and FR2, respectively, over 4G radio interface. In addition, the maximum number of CCs supported in carrier aggregation (CA) is increased from 5 in 4G to 16 in 5G. As an example, 400 MHz (or 1.6 GHz) of instantaneous bandwidth can be achieved by using CA with four 100-MHz (or 400-MHz) channels in FR1 (or FR2). More spectrally efficient modulation formats such as 256-QAM and 1024-QAM are supported by 5 G. These enhancements directly lead to a substantial increase in transmission link capacity.

In terms of the waveform, 5G NR uses cyclic-prefix orthogonal frequency division multiplexing (CP-OFDM) for downlink transmission and discrete fourier transform-spread-orthogonal frequency division multiplexing (DFT-s-OFDM) or CP-OFDM for uplink transmission. Owing to its relatively low peak-to-average-power ratio, DFT-s-OFDM is often used in power-limited scenarios. Unlike 4G, where the subcarrier spacing is fixed to 15 kHz, 5G NR adopts a set of flexible subcarrier spacings to make the frame design more scalable and flexible, with the aim to better support a wide range of applications with different bandwidth and latency requirements.

Table 2.3 shows the newly introduced μ-parameter and its corresponding subcarrier spacing (Δf_{sc}), slot duration (T_{slot}), number of slots per 1-ms subframe ($N_{slots\text{-}per\text{-}ms}$), number of OFDM symbols per slot ($N_{OFDM\text{-}per\text{-}slot}$), and number of OFDM symbols per 1-ms subframe ($N_{OFDM\text{-}per\text{-}ms}$).

Table 2.2 Comparison between 4G and 5G radio interfaces.

	4G Radio interface	5G NR
Maximum bandwidth per CC	20 MHz	100 MHz in FR1; 400 MHz in FR2.
Maximum number of CCs	5	16
Modulation formats	BPSK, π/2-BPSK, QPSK, 16/64-QAM	BPSK, π/2-BPSK, QPSK, 16/64/256/1024-QAM
Waveform	CP-OFDM for DL; SC-FDM for UL.	CP-OFDM for DL; DFT-s-OFDM or CP-OFDM for UL
Subcarrier spacing	15 kHz	$2^n \cdot 15$ kHz
Channel coding	Turbo codes for data channel; TBCC for control channel	LDPC for data channel; Polar codes for control channel.
MIMO	Up to 8 × 8	Typically 64 × 64
Multiuser beamforming	Typically one beam	Typically 8−16 beams
Air interface latency	10 ms	1 ms
Duplexing	FDD, Static TDD	FDD, Static and dynamic TDD

CC, Component carrier; *CP-OFDM*, cyclic-prefix orthogonal frequency division multiplexing; *FDD*, frequency division duplex; *TBCC*, tail-biting convolutional code; *TDD*, time division duplex.

Table 2.3 Scalable 5G NR numerology with flexible subcarrier spacing.

μ	Δf_{sc} ($=2^{\mu} \cdot 15$ kHz)	T_{slot} ($=2^{-\mu}$ ms)	$N_{slots\text{-}per\text{-}ms}$ ($=2^{\mu}$)	FR	$N_{OFDM\text{-}per\text{-}slot}$	$N_{OFDM\text{-}per\text{-}ms}$ ($=2^{\mu} \cdot N_{OFDM\text{-}per\text{-}slot}$)
0	15 kHz	1 ms	1	FR1	14	14
1	30 kHz	500 μs	2	FR1	14	28
2	60 kHz	250 μs	4	FR1 and FR2	14 (normal CP) 12 (extended CP)	56 48
3	120 kHz	125 μs	8	FR2	14	112
4	240 kHz	62.5 μs	16	FR2	14	224

With the increase of the μ-parameter, Δf_{sc} exponentially increases while T_{slot} exponentially decreases, enabling faster slot-based scheduling to support low-latency applications. Subcarrier spacings of below and above 60 kHz are available in FR1 and FR2, respectively, while subcarrier spacing of 60 kHz is available in both FR1 and FR2. A benefit of the scalable OFDM numerology is that the number of subcarriers for wider bandwidth signals is reasonably contained such that the digital signal processing (DSP) complexity does not increase unnecessarily. Note that the normal CP duration is inversely proportional to the subcarrier spacing, for example, 4.7 μs at $\Delta f_{sc} = 15$ kHz and 1.2 μs at $\Delta f_{sc} = 60$ kHz. At 60-kHz subcarrier spacing, both the normal CP and the extended CP are supported, with the latter intended for increasing the signal tolerance to intersymbol interference, which results from transmission impairments such as multipath interference.

To better support low-latency applications, mini-slots have been introduced in 5G NR. Unlike a standard slot with 14 or 12 OFDM symbols, a mini-slot can contain only 7, 4, or 2 OFDM symbols. Mini-slots can be transmitted without the need to wait for slot boundaries, thereby enabling quick delivery of payloads with low latency. The flexible 5G NR slot structure allows for dynamic assignment of bidirectional downlink and uplink slots to support dynamic time division duplex (TDD). In 4G mobile networks, frequency division duplex (FDD) and static TDD were adopted. In FDD, downlink and uplink signals are transmitted over different frequency bands. In TDD, downlink and uplink signals are transmitted at different time slots over the same frequency band. In today's mobile networks, traffic patterns are varying and more uplink bandwidth-hungry applications, such as cloud storage and personal broadcasting, are emerging. In response to this situation, dynamic TDD has been introduced in 5G NR to provide dynamically adjustable uplink and downlink bandwidth allocations such that the network capacity and performance are tailored to satisfactorily meet the mobile communication demands at any given time.

Regarding the channel coding, 4G adopted turbo codes for data channels and tail-biting convolutional code for control channels. 5G NR has adopted low-density parity check (LDPC) codes [16] for data channels and polar codes [17] for control channels [18]. LDPC offers inherent parallelism and thus provides high decoding speeds needed for 5G eMBB applications. Polar codes offer capacity-approaching performance with modest encoding and decoding complexities.

Another key feature of 5G NR is the use of m-MIMO with multiuser beamforming. 5G m-MIMO is capable of forming eight or more radio beams to communicate with multiple users simultaneously, thereby dramatically increasing the communication capacity. More in-depth description on m-MIMO will be provided in the following section.

2.4 Massive multiple-input and multiple-output

In mobile communication, MIMO is a well-established technology that exploits the space dimension to achieve increased network capacity and coverage by using multiple transmitting antennas, propagation paths, and receiving antennas [19,20]. MIMO provides the following three main functions.

- Beamforming and beam steering—MIMO enables the formation of a radio beam with an adjustable direction by controlling the phase relationship of the signals emitting from multiple transmitting antennas. Beamforming directionally focuses the RF energy on an intended user equipment (UE) and provides two direct benefits:
  - Increased radio power to the intended UE
  - Reduced interference to other UEs

  Similarly, beamforming enables a base station receiver to receive an uplink signal from a UE along a specific direction that has low noise and interference. With the channel information being available, beam steering can be applied to dynamically track the UE, increasing the receiver SNIR wherever the user is moving. In a multipath environment, beam steering can dynamically direct the RF energy toward the best path to constantly maintain the best connectivity and thus increase the network coverage.
- Spatial diversity—MIMO offers the ability to exploit spatial diversity, which improves the reliability of the communication system by sending the same data across different spatial paths. Techniques such as space-time coding (STC) can be applied. STC may be subdivided into three categories according to whether the receiver knows the channel conditions:
  - Coherent STC, where the receiver knows the channel conditions through training or estimation
  - Noncoherent STC, where the receiver does not know the channel conditions but knows the statistics of the channel
  - Differential STC, where the receiver neither knows the channel conditions nor the statistics of the channel

- Spatial multiplexing—MIMO naturally supports spatial multiplexing to multiply communication capacity without requiring additional spectral bandwidth. In spatial multiplexing, multiple messages can be simultaneously transmitted over different spatial paths. Note that the channel conditions are often time-varying because of the continuous movement of the mobile user and/or change in the surrounding environment; thus both precoding and postprocessing are desired to dynamically optimize the communication performance.

In 4G, the maximum number of MIMO transmitting antennas (or receiving antennas) at each base station is limited to 8, resulting in up to 8×8 MIMO. In 5G, m-MIMO with many transmitting and receiving antennas, for example, 64×64 MIMO and 256×256 MIMO are used. The use of "massive" number of antenna elements helps form beams with tighter beam sizes, thus further increasing the SNIR and the signal range. In addition, m-MIMO beamforming is capable of the so-called three-dimensional beamforming, creating horizontal and vertical beams toward users at different geographic locations to further increase the coverage. Moreover, m-MIMO allows for more users to be simultaneously supported via multiuser MIMO (MU-MIMO) or multiuser beamforming, thus further enhancing the system capacity.

Fig. 2.4 illustrates the use of m-MIMO for multiuser beamforming where a 5G base station antenna unit is forming multiple RF beams to simultaneously communicate with multiple mobile UEs.

The implementation of an m-MIMO system requires not only more antenna elements at the base station but also more sophisticated algorithms, at the center of which is the precoding algorithm. For a downlink MU-MIMO system with N transmitting antennas at the base station and K single-antenna users, the signal received by the kth user can be expressed as:

$$y_k = \boldsymbol{h}_k^H \boldsymbol{x} + n_k \tag{2.2}$$

where $\boldsymbol{h}_k$ is an $N \times 1$ vector describing the channel response for the kth user, $\boldsymbol{x}$ is an $N \times 1$ vector describing the N signals emitted by the N transmitting antennas, n_k is

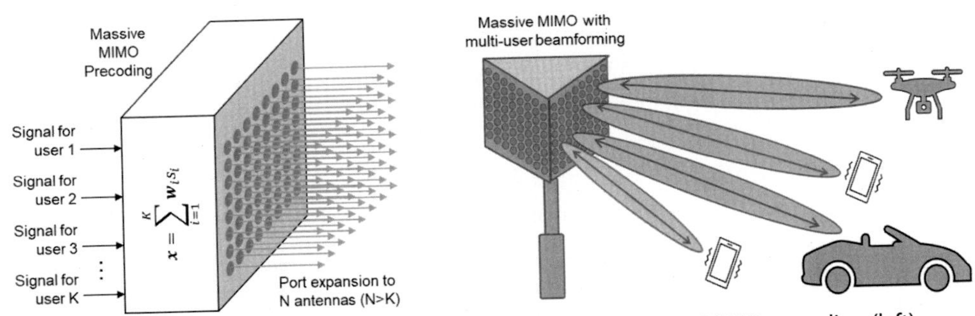

Figure 2.4 Illustration of multiuser beamforming (right) enabled by m-MIMO precoding (left).

the noise added to the signal received by the kth user. With precoding to precompensate the channel response, the signals to be emitted by N transmitting antennas become

$$\boldsymbol{x} = \sum_{i=1}^{K} \boldsymbol{w}_i s_i \quad (2.3)$$

where $\boldsymbol{w}_i$ is the $N \times 1$ precoding vector for the ith signal and s_i is the signal intended for the ith user. The precoding vectors for the K users can be obtained by using the uplink–downlink duality, which states that the precoding vector for the ith downlink signal is the same as the optimal receiver filter for the uplink signal coming from the ith user, up to a scaling factor. In effect, the base station MIMO receiver that receives the uplink signals from the K users can obtain the channel state information (CSI) and pass it to the MIMO transmitter for precoding of the downstream signals to the same K users. It is worth noting that the scalable 5G NR TDD slot structure nicely supports fast channel feedback to accurately obtain the CSI. In real-world implementations, much effort has been devoted to efficiently achieving accurate channel estimation under a diverse set of channel conditions.

As shown in Fig. 2.4, precoding may be performed at the radio equipment (RE) so that the data interface between the RE and the RE control (REC) only needs to transmit the K independent user data streams. This makes the data interface, often referred to as the front-haul interface, bandwidth-efficient. The front-haul interface is specified by in the evolved Common Public Radio Interface [21] and the next-generation front-haul interface [22]. More descriptions on the front-haul interface will be provided in the next chapter.

In 4G radio access network (RAN), each antenna is a passive unit and is driven by a RF signal sent from a remote radio unit (RRU) via an RF cable. For m-MIMO with ~100 antennas, it becomes difficult to have ~100 RF cables to connect all the antennas with an RRU, especially when high-frequency bands are used. To address this challenge, active antenna unit (AAU) has been introduced to 5G RAN. An AAU consists of m-MIMO antennas, the hardware and software required for the radio signals to be transmitted and received, as well as some needed DSP functions. The key hardware components include DSP chip, digital-to-analog converter, analog-to-digital converter, RF carrier frequency generators, RF mixers, RF power amplifiers, RF preamplifiers, RF filters, duplexers, and antennas. With the antenna–radio integration in AAU, the deployment of 5G antenna sites becomes simpler.

Fig. 2.5 illustrates the configuration of a 5G m-MIMO antenna site that consists of an AAU that is connected to a distributed unit (DU) by a lightweight optical fiber cable. For comparison, a typical 4G antenna site that consists of a passive MIMO antenna array, an RRU, and a baseband processing unit (BBU) is also shown, in both distributed radio access

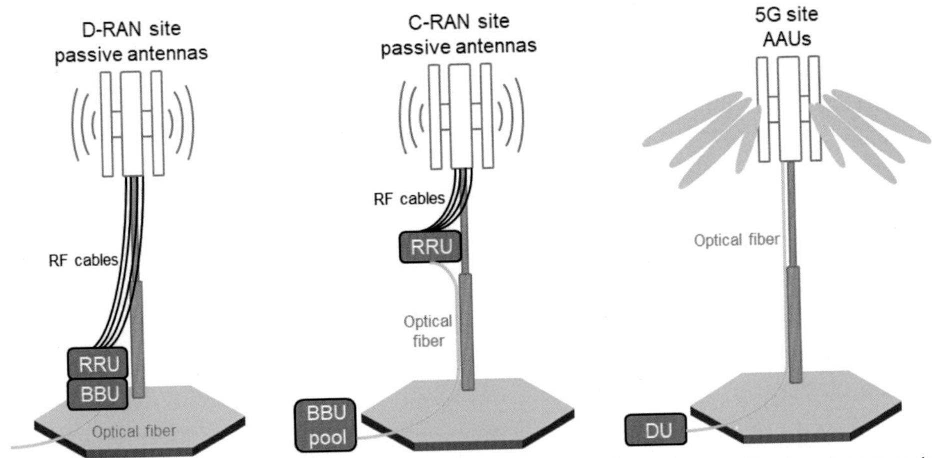

Figure 2.5 Illustration of typical 4G and 5G antenna site configurations with signal connections shown and power connections omitted. *AAU*, Active antenna unit; *C-RAN*, cloud radio access network; *D-RAN*, distributed radio access network.

network (D-RAN) and C-RAN configurations. The connection between the passive antenna array and the RRU is via multiple RF cables, each connecting to one antenna element. Evidently, the use of AAU drastically simplifies the construction of antenna sites.

In addition, the beamforming associated processing and waveform generation can be done inside the AAU, so the connection between the AAU and the DU only needs to provide the independent data streams to be transmitted and received. This facilitates the use of m-MIMO with many antenna elements (e.g., 256×256 MIMO). As a larger number of antenna elements leads to a finer spatial resolution of the radio beams formed, the performance of the multiuser beamforming increases with the number of antennas used. Thus the use of AAU enables better beamforming without adding extra load on the connection between the DU and the AAU, which makes the solution highly scalable.

In summary, m-MIMO is a key 5G technology that offers the following important benefits.

- Increased network capacity via MU-MIMO, where multiple users are simultaneously served with the same time and frequency resources
- Improved coverage via three-dimensional beamforming that improves the SNIR for users at any location inside each cell, even when the user is at a boundary between cells
- User experience via dynamic beamforming that tracks each user to maintain good connectivity, even when the user is moving quickly and/or the surrounding environment is changing.

m-MIMO is an outstanding example of an elegant scientific concept turning into a successful industry solution after decades of research and development to better serve our global community.

2.5 Cloud radio access network

In the wireless network evolution from 2G to 5G, there have been important architecture changes, as illustrated in Fig. 2.6. The key features of these generations of networks are summarized below.

In 2G, base transceiver station (BTS) and base station controller (BSC) are separated. The D-RAN architecture, in which the baseband processing is performed at the antenna site, is used. The supporting optical network is primarily based on synchronous optical networking and synchronous digital hierarchy. In 3G, BTS and BSC evolved to Node B (NB) and radio network controller, respectively, which remained to be separated. The supporting optical network is primarily based on multiservice transport platform.

In 4G, BSC disappeared because of the emergence of C-RAN [3,4]. C-RAN is a centralized, cloud computing-based architecture for radio access networks that supports 2G, 3G, 4G and future wireless communication standards. It supports centralized processing, collaborative radio, and simplified (clean) RRUs. C-RAN is widely regarded to bring not only network performance improvements but also substantial reduction in the total cost of ownership (TCO) of mobile networks. The supporting optical network is primarily internet protocol–based radio access network.

In 5G, the wireless architecture is based on the cloud architecture, with the addition of mobile cloud engine (MCE) and MEC in distributed data centers. With MCE and MEC, multiple RATs can be readily supported. Connection between multiple RATs can be realized for resource integration and superior user experience. Moreover, C-RAN is further evolved to include AAUs for simplified antenna sites,

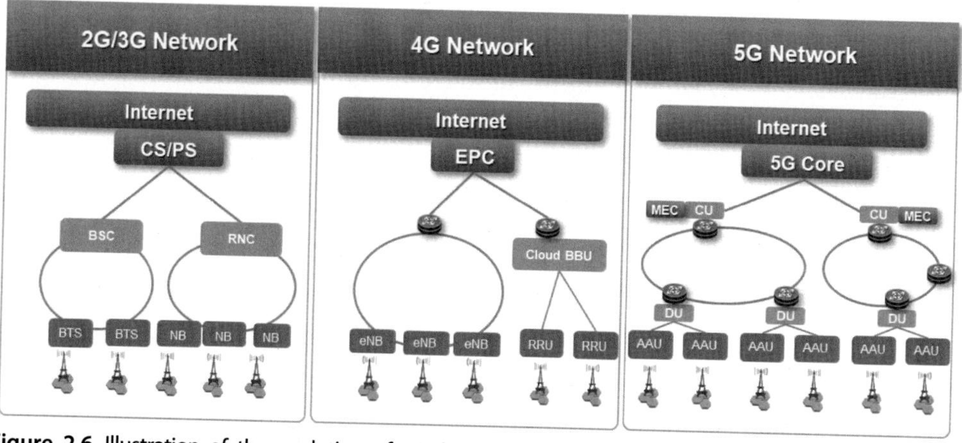

Figure 2.6 Illustration of the evolution of wireless network architectures from 2G/3G to 4G and 5G. *CS/PS*, Circuit switched core/packet switched core; *CU(DU)*, central unit (distributed unit) for 5G next-generation node B (gNB); *MEC*, mobile edge computing; *NB*, node B; *eNB*, evolved node B for 4G;.

DUs for distributed processing (especially for processing of low-latency real-time functions), and CUs for centralized processing. Thus elastic network architecture with both distributed and centralized controls can be formed, allowing for on-demand deployment of new network functions. The supporting optical network can be the aforementioned 5G-oriented optical network.

Owing to the widespread deployment of cloud data centers, C-RAN is playing an increasingly vital role in 5G. The key benefits of 5G C-RAN can be summarized as follows [23].

- Improved cell-edge performance by avoiding intercell interference via CoMP.
- Reduced processing latency by suitably partitioning the real-time functions and nonreal-time functions in DUs and CUs.
- Simplified antenna sites by moving most of the processing functions to nearby cloud data centers.
- Increased scalability and flexibility as upgrades in both cloud processing and AAU equipment can be more easily implemented.
- Reduced capital expenditure CAPEX and operational expenditure (OPEX) by greatly simplifying the antenna sites, reducing real estate and rental fees, and sharing the computing resources at the DUs and CUs.

Fig. 2.7 illustrates some of these key benefits of 5G C-RAN. To support C-RAN, optical transmission links are needed to connect AAUs with DUs, and DUs with CUs. The interfaces are based on Common Public Radio Interface and evolved Common Public Radio Interface, which will be described in the next chapter.

2.6 Progresses made in 5G deployment

The year 2020 has witnessed a rapid global deployment of 5G networks. Since most of the 5G specifications have been well standardized, the deployed 5G networks satisfactorily delivered the promised eMBB, uRLLC, and mMTC features, such as peak downlink speed of 10.3 Gb/s in the sub-6-GHz band [24], user experienced downlink speed of 100 Mb/s, capacity density of 10 Mb/s/m^2, end-to-end latency of as low as 1 ms, mobility of 500 km/hour, and massive connection density of 1 million/km^2.

The rapid deployment of 5G networks benefits from the collaboration between 4G and 5G networks. To facilitate a quick and seamless transition to 5G, 3GPP presented mobile network operators with an evolution path from 4G to nonstandalone (NSA) 5G and then to standalone (SA) 5G, as shown in Fig. 2.8.

In the NSA 5G configuration, the control signaling is supported by the existing 4G network for control functions, while the high-speed 5G user data transfer is supported by the 5G NR. This NSA option offers a seamless transition to 5G as both 4G and 5G users can be simultaneously supported. In addition, in areas where the 5G coverage is lacking, the user data transfer can be continuously supported by the

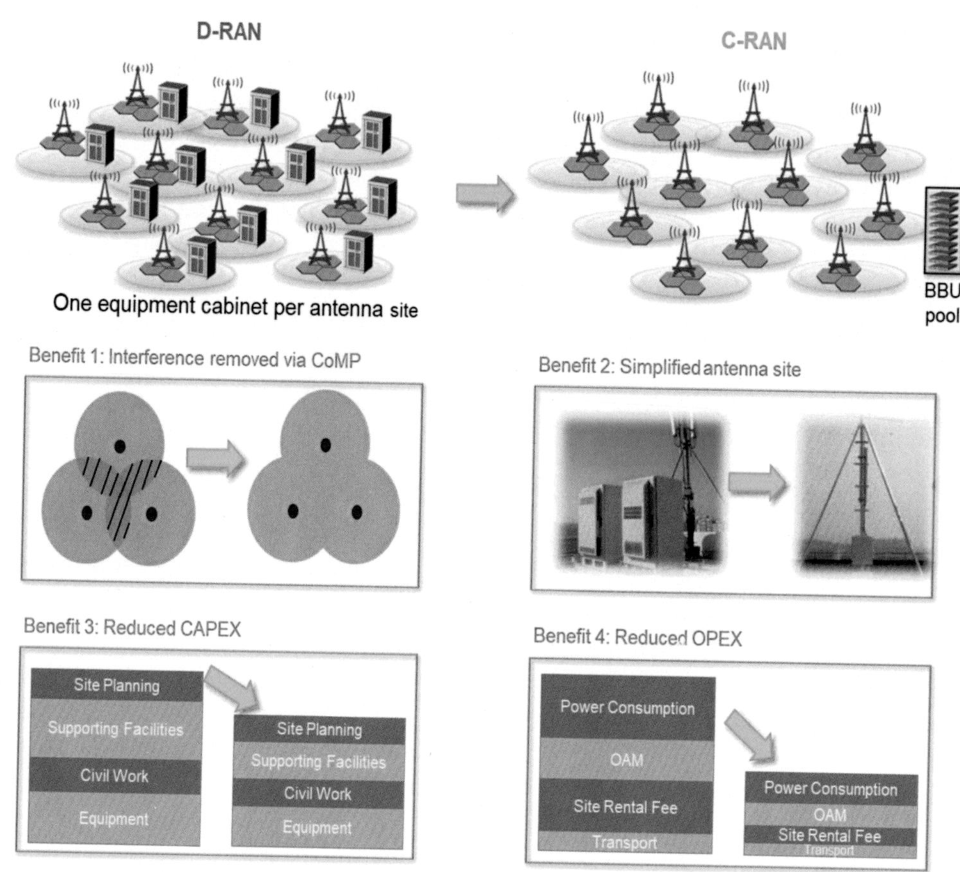

Figure 2.7 Illustration of the key benefits of C-RAN as compared to D-RAN.

4G network. In effect, the 5G network is complementing the existing 4G network to provide network performance enhancements wherever possible.

In the SA 5G configuration, both the control signaling and the high-speed 5G user data transfer are supported by a completed 5G network that also includes the new 5G packet core to handle the control functions. This SA 5G configuration does not rely on the 4G evolved packet core for the control signaling and can fully unleash the promised capabilities of 5G.

In addition to 4G and 5G networks, there are legacy 2G and 3G networks in many areas globally. Owing to limited network coverage of current 4G and 5G networks, voice services in some regions have to fall back to 2G/3G networks. In addition, many enhanced machine type communication services such as narrowband IoT require large coverage and low power consumption, which may be better supported by 2G/3G. All these considerations hinder the phaseout of legacy 2G/3G networks. However, an antenna site with multiple generations of RATs including 2G, 3G, 4G, and 5G are

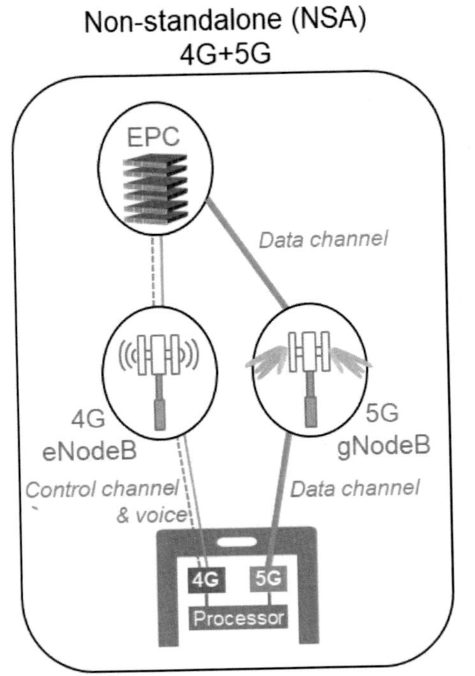

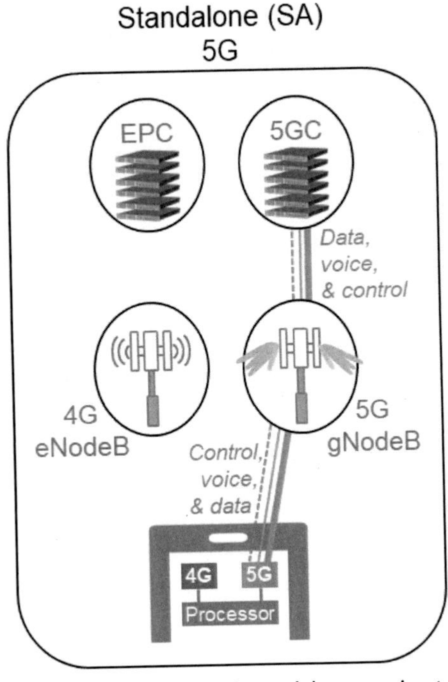

Figure 2.8 Illustration of an evolution path from nonstandalone service and standalone service to support fast and seamless transition to 5G. *EPC*, Evolved packet core; *5GC*, 5G core.

complex and difficult to maintain, resulting in high TCO. Thus how to cost-effectively manage a multi-RAT (2G/3G/4G/5G) site presents a major challenge.

To address the abovementioned challenge, the single-RAN approach [25] was introduced with the following three objectives:

- Site simplification, enabled by compact full-outdoor base stations to reduce site deployment difficulty and cost;
- RAT simplification, enabled by smooth RAT migration and consolidation;
- Network intelligence, enabled by adding intelligence into sites, networks, and the cloud.

Site simplification is important as the costs associated with site engineering, construction, and rental fees account for a large portion of the TCO. It is desirable to make each multi-RAT base station to be compact enough to be installed on a single pole, tower, wall, or rooftop. The start-of-the-art solution is the "1 + 1" antenna, in which all the passive antennas needed for 2G/3G/4G, typically in the sub-3-GHz bands, are aggregated into one passive antenna unit, while the 5G AAU is installed next to it, as shown in Fig. 2.9. As also illustrated in the same figure, the RRUs and BBUs used for 2G/3G/4G can be made with full-outdoor modules that can be installed with the passive and active antenna units on a single pole. This eliminates the

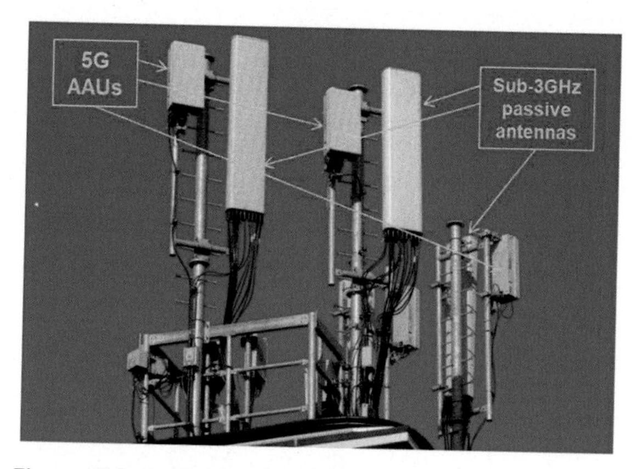

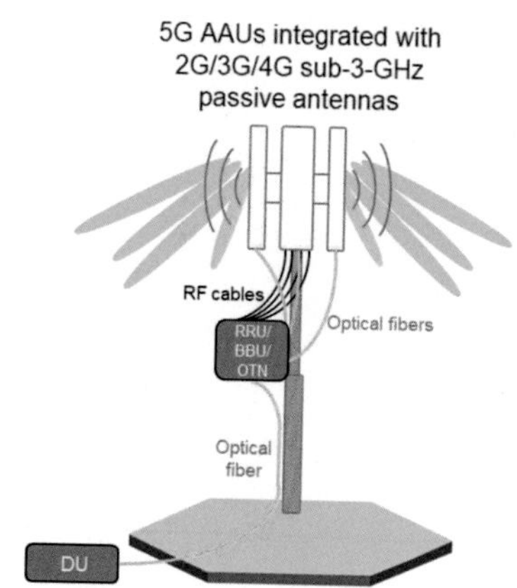

Figure 2.9 A picture showing the coexistence of 5G AAUs and sub-3-GHz passive antennas for 2G/3G/4G at a cell site (left) and an illustration of a simplified multi-RAT base station supporting 2G/3G/4G/5G with compact outdoor modules (right). *OTN*, optical transport network module for aggregating all the mobile signals in one high-speed optical wavelength signal.

need for an equipment room or cabinet, which substantially reduces site rental and maintenance costs. Moreover, all the wireless signals can be aggregated by an optical module, for example, an optical transport network module, to form a high-speed optical wavelength signal (e.g., at 100 Gb/s) for communication with the DC via a single optical fiber. This reduces the needed fiber resource and the associated transport cost.

RAT simplification aims to enable a smooth RAT migration and consolidation by continued 4G LTE during 5G deployment. 4G LTE is evolved into a full-service fundamental network that is capable of fully replacing the coverage provided by 2G and 3G networks. As an increasing number of 2G/3G users are upgraded to 4G/5G, the legacy 2G and 3G networks can be gradually retired. DSS is a key technology that enables the smooth RAT migration and consolidation [26]. With triple-mode DSS, three network modes, for example, 2G/4G/5G, 3G/4G/5G, or NB/4G/5G, can dynamically share the spectral resources. In effect, DSS frees up valuable low-frequency spectral bands for 4G LTE so that universal 4G coverage can be provided at a low cost.

The evolution from 4G to 5G is also readily supported by DSS. China Mobile and Huawei demonstrated DSS between 4G and 5G in a live mobile network in late 2019 [26]. This field trial used 2.6 GHz 4G/5G integrated base stations and 5G-ready consumer equipment. Fig. 2.10 shows the DSS configuration in which up to 40 MHz of spectrum can be dynamically shared between 4G and 5G. Depending on the traffic loads of the 4G and 5G networks, the shared spectrum can be allocated to support the network that demands more capacity. Using such a load-based spectrum resource allocation, the performances of both 4G and 5G networks can be jointly optimized to provide the best user experience.

Network intelligence is becoming more and more important as mobile networks become increasingly sophisticated. To realize the full potential of network intelligence,

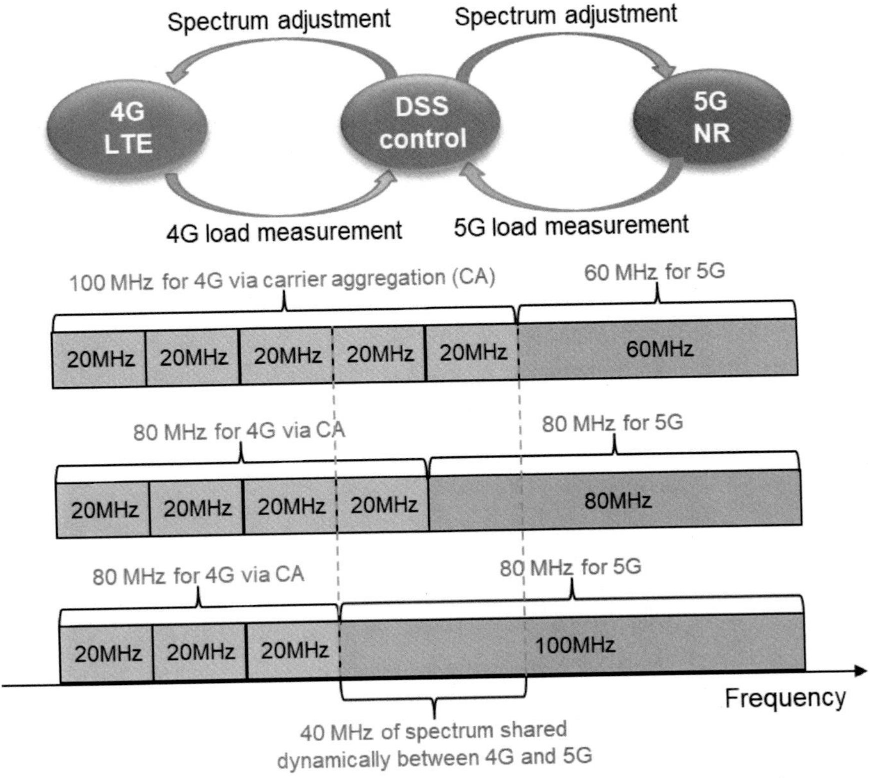

Figure 2.10 Illustration of load-based dynamic spectrum sharing (DSS) between 4G long-term evolution (LTE) and 5G new radio (NR) to optimize the overall network performance [26].

intelligence needs to be introduced into base station sites, RAN and core networks, and the cloud. At base station sites, real-time analysis and fast decision making can be made for tasks that require extremely low latency. In mobile networks, intelligent operations and maintenance are introduced to provide capabilities such as automation, prediction, and analytics. In the cloud, artificial intelligence (AI) and machine learning can be introduced to continuously improve the overall performance of the mobile networks. For example, energy savings can be obtained by dynamically optimizing radio resource utilization and network efficiency, thereby lowering the OPEX. The newly established Open RAN Alliance also highlights network intelligence as a key objective for next-generation wireless networks [27].

Substantial advances have been made toward the abovementioned three objectives, site simplification, RAT simplification, and network intelligence [28], as highlighted below:

- All-in-one blade AAU with active and passive integration of all sub-6-GHz frequency bands from 700 MHz to 4.9 GHz to support all mobile network generations from 2G through 5G, enabling site simplification and RAT simplification

- Lightweight 5G AAU that weighs only 25 kg and supports 64×64 m-MIMO with 400-MHz bandwidth, enabling site simplification and better support of fragmented spectrum bands
- AI-enabled multistandard, multifrequency-band coordination to save energy consumption by up to 20%, showing the value of network intelligence

Significant progress has been made in the early deployment of 5G, especially via technological innovations to realize site simplification, RAT simplification, and network intelligence such that the deployment of 5G has become economically viable. Going forward, continued technological innovations to further reduce the CAPEX and OPEX associated with the 5G deployment are necessary for a widespread adoption of 5G. Global cooperation and collaboration to foster such innovations are deemed imperative.

References

[1] ITU-R M.2083, IMT vision—framework and overall objectives of the future development of imt for 2020 and beyond; September 2015.
[2] ITU report on Minimum requirements related to technical performance for IMT-2020 radio interface; November 2017.
[3] China Mobile Research Institute, C-RAN: the road towards green RAN, White paper v. 2.6; September 2013.
[4] I C, Huang J, Duan R, Cui C, Jiang J, Li L. Recent progress on C-RAN centralization and cloudification. IEEE Access 2014;2:1030–9.
[5] ETSI GS MEC 010-1, Mobile edge computing (MEC); mobile edge management; part 1: system, host and platform management; October 2017.
[6] A Reznik, LMC Murillo, Y Fang, W Featherstone, M Filippou, F Fontes, et al. Cloud RAN and MEC: a perfect pairing, ETSI white paper, no. 22; February 2018, p. 1–24.
[7] The Next Generation Mobile Networks (NGMN) Alliance, 5G white paper; February 2015.
[8] Foukas X, et al. Network slicing in 5G: survey and challenges. IEEE Commun Mag 2017;55(5):94–100.
[9] Popovski P, Trillingsgaard KF, Simeone O, Durisi G. 5G Wireless network slicing for eMBB, URLLC, and mMTC: a communication-theoretic view. IEEE access 2018;6:55765–79.
[10] Lei W, Soong ACK, Liu J, Wu Y, Classon B, Xiao W, et al. 5G System design – an end to end perspective. Springer International Publishingr; 2020.
[11] Al-Dulaimi A, Wang X, Chih-Lin I. 5G Networks: fundamental requirements enabling technologies and operations management. Hoboken, NJ: Wiley; 2018.
[12] Global System for Mobile Communications Association (GSMA) position paper on 5G Spectrum; March 2020.
[13] Shannon CE. A mathematical theory of communication. Bell Syst Tech J 1948;27(3):379–423.
[14] Bi Q. On 5G evolution – challenges and opportunities, Invited talk at the 1st OSA 5G Summit, Oct. 16–17, Washington DC; 2019.
[15] 3GPP specification series: 38 series.
[16] Gallager RG. Low-density parity-check codes. MA, Cambridge: MIT Press; 1963.
[17] Arıkan E. Channel polarization: a method for constructing capacity-achieving codes for symmetric binary-input memoryless channels. IEEE Trans Inf Theory 2009;55(7):3051–73.
[18] 3GPP RAN1 meeting #87 final report.
[19] Foschini GJ, Gans MJ. On limits of wireless communications in a fading environment when using multiple antennas. Wirel Pers Commun 1998;6:311–35.

[20] Larsson EG, Edfors O, Tufvesson F, Marzetta TL. Massive MIMO for next generation wireless systems. IEEE Commun Mag 2014;52(2):186–95.
[21] The CPRI Specification eCPRI 2.0. Common public radio interface: eCPRI interface specification; May 10, 2019.
[22] I C, Li H, Korhonen J, Huang J, Han L. RAN revolution with NGFI (xhaul) for 5G. J Lightwave Technol 2018;36(2):541–50.
[23] Liu X. Optical communication technologies for 5G wireless. In: Optical fiber communication conference (OFC), Short course 444; 2019.
[24] Bi Q. Ten trends in the cellular industry and an outlook on 6G. IEEE Commun Mag 2019;57(12):31–6.
[25] See for example, <https://www.zdnet.com/article/huawei-launches-5g-singleran-pro/>.
[26] See for example, <https://www.lightreading.com/partner-perspectives-(sponsored-content)/first-cloudair-lte-tdd-and-nr-dynamic-spectrum-sharing-commercial-verified-by-china-mobile-and-huawei/a/d-id/756453>.
[27] O-RAN Alliance, <https://www.o-ran.org/>.
[28] See for example, <https://www.rcrwireless.com/20200220/wireless/huawei-shows-off-new-5g-gear-restates-tech-leadership>.

CHAPTER 3

Optical interfaces for 5G radio access network

3.1 Overview of mobile front-, mid-, and back-haul

Optical networks form the underlying infrastructure that transports mobile data globally. Wireless base stations in radio access network (RAN) communicate with mobile core networks via the so-called mobile X-haul networks, which include mobile front-, mid-, and back-haul networks [1–11]. In 4G wireless networks, distributed RAN (D-RAN) and Cloud RAN (C-RAN) are two main RAN architectures. In comparison to the former, the latter offers the several advantages such as improved network performance via coordinated multipoint (CoMP), simplified antenna sites, and reduced capital expenditures and operating expenditures (OPEX), as described in Chapter 2. In 4G D-RAN, each base station consists of a passive antenna, a remote radio unit (RRU), and a colocated baseband unit (BBU), and communicates with a 4G evolved packet core (EPC) via a back-haul link, as shown in Fig. 3.1A. In 4G C-RAN, the BBUs from multiple base stations are centralized as a BBU pool in a central office (CO) with optical front-haul links connecting the RRUs with the BBU pool, as shown in Fig. 3.1B.

Fig. 3.2 shows the typical 4G D-RAN and C-RAN configurations. For the uplink direction, the wireless signal from a user equipment (UE) is first received by an antenna and then converted to the digital domain by an RRU, which is connected to a BBU that sequentially performs physical layer (PHY) processing, media access control (MAC) layer processing, radio link control (RLC), packet data convergence protocol (PDCP) processing, and radio resource control (RRC), before sending the mobile payload data to the EPC via a back-haul link. In the downlink direction, the EPC sends mobile payload data to the BBU via the back-haul link, and the BBU then performs the above baseband processing functions in the reserved order. The connection interface between the BBU and the RRU is typically the Common Public Radio Interface (CPRI) [5], which supports both electrical and optical transmissions [12–16]. More details on the CPRI will be provided in the following section. In the D-RAN case, the BBU and RRU of a base station are colocated and electrical CPRI connection may be used. In the C-RAN case, on the other hand, the BBU and RRU are often separated by up to 20 km, so an optical front-haul link is commonly used for the CPRI connection.

Optical Communications in the 5G Era
DOI: https://doi.org/10.1016/B978-0-12-821627-9.00011-5

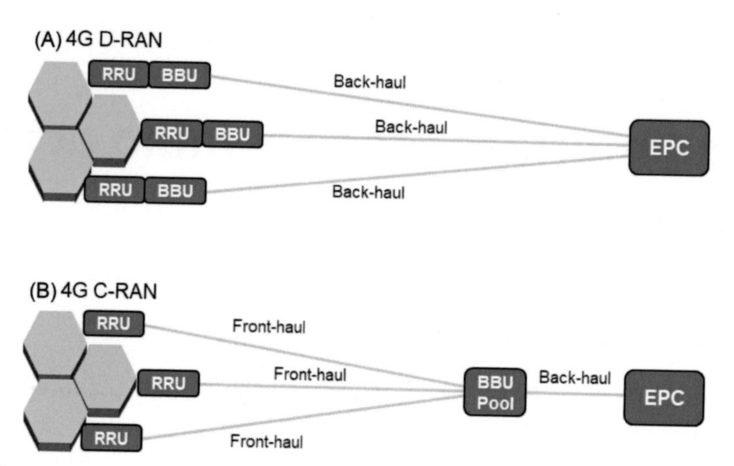

Figure 3.1 Illustrations of 4G D-RAN with only back-haul optical links (A) and C-RAN with both back-haul and front-haul optical links (B).

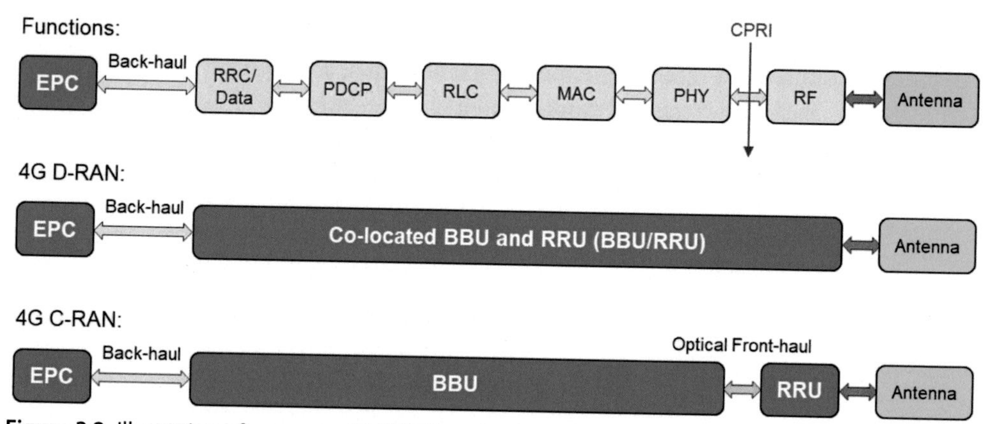

Figure 3.2 Illustration of common 4G D-RAN and C-RAN configurations.

In 5G wireless networks, RAN has evolved to better support enhanced mobile broadband, ultra-reliable and low-latency communications, and massive machine machine-type communications by a three-level architecture consisting of centralized units (CUs), distributed units (DUs), and radio units (RUs) that are typically integrated with antennas in active antenna units (AAUs), and X-haul network segments [6–11], which include:

- The front-haul network segment connecting RUs and DUs
- The mid-haul network segment connecting DUs and CUs
- The back-haul network segment connecting CUs and the core networks

The partition among the three network segments can be based on different functional split options. The 3rd Generation Partnership Project (3GPP) has defined eight functional slit options from Option 1 to Option 8, respectively between two adjacent functions in a nine-function sequence consisting of RRC, PDCP, high-level RLC, low-level RLC, high-level MAC, low-level MAC, high-level PHY, low-level PHY, and radio frequency (RF) functions [6–11]. Fig. 3.3 shows four common 5G RAN configurations enumerated as follows:

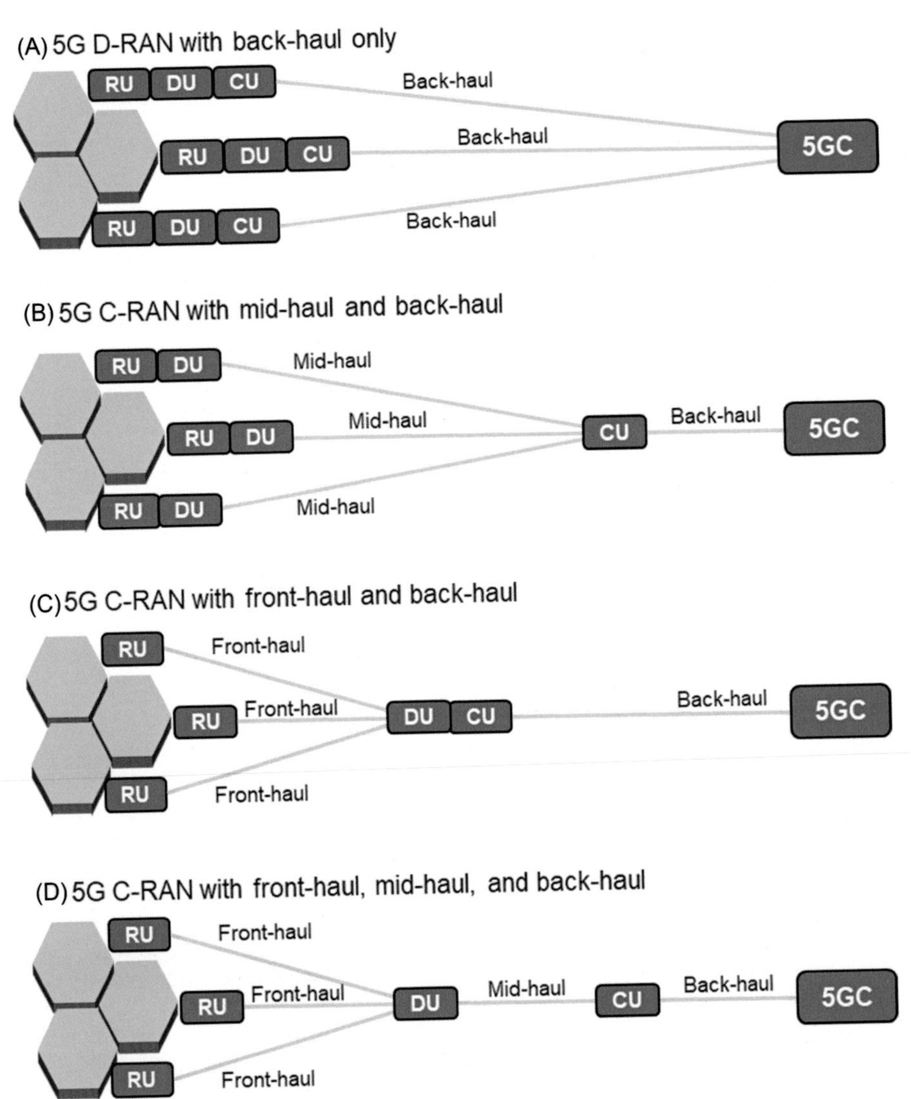

Figure 3.3 Illustrations of four common 5G RAN configurations: (A) D-RAN with back-haul network segment only, (B) C-RAN with both mid-haul and back-haul network segments, (C) C-RAN with both front-haul and back-haul network segments, and (D) C-RAN with front-haul, mid-haul, and back-haul.

- 5G D-RAN with back-haul network segment only
- 5G C-RAN with mid- and back-haul network segments
- 5G C-RAN with front- and back-haul network segments
- 5G C-RAN with front-, mid-, and back-haul network segments

Fig. 3.4 shows the common functional slit partitions corresponding to the above four 5G RAN configurations. Each configuration has concomitant benefits and drawbacks. Table 3.1 compares the same for these four configurations.

As shown in Figs. 3.3 and 3.4, as well as Table 3.1, the three-level RAN architecture is well supported by configurations (c) and (d), in which optical front-haul enables dramatically simplified antenna site with compact AAUs. In addition, full performance benefit of CoMP can be obtained. Moreover, real-time functions can be processed at nearby DUs with low latency. In configuration (c), the DU and CU of a given 5G next-generation node B (gNB) are colocated in one CO, which simplifies network connections by not having the mid-haul segment. In addition, the already-deployed COs used for 4G baseband processing may be rearchitected to efficiently support 5G DU and CU processing as well. In configuration (d), the DU and CU of a given gNB are physically separated in two locations, which allows for more flexible distribution of DUs and CUs. This configuration may facilitate DUs to be placed closer to AAUs for ultralow-latency applications.

Fig. 3.5 illustrates a common 5G C-RAN architecture with front-, mid-, and back-haul network segments. Table 3.2 shows the typical features of these three

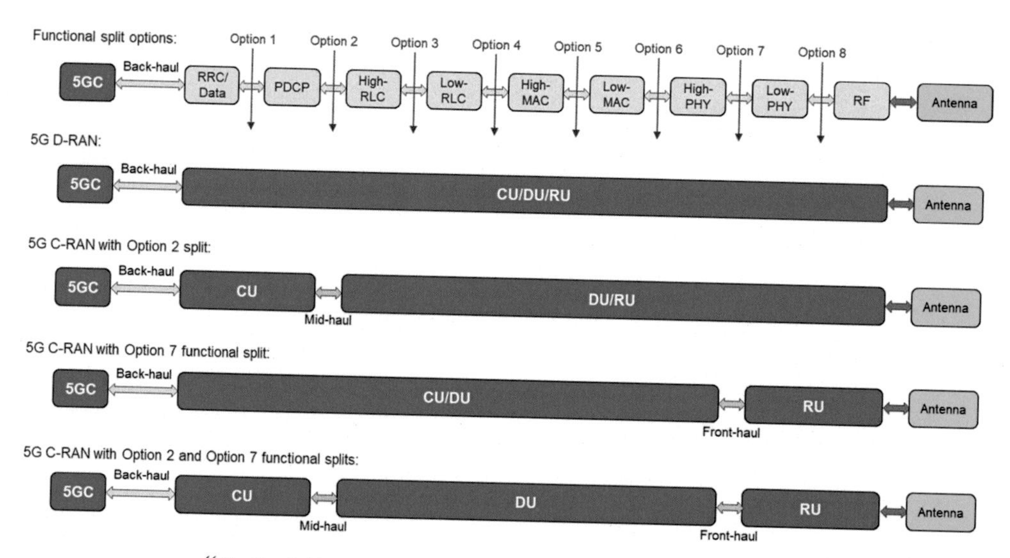

Figure 3.4 Common functional slit partitions for the four 5G RAN configurations illustrated in Fig. 3.3.

Table 3.1 Comparison among four common 5G RAN configurations.

Configuration	Benefits	Drawbacks/limitations
(a) 5G D-RAN with back-haul only • RU/DU/CU together at the antenna site	• Low optical transport data rate requirements for the back-haul links	• High complexity at antenna site • No support for CoMP
(b) 5G C-RAN with mid- and back-haul • RU/DU at antenna site • CU in a CO	• Low optical transport data rate requirements for the mid-haul links	• Medium complexity at antenna site • Limited CoMP gain
(c) 5G C-RAN with front- and back-haul • RU at the antenna site • DU and CU colocated in a single CO	• Low complexity at the antenna site, enabling compact AAUs • Full support for CoMP • Simplified connections by not having mid-haul	• Medium optical transport data rate requirements for the front-haul links
(d) 5G C-RAN with front-, mid-, and back-haul • RU at the antenna site • DU and CU separated in two locations	• Low complexity at the antenna site enabling compact AAUs • Full support for CoMP • More flexible distribution of DUs and CUs	• Medium optical transport data rate requirements for the front-haul links

CO, Central office; *CoMP*, coordinated multipoint; *C-RAN*, cloud RAN; *CU*, centralized unit; *D-RAN*, distributed RAN; *DU*, distributed unit; *RAN*, radio access network; *RU*, radio unit.

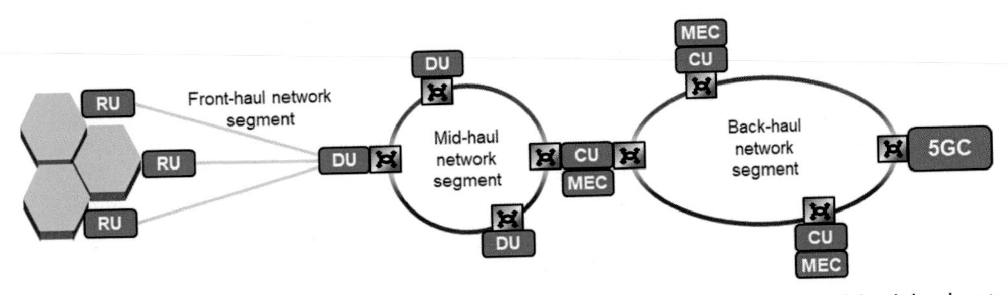

Figure 3.5 Illustration of a common 5G C-RAN architecture with front-, mid-, and back-haul network segments.

network segments. Evidently, the three-level RAN architecture effectively supports (1) highly synchronized low-latency real-time processing in the front-haul network segment, (2) traffic aggregation and mobile edge computing (MEC) in the mid-haul

Table 3.2 Typical features of the front-, mid-, and back-haul network segments of 5G RAN.

	Front-haul network segment	Mid-haul network segment	Back-haul network segment
One-way latency	<100 μs	<0.4 ms	<10 ms
Synchronization	±65 ns to ±260 ns (relative)	±1.5 μs (absolute)	±5 μs (absolute)
One-way distance	<20 km	<80 km	<200 km
Node–node connection	Point-to-point	Point-to-multipoint	Point-to-multipoint

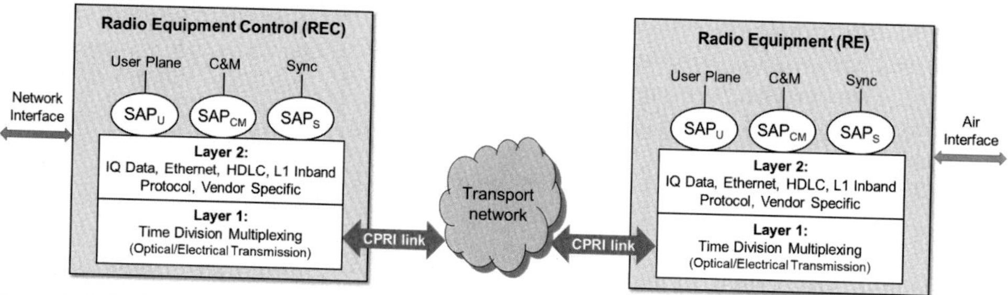

Figure 3.6 Basic system architecture of REC and RE that are connected via CPRI.

network segment, and (3) bandwidth-efficient transport in the back-haul network segment. In an exemplary metropolitan (metro) deployment scenario [8], there may be four 5G Core nodes, each supporting five CU nodes. Each CU node may then support 20 DUs, each connecting to about 30 RUs. Thus this illustrative metro network can support around 12,000 AAUs. More details on the optical connections and networks for front-, mid-, and back-haul network segments will be presented in Chapter 4.

3.2 Common public radio interface

To provide deployment flexibility for mobile networks, radio base station is separated into two basic building blocks, the radio equipment control (REC) and the radio equipment (RE). In D-RAN, the RE and REC are colocated at the antenna site. In C-RAN, the RE is located at the antenna site, while the REC is located in a centralized site. CPRI is a widely used digital interface for transporting antenna samples between an REC and an RE. Fig. 3.6 illustrates the basic system architecture of REC and RE that are connected via CPRI [5].

The REC communicates with the core network via a network interface and the RE via CPRI. Inside REC, there are three service access points, respectively for the user plane, control and management (C&M), and synchronization (Sync). The user plane contains the in-phase and quadrature (IQ) data of the antenna signals. The IQ data of different antenna carriers are multiplexed via time division multiplexing (TDM). The C&M data are either sent by in-band protocol for time critical signaling or by layer 3 protocols that are not defined by CPRI. High-level data link control and Ethernet are also supported by CPRI. In the downlink direction, these additional C&M data, as well as Sync signals and any vendor specific information, are multiplexed with the IQ data via TDM to form a CPRI signal. The CPRI signal is then sent to the RE over an electrical transmission line or an optical transmission line. At the RE, all the information carried by the CPRI signal is retrieved such that the needed analog RF functions can be performed accordingly. The RE then communicates with each UE through the antenna and the air interface. In the uplink direction, the above processes are reversed.

Fig. 3.7 shows the basic frame structure of CPRI. Each basis frame contains digitized I/Q bits of multiple antennas and carriers, as well the control words (CWs) that carry the C&M and Sync information. The ratio between the bits used for CWs and the I/Q bits is 1:15. Each CPRI basic frame lasts about 260 ns (1/3.84 MHz) and contains $(Y+1)$ containers each having 128 bits. The data rate of CPRI frame with $(Y+1)$ containers is $(Y+1)\cdot 491.52$ Mb/s.

For high-speed transmission, binary on-off-keying (OOK) is typically used. 8b/10b or 64b/66b line coding is often used to facilitate OOK clock recovery and error detection. Thus the required CPRI line bit rate can be expressed by Eq. (3.1):

$$R_{\mathrm{CPRI}} = A\cdot f_s\cdot b_s\cdot 2\cdot (16/15)\cdot \varepsilon_{\mathrm{line}} \tag{3.1}$$

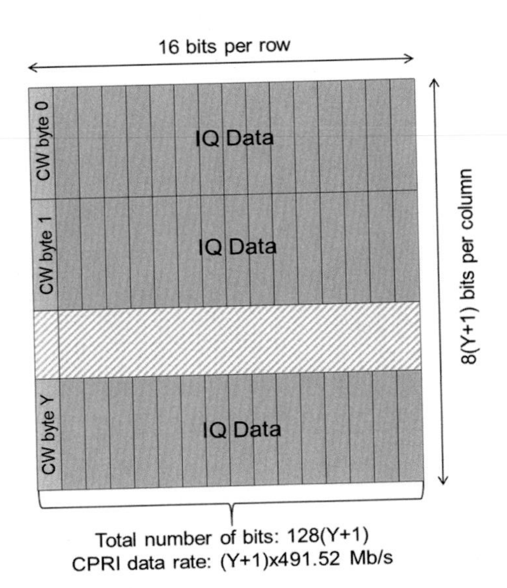

Figure 3.7 CPRI basic frame structure.

where A is the number of antennas, f_s is the sampling frequency (e.g., 30.72 MHz for 20-MHz RF bandwidth), b_s is the quantized bits per sample per quadrature component (e.g., 15 for Long-Term Evolution [LTE]), the factor 2 accounts for both I and Q quadrature components, the factor 16/15 accounts for the overhead of CW, and $\varepsilon_{\text{line}}$ is the rate expansion factor due to the line coding used, for example, 10/8 and 66/64 for 8b/10b and 64b/66b line coding, respectively. As an example, the CPRI optical line rate for 8 × 8 MIMO with 20-MHz LTE signal that is sampled at 30.72 MHz with 15 bits per quadrature component is $8 \cdot 30.72(\text{MHz}) \cdot 15(\text{bits}) \cdot 2 \cdot 16/15 \cdot 10/8$ or 9.8304 Gb/s.

Fig. 3.8 shows the CPRI overall frame structure, which consists of basic frames, hyperframes, and node B frames with the following hierarchy.

- Basic Frame (260.416 ns)

  The duration of each basic frame is 1x UMTS (Universal Mobile Telecommunications System) chip, or 260.416 ns (=1/3.84 MHz)
- Hyperframe (66.66 μs)

  Each hyperframe contains 256 basic frames, indexed from 0 to 255. The frame header is K28.5 (an 8b/10b keyword) for frame synchronization. Hyperframe readily supports TDMA with help from control words
- Node B Frame (10 ms)

  Each 10-ms node B frame contains 150 hyperframes, indexed from 0 to 149.
- 40.96-s UMTS frame cycle period

Each 40.96-s UMTS frame cycle period contains 4096 node B frames, indexed from 0 to 4095.

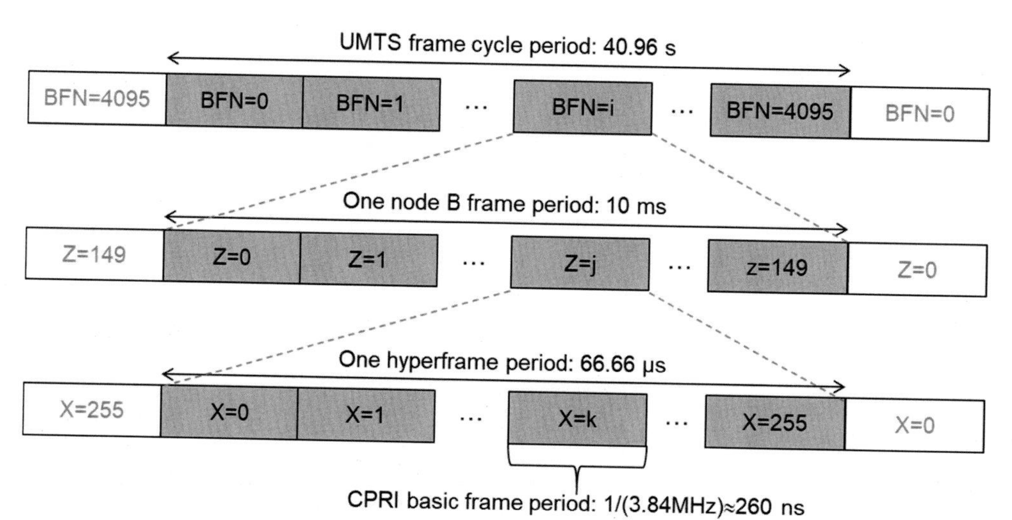

Figure 3.8 CPRI overall frame structure. *BFN*, Node B frame number.

Table 3.3 Typical line bit rate options defined by CPRI.

CPRI option	CPRI line bit rate[a] (Gb/s)	No. of CPRI containers	Typical application scenario	Payload data rate[b] (Gb/s)
1	0.6144	1	10-MHz channel	0.1
2	1.229	2	20-MHz channel	0.2
3	2.458	4	20-MHz, 2 × 2 MIMO	0.2
4	3.072	5	25-MHz, 2 × 2 MIMO	0.25
5	4.915	8	20-MHz, 4 × 4 MIMO	0.2
6	6.144	10	5 × 10-MHz, 2 × 2 MIMO	0.5
7	9.830	16	20-MHz, 8 × 8 MIMO	0.2
7A	8.110	16	20-MHz, 8 × 8 MIMO	0.2
8	10.14	20	5 × 20-MHz, 2 × 2 MIMO	1
9	12.16	24	3 × 20-MHz, 4 × 4 MIMO	0.6
10	24.33	48	3 × 20-MHz, 8 × 8 MIMO	0.6

CPRI, Common public radio interface; *MIMO*, multiple-input multiple-output.
[a]Options 1–7 use 8b/10b line coding, while Options 7A–10 use 64b/66b line coding.
[b]For simplicity, the spectral efficiency is assumed to be 10 b/s/Hz and no multiuser beamforming is assumed.

Table 3.3 shows the typical line bit rate options defined by CPRI. From Option 1 to Option 7, 8b/10b line coding is used. For a commonly used Option 7, the CPRI line bit rate needed for aggregating eight 20-MHz LTE signals (with 30.72-MHz sampling rate) in the 8 × 8 MIMO case is as high as 9.8304 Gb/s, although the wireless payload data rate is only 0.2 Gb/s even assuming that the spectral efficiency (SE) is 10 b/s/Hz. This indicates that CPRI is not bandwidth-efficient. From Option 7A to Option 10, 64b/66b line coding is adopted. For Option 10, three sectors of 8 × 8 MIMO with 20-MHz RF bandwidth can be supported at a CPRI line bit rate of 24.33 Gb/s. The wireless payload data rate is only 0.6 Gb/s assuming that the SE is 10 b/s/Hz. This again indicates that CPRI is not bandwidth-efficient. One key reason for the bandwidth inefficiency is that the CPRI line bit rate is proportional to the number of antennas rather than the actual payload data rate.

This bandwidth inefficiency becomes more pronounced in 5G use cases where massive MIMO (m-MIMO) is used. Table 3.4 shows the CPRI line bit rates for some typical 5G use cases based on Eq. (3.1) with the assumption of the same oversampling ratio as that in LTE, 15 for b_s, and the use of 64b/66b line coding. For a typical 5G use case with 64 × 64 m-MIMO and 200-MHz RF bandwidth, the needed CPRI data rate would become as high as 648.8 Gb/s, which is too high to be supported cost-effectively.

One way to increase the bandwidth efficiency of CPRI is to perform CPRI compression [17]. It was found that a compression ratio of 3:1 can be realized with small degradation of the overall CPRI performance. Even with 3:1 CPRI compression, the required interface data rates for most 5G use cases are still too high to be realized cost-effectively. This leads the industry to consider new functional split options, such as

Table 3.4 CPRI line bit rate requirements for some typical 5G use cases.

m-MIMO type	Radio channel bandwidth			
	40 MHz	100 MHz	200 MHz	1 GHz
32 × 32	64.88 Gb/s	162.2 Gb/s	324.4 Gb/s	1.622 Tb/s
64 × 64	129.8 Gb/s	324.4 Gb/s	648.8 Gb/s	3.244 Tb/s
128 × 128	259.5 Gb/s	648.8 Gb/s	1.298 Tb/s	6.488 Tb/s
256 × 256	519.1 Gb/s	1.298 Tb/s	2.595 Tb/s	12.98 Tb/s

CPRI, Common public radio interface; *m-MIMO*, massive multiple-input multiple-output.

evolved CPRI (eCPRI) [7] and next-generation front-haul interface [11], to substantially reduce the interface data rates, albeit at the expense of somewhat increased complexity at the RUs. We will describe eCPRI and compare its bandwidth efficiency with CPRI in the following section.

3.3 Evolved common public radio interface

To address the aforementioned bandwidth inefficiency issue of CPRI, the CPRI industry association, consisting of four member companies, Ericsson, Huawei, NEC, and Nokia, released the first version of eCPRI specification on August 31, 2017 [7]. The new specification is designed to support the 5G front-haul use cases and provide enhancements to meet the increased requirements of 5G. The eCPRI specification offers several advantages to the base station design enumerated as follows [7].

1. The new interface may enable about 10 times reduction of the required bandwidth.
2. Required bandwidth can scale flexibly according to the user plane traffic.
3. Use of packet-based transport technologies will be enabled. Mainstream technologies, such as Ethernet, open the possibility to carry eCPRI traffic and other traffic simultaneously, in the same switched network. In addition, the use of well-established Ethernet and IP protocols is possible for operation, administration, maintenance, provisioning, and troubleshooting of the network.
4. The new interface is a real-time traffic interface that enables the use of sophisticated coordination algorithms to guarantee best possible radio performance.
5. The interface is future-proof, allowing new feature introductions via software updates in the REC without changing the RE.

eCPRI enables flexible functional decomposition while limiting the complexity of the RE, as shown in Fig. 3.9. Split points located inside the PHY layer is one set of examples covered in the eCPRI specification. The traditional CPRI is at split position *E*, which is between the RF equipment and the PHY processing layer and requires the highest interface line bit rate. On the other hand, the gain from CoMP is

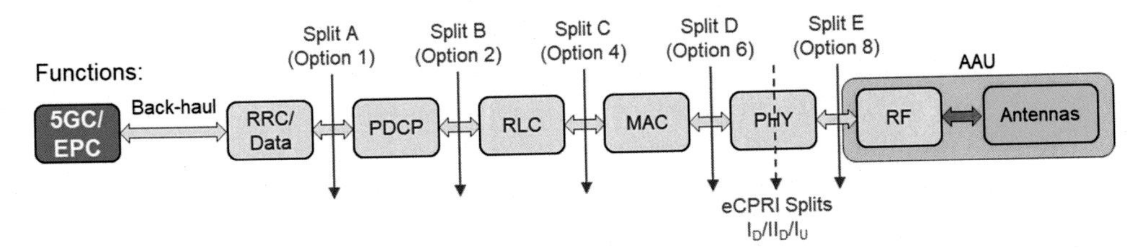

Figure 3.9 eCPRI functional split options.

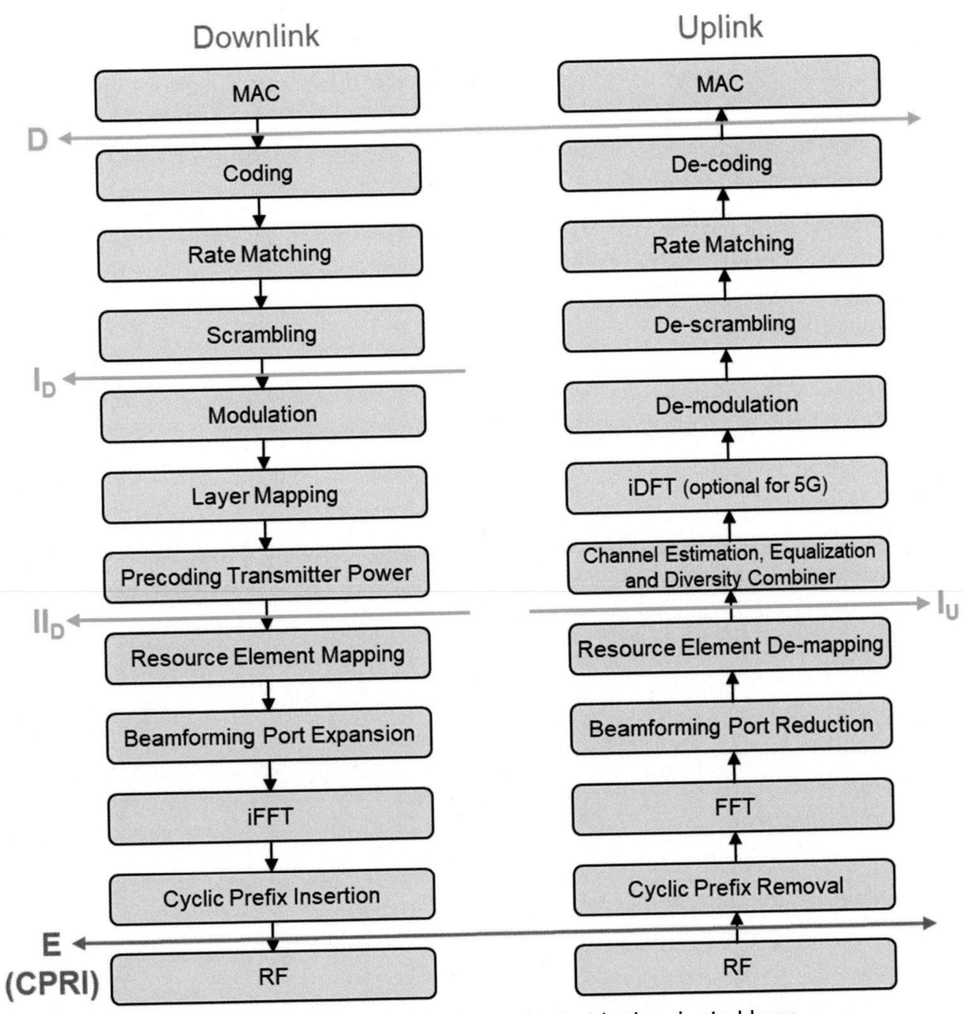

Figure 3.10 eCPRI functional split positions I_D, II_D, and I_U inside the physical layer.

diminished when the split position is close to the MAC. It is thus important to find appropriate functional split positions inside the PHY layer to reduce interface line bit rate while keeping the support of real-time coordination functions such as CoMP.

Fig. 3.10 shows the eCPRI functional split positions inside the physical layer. For the downlink direction, there are two split positions, labeled as I_D and II_D. For the uplink direction, there is one split position, labeled as I_U. At these slit positions, the interface bandwidth efficiency is improved without sacrificing the support of sophisticated coordination functions. The key reason for the improved interface bandwidth efficiency is that the eCPRI line bit rate is traffic-dependent, rather than traffic-independent as in the case of CPRI. Resource element mapping (for the downlink direction) and resource element de-mapping (for the uplink direction) are performed at the RE, so that only the actual traffic is communicated between the RE and the REC. In addition, antenna port expansion (for the downlink direction) and antenna port reduction (for the uplink direction) are performed at the RU, thus making eCPRI bit rate independent of the scale of the antenna to support m-MIMO with higher bandwidth efficiency.

When going to a split that is above the split *E*, there are many factors that will have an impact on the interface data rate, such as:

- Throughput (closely related to the wireless signal bandwidth)
- Number of antennas
- Number of MIMO layers
- Multiuser MIMO support
- Selection of beamforming algorithm
- Modulation scheme
- Code rate

In the specification of eCPRI 1.0 [7], the data rate needed by eCPRI is compared with that needed by CPRI for a typical 5G use case. The following values are used for an illustrative interface data rate calculation:

- Throughput*: 3 Gb/s for downlink and 1.5 Gb/s for uplink
  (*: the throughput here is the end-user data rate at the interface for split *D*)
- Wireless signal bandwidth: 100 MHz
- Sampling frequency: 153.6 MSa/s (assuming the same oversampling ratio as in LTE)
- b_s: 15 bits per I(Q) sample
- Number of antennas: 64
- Number of downlink MIMO-layers: 8
- Number of uplink MIMO-layers: 4 (with 2 diversity streams per uplink MIMO layer)
- Transmission time interval: 1 ms
- Digital beamforming with beamforming coefficients calculated in the REC

Table 3.5 Comparison of interface line bit rates at key split positions for an illustrative 5G use case with 64 × 64 MIMO and 100-MHz RF bandwidth.

Direction	Split *E* (CPRI)		eCPRI split II_D/I_U for downlink/uplink		Split *D*	
	User data rate (Gb/s)	Interface bit rate (Gb/s)	User data rate (Gb/s)	Interface bit rate (Gb/s)	User data rate (Gb/s)	Interface bit rate (Gb/s)
Downlink	295	324	~20	~25	3	~3
Uplink	295	324	~20	~25	1.5	~1.5

- Modulation scheme: 256-quadrature amplitude modulation
- Subcarrier spacing: 15 kHz
- Code rate: ~0.80

The user data rates and the overall interface line bit rates (including the overheads needed for control signaling and line coding) for three key split positions, split *E*, eCPRI split II_D/I_U, and split *D*, are summarized in Table 3.5.

CPRI, Common public radio interface; *eCPRI*, evolved common public radio interface; *MIMO*, multiple-input multiple-output; *RF*, radio frequency.

For split *E* (or CPRI), a high interface data rate of 324 Gb/s is needed for either the downlink or the uplink direction. Remarkably, the overall interface line bit rate at the eCPRI split II_D/I_U is substantially reduced to only ~25 Gb/s, representing over 10-fold reduction in interface line bit rate as compared to CPRI. It is worth noting that eCPRI split II_D/I_U provides this substantial bandwidth efficiency improvement over CPRI without sacrificing the support of real-time coordination functions such as CoMP. Thus eCPRI split II_D/I_U is well suited for serving as the front-haul interface.

For split *D*, the control signal bit rate is negligible as compared to the user data rate as most of the real-time control functions have been conducted under this split position. Thus the overall interface bit rates at split *D* are just slightly higher than the user data rates, that is, 3 Gb/s for the downlink and 1.5 Gb/s for the uplink as assumed in this illustrative 5G use case. Note also that although split *D* is more bandwidth-efficient than eCPRI split II_D/I_U, it sacrifices the support of real-time coordination functions and is thus not suitable to be the front-haul interface. The improved interface bandwidth efficiency of eCPRI as compared to CPRI is further illustrated in Fig. 3.11.

Fig. 3.12 shows the protocol stack layouts in an eCPRI REC (eREC) unit and an eCPRI RE (eRE) unit that are connected via an eCPRI link [18]. As compared to the basic system architecture of REC and RE that are connected via CPRI, several enhancements have been made. Essentially, eCPRI is a packet-based front-haul interface that leverages advanced Ethernet/IP networking features such as Ethernet

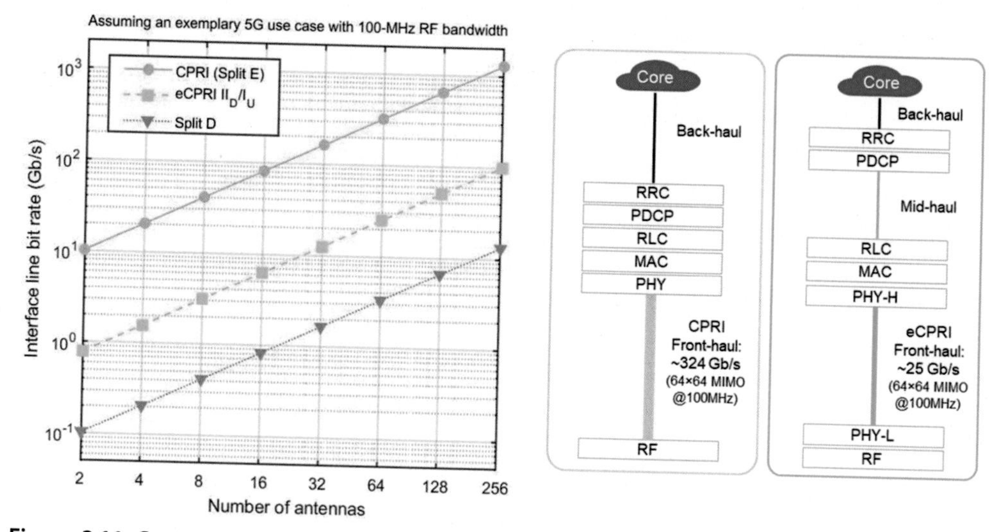

Figure 3.11 Comparison of interface bit rates of CPRI and eCPRI as a function of the number of antennas for an illustrative 5G use case with 100-MHz RF bandwidth.

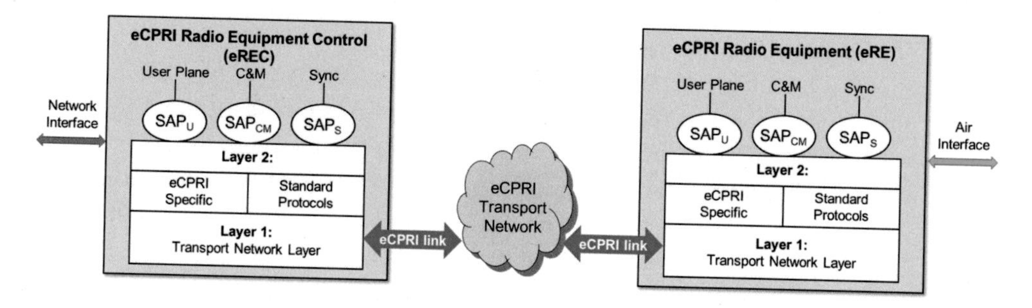

Figure 3.12 Basic system architecture of eREC and eRE that are connected via eCPRI.

Operations, Administration, and Maintenance. In addition, the existing standard protocols for synchronization and security are adopted.

A set of requirements for the eCPRI transport network are specified to ensure that eCPRI systems can comply with the requirements associated with the more stringent radio technologies' features in terms of timing and frequency accuracy, bandwidth capacity, latency, packet loss, etc.

For requirements related to timing accuracy, there are four categories, *A+*, *A*, *B*, and *C*, for representative applications such as MIMO with transmitter diversity, contiguous carrier aggregation, noncontiguous carrier aggregation, and LTE Time-Division Duplex, respectively. Table 3.6 list the key time alignment error (TAE) and time error (TE) requirements for these

Table 3.6 Timing accuracy requirements in eCPRI transport network.

Category	*A*+ (ns)	*A* (ns)	*B* (ns)	*C* (ns)
3GPP TAE at antenna ports	65	130	260	3000
|TE| at UNI	20 (relative)	70 (relative)	200 (relative)	1100 (absolute)

eCPRI, Evolved common public radio interface; *3GPP*, the 3rd generation partnership project; *TAE*, time alignment error; *UNI*, user network interface.

four categories [18]. For categories *A*+ , *A*, and *B*, the requirements are expressed as relative requirements between two user network interfaces (UNIs), instead of relative to a common clock reference, for example, a primary reference time clock. For category C, the requirement is expressed as an absolute requirement at the UNI as in ITU-T G.8271.1. In most cases, categories *A*+ , *A*, and *B* also need to meet the absolute TE requirement of Category *C*.

In the above table, the TE at UNI is for a typical case where the telecom time slave clock (T-TSC) is not integrated in eRE. The termination of precision time protocol is in the T-TSC at the edge of the transport network, and the phase and time references are delivered from the T-TSC to the colocated eRE via a synchronization distribution interface. This case can be referred to the "deployment case 2" in Figure 7-1 of ITU-T G.8271.1 [19]. Evidently, the timing accuracy requirements in eRE are very stringent, especially when m-MIMO is used, and the eCPRI transport network has to provide the needed synchronization accuracy. Therefore accurate synchronization is one of the key features of 5G-oriented optical transport networks [20,21].

For requirements related to latency, there are several classes of maximum one-way frame delay requirements ranging from 25 μs to 100 ms, as summarized in Table 3.7. The maximum one-way frame delay requirement value includes fiber propagation delay and switching/processing delay. The low requirement class is used for transporting nontime-critical C&M traffic. The medium requirement class is used for transporting time-critical C&M traffic and user plane data with relaxed latency requirements. The high requirement class is used for transporting user plane data with stringent latency requirements. Certain illustrative use cases are also listed. The most stringent class is the High25 class, which requires the maximum one-way frame delay to be less than 25 μs, for ultralow-latency applications. To achieve this, the maximum front-haul fiber distance needs to be less than 5 km as the speed of light in fiber is 0.2 km/μs. In addition, the switching/processing delay is desired to be much less than 5 μs. Thus low-latency transmission is another key feature of 5G-oriented optical transport networks [20,21].

It is worth noting that although eCPRI is more bandwidth-efficient than CPRI, the latter offers the benefit of more simplified remote radio equipment. Also, CPRI-supported REC and RE have been extensively deployed. The CPRI industry association thus noted that "in addition to the new eCPRI specification, the work continues

Table 3.7 Latency requirements in eCPRI transport network.

Class		Maximum one-way frame delay	Use case example
High (for fast user plane)	High25	25 μs	Ultra-low latency applications
	High100	100 μs	Full new radio performances
	High200	200 μs	Installations with fiber length up to 40 km
	High500	500 μs	Large latency installations
Medium		1 ms	Slow user plane and fast C&M plane
Low		100 ms	Slow C&M plane communication

C&M, Control and management; *eCPRI*, evolved common public radio interface.

to further develop the existing CPRI specifications to keep it as a competitive option for all deployments with dedicated fiber connections in front-haul including 5G." On May 10, 2019, the CPRI association had released eCPRI 2.0, a new specification that enhances the support for the 5G front-haul by providing functionality to support CPRI (version 7.0) over Ethernet [22]. Moreover, the eCPRI 2.0 specification offers eCPRI−CPRI interworking features by

- Extending eCPRI flexibility for front-haul transport between eCPRI nodes and CPRI nodes
- Enabling the coexistence of new and legacy equipment in the same Ethernet-based front-haul networks

A main objective of eCPRI 2.0 is to enable the reuse of existing RE and REC units with an eCPRI transport network. This is achieved by introducing a new interworking function (IWF) to bridge CPRI and eCPRI nodes. There are three types of IWFs defined as follows:

- IWF type 0, which connects an eREC with a RE in the "eREC ⇔ eCPRI transport network ⇔ IWF type 0 ⇔ RE" configuration
- IWF type 1, which connects an REC with an eCPRI transport network in the "REC ⇔ IWF type 1 ⇔ eCPRI transport network ⇔ IWF type 2 ⇔ RE" configuration
- IWF type 2, which connects a RE with an eCPRI transport network in the "REC ⇔ IWF type 1 ⇔ eCPRI transport network ⇔ IWF type 2 ⇔RE" configuration

Fig. 3.13 illustrates the interworking between RE and REC/eREC with an eCPRI transport network via IWF type 0, 1, and 2. Effectively, these IWFs enable

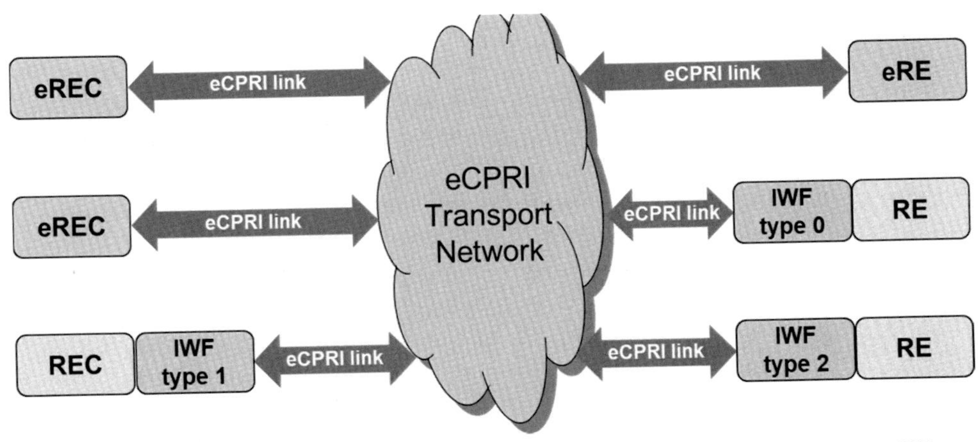

Figure 3.13 Interworking between RE and eREC/REC with an eCPRI transport network via IWF type 0, 1, and 2, enabling the coexistence of CPRI and eCPRI with the same eCPRI transport network.

legacy RE to communicate with legacy REC/eREC over an eCPRI transport network. The generic connection between eRE and eREC via eCPRI is also shown for comparison. This new development enables both CPRI and eCPRI to coexist in 5G radio access networks to provide the optimal interfaces for a diverse set of application scenarios (Fig. 3.13).

3.4 O-RAN interfaces

As shown in the previous sections, RANs are continuously evolving in the 5G era. To aid the mobile communication industry to have a collective effort to further improve RANs, the Open RAN (O-RAN) Alliance was founded by AT&T, China Mobile, Deutsche Telekom, NTT DOCOMO, and Orange in February 2018 [23]. As of 2020, the O-RAN Alliance has over 200 members ranging from mobile network operators, system vendors, and research and academic institutions. The O-RAN Alliance is in pursuit of the vision of openness and intelligence for the next-generation wireless networks. The vision of openness aims to bring service agility and cloud scale economics, particularly via open interfaces to enable multivendor deployments and a more competitive and vibrant supplier ecosystem. The vision of intelligence aims to enable self-driving networks with optimized network-wide efficiency and reduced OPEX by embedding intelligence in every layer of the RAN architecture.

To realize the vision of openness and intelligence for the next-generation wireless networks, the following efforts are made by the O-RAN Alliance:

- Openness
- Leading the industry toward open, interoperable interfaces
- Specifying open interfaces
- Driving standards to adopt the open interfaces as appropriate

- Exploring open source software where appropriate
- Intelligence
- Software-defined self-driving RAN
- RAN virtualization
- Big data—enabled RAN intelligence
- Artificial intelligence (AI)—enabled RAN intelligence.

Regarding open interfaces, the O-RAN Alliance introduced a reference O-RAN architecture in 2018 [24], as shown in Fig. 3.14. O-RAN aims to enhance the openness of multiple interfaces such as the A1, E1, E2, F1, and fronthaul interfaces (see Fig. 3.14). These interfaces are connected to the following segments of the reference architecture.

- *Nonreal-time (non-RT) layer of RAN intelligent controller (RIC)*

  In this O-RAN reference architecture, non-RT control functions (with required response time longer than 1 s) and near-real-time (near-RT) control functions (with required response time less than 1 s) are decoupled into two layers in the RIC. The non-RT layer includes functions such as service and policy management, RAN analytics, and model training for the near-RT layer. The trained models and real-time control functions produced in the non-RT layer are passed to the near-RT layer for runtime execution. The interface between the non-RT layer and the near-RT layer is referred to as the A1 interface.

  According to the 2018 O-RAN white paper [24], "With the introduction of A1, network management applications in RIC non-RT are able to receive and act on highly reliable data from the modular CU and DU in a standardized format. Messages generated from AI-enabled policies and ML based training models in RIC non-RT are conveyed to RIC near-RT. The core algorithm of RIC non-RT is developed and owned by operators. It provides the capability to modify the RAN behaviors by deployment of different models optimized to individual operator policies and optimization objectives."

- *Near-RT layer of the RIC*

  The near-RT layer of the RIC provides enhanced radio resource management (RRM) with embedded intelligence. It enhances operational challenging functions such as per-UE controlled load-balancing, resource block management, and interference detection and mitigation. In addition, it provides new functions such as Quality of Service management, connectivity management, and seamless handover control. The near-RT layer utilizes a common database called the Radio Network Information Base. The interface between the near-RT layer of the RIC and the underlying network is referred to as the E2 interface.

  According to the 2018 O-RAN white paper [24], "E2 is the interface between the RIC near-RT and the Multi-Radio Access Technology (RAT) CU protocol

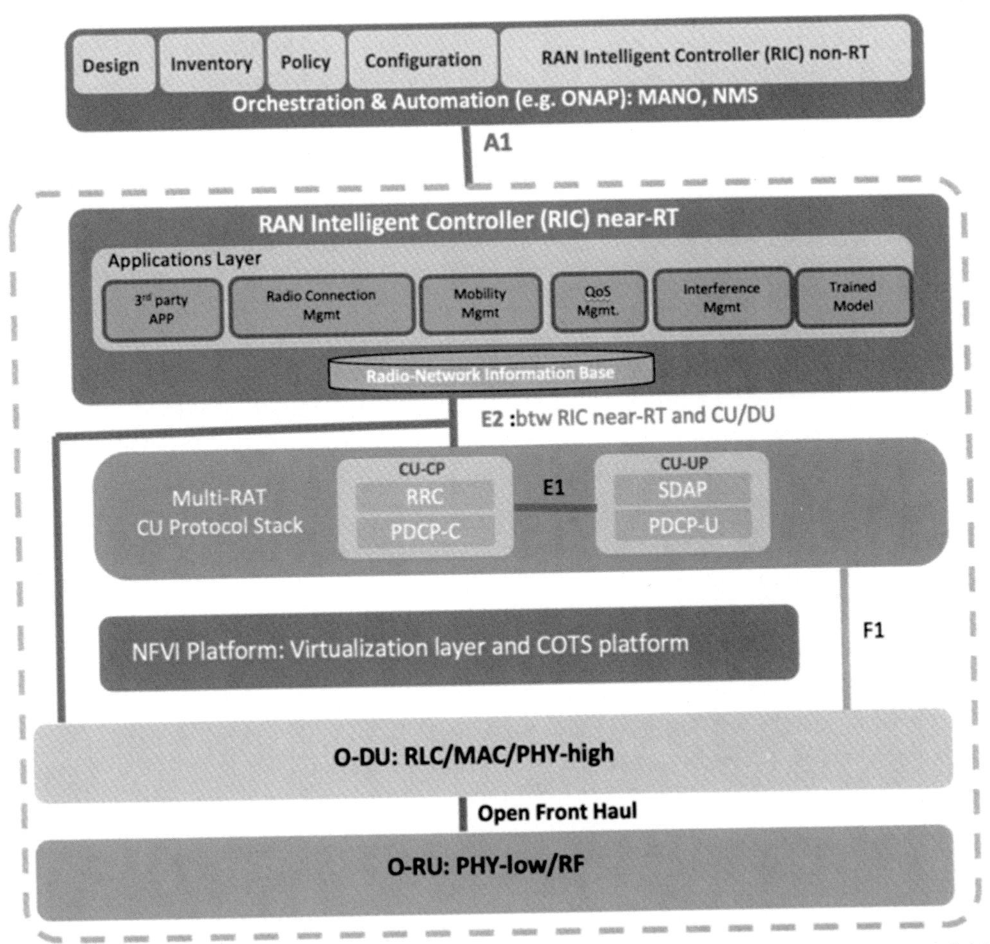

Figure 3.14 Illustration of the O-RAN reference architecture with open interfaces F1, E1, E2, and A1 (after [24]). *COTS*, Commercial-off-the-shelf; *MANO*, management and orchestration; *NFV*, network function virtualization; *NFVI*, NFV infrastructure; *NMS*, network management system; *ONAP*, open network automation platform; *RT*, real-time; *SDAP*, service data adaptation protocol; *SDR*, software-defined radio.

stack and the underlying RAN DU. Originated from the interface between legacy RRM and RRC in traditional systems, the E2 delivers a standard interface between the RIC near-RT and CU/DU in the context of an O-RAN architecture. While the E2 interface feeds data, including various RAN measurements, to the RIC near-RT to facilitate RRM, it is also the interface through which the RIC near-RT may initiate configuration commands directly to CU/DU."

- *Multi-RAT CU*

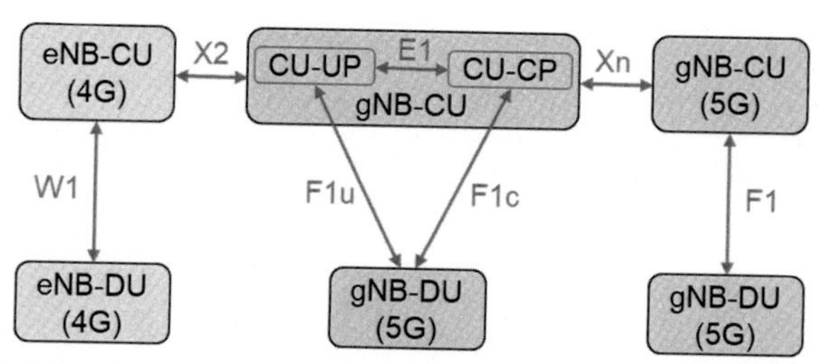

Figure 3.15 The interfaces between CU and CU/DU in eNB and gNB defined and specified by 3GPP and O-RAN (after [24]).

The Multi-RAT CU protocol stack supports multiple generations of protocol processing. The CU protocol stack can be divided into CU-CP and CU-UP, where CU-UP consists of RRC and the control plane of PDCP (PDCP-C) and CU-UP consists of service data adaptation protocol (SDAP) and the user plane of PDCP (PDCP-U). The interface between the CU-CP and CU-UP is referred to as the E1 interface. In addition, there are four interfaces between CU and CU/DU in eNB and gNB specified by 3GPP:

- F1, which connects a 5G CU (gNB-CU) with a 5G DU (gNB-DU) for both the user plane (F1u) and the control plane (F1c)
- W1, which connects a 4G CU (eNB-CU) with a 4G DU (eNB-DU)
  - X2, which connects a gNB-CU with an eNB-CU
  - Xn, which connects a gNB-CU with another gNB-CU

All the interfaces, E1/F1/F1u/F1c/W1/X2/Xn, are illustrated in Fig. 3.15. These interfaces have been defined by 3GPP, and O-RAN aims to enhance their multivendor interoperation capabilities.

- *DU and RU*

As described previously, 5G DU and RU perform real-time L2 functions, baseband processing, and RF processing. The interface that connects the DU and the RU is the front-haul interface, which is commonly based on CPRI and eCPRI. Like eCPRI, the O-RAN front-haul interface is located between the High-PHY and Low-PHY, and O-RAN aims to enhance the interoperability of the interface.

It can be seen that multiple global organizations such as 3GPP, CPRI, IEEE, ITU, and O-RAN have been defining and specifying various RAN interfaces. In some use cases, end-to-end integration of multiple network architecture segments is desirable for achieving high network performance and/or energy efficiency. The end-to-end integration may also prefer a tighter integration between software and hardware. The collective efforts made by the mobile network community toward network intelligence via both open and innovative network architectural designs are beneficial to our society.

References

[1] China Mobile Research Institute. C-RAN: the road towards green RAN, White paper v. 2.6; September 2013.
[2] I C-L, Rowell C, Han S, Xu Z, Li G, Pan Z. Towards green & soft: a 5G perspective. IEEE Commun Mag 2014;52(2):66–73.
[3] I C-L, Huang J, Duan R, Cui C, Jiang J, Li L. Recent progress C-RAN centralization and cloudification. IEEE Access 2014;2:1030–9.
[4] I C-L, Yuan Y, Huang J, Ma S, Duan R, Cui C. Rethink fronthaul for Soft RAN. IEEE Commun Mag 2015;53(9):82–8.
[5] CPRI 7.0 Specification [Online], <http://cpri.info/spec.html>; October 9, 2015.
[6] 3GPP TR 38.801, Technical specification group radio access network; study on new radio access technology: radio access architecture and interfaces; March 2017.
[7] eCPRI 1.0 Specification [Online], <http://cpri.info/spec.html>; August 22, 2017.
[8] China Telecom. 5G-Ready OTN technical white paper [Online], <http://www.ngof.net/en/download/5G-Ready_OTN_Technical_White_Paper.pdf>; September 2017.
[9] I C, Li H, Korhonen J, Huang J, Han L. RAN revolution with NGFI (xhaul) for 5G. J Lightwave Technol 2018;36(2):541–50.
[10] ITU-T Technical Report GSTR-TN5G, Transport network support of IMT-2020/5G; October 19, 2018.
[11] IEEE 1914.1-2019, IEEE standard for packet-based fronthaul transport networks; 2019.
[12] Pizzinat A, Chanclou P, Saliou F, Diallo T. Things you should know about fronthaul. J Lightwave Technol 2015;33(5):1077–83.
[13] Kani J, Kuwano S, Terada J. Options for future mobile backhaul and fronthaul. Opt Fiber Technol 2015;26:42–9.
[14] Liu X, Zeng H, Chand N, Effenberger F. Efficient mobile fronthaul via DSP-based channel aggregation. J Lightwave Technol 2016;34(6):1556–64.
[15] J Li. Photonics for 5G in China. In: 2017 Asia communications and photonics conference, paper S4E.1; 2017.
[16] Alimi IA, Teixeira AL, Monteiro PP. Toward an efficient C-RAN optical fronthaul for the future networks: a tutorial on technologies, requirements, challenges, and solutions. IEEE Commun Surv Tutor 2018;20(1):708–69 First Quarter.
[17] Guo B, Cao W, Tao A, Samardzija D. LTE/LTE-A signal compression on the CPRI interface. Bell Labs Tech J 2013;18(2):117–33.
[18] eCPRI 1.2 Specification [Online], <http://cpri.info/spec.html>; June 25, 2018.
[19] ITU-T G.8271.1/Y.1366.1, Network limits for time synchronization in Packet networks with full timing support from the network. [Online]. <https://www.itu.int/rec/T-REC-G.8271.1-202003-I/en>.
[20] X Liu, N Deng. Chapter 17: Emerging optical communication technologies for 5G. In: Willner A, editor. Optical fiber telecommunications VII; 2019.
[21] Liu X. Evolution of fiber-optic transmission and networking toward the 5G era. iScience 2019;22:489–506. Available from: https://www.cell.com/iscience/pdf/S2589-0042(19)30476-6.pdf.
[22] eCPRI 2.0 Specification [Online], <http://cpri.info/spec.html>; May 10, 2019.
[23] O-RAN Alliance, <https://www.o-ran.org/>.
[24] O-RAN 2018 white paper, Building the next generation RAN [Online], <https://www.o-ran.org/resources>; October 2018.

CHAPTER 4

Optical technologies for 5G X-haul

4.1 Overview on 5G X-haul

As described in Chapter 3, Optical Interfaces for 5G Radio Access Network, a common 5G cloud radio access network (C-RAN) consists of front-, mid-, and back-haul network segments, which are referred to as X-haul network segments. Fig. 4.1 illustrates a metro-area 5G C-RAN with the three X-haul network segments where

- The front-haul network segment connects radio units and their corresponding distributed units (DUs).
- The mid-haul network segment connects DUs and their corresponding centralized units (CUs).
- The back-haul network segment connects CUs and their corresponding 5G Core (5GC) nodes.

According to the 3rd Generation Partnership Project (3GPP) [1], the one-way latencies for Common Public Radio Interface (CPRI)/evolved CPRI (eCPRI), ultra-reliable and low-latency communication (uRLLC), enhanced mobile broadband (eMBB) user plane (UP), eMBB control plane (CP), and massive machine-type communications (mMTC) are 100 μs, 0.5 ms, 4 ms, 10 ms, and 10 s, respectively. As shown in Fig. 4.1, uRLLC, eMBB-UP, and eMBB-CP are, respectively, processed at an access data center (DC), an aggregation DC, and a core DC. With the consideration of the latencies due to fiber propagation and DU/CU processing, as well as the typical metro-area network configurations, the typical distances of front-, mid-, and back-haul segments are limited to 10, 40, and 80 km, respectively. In addition, there are usually multiple core DCs that serve one metro-area to provide the needed processing and protection capabilities. The data center interconnection (DCI) distances are usually less than 120 km.

In the first phase 5G deployment in China, approximately 600,000 base stations were deployed in 50 major cities by the end of 2020 [2], indicating an average of about 12,000 active antenna units (AAUs) for a typical major city. In an illustrative metro network deployment scenario, 12,000 AAUs can be supported by 400 DUs if each DU serves 30 AAUs [3]. These 400 DUs can then be supported by 20 CUs if each CU serves 20 DUs. Finally, these 20 CUs can be served by four 5GC nodes that reside in four core DCs, each supporting five CUs. The four core DCs are interconnected via a mesh network for low-latency communication and high-reliability

Optical Communications in the 5G Era
DOI: https://doi.org/10.1016/B978-0-12-821627-9.00007-3

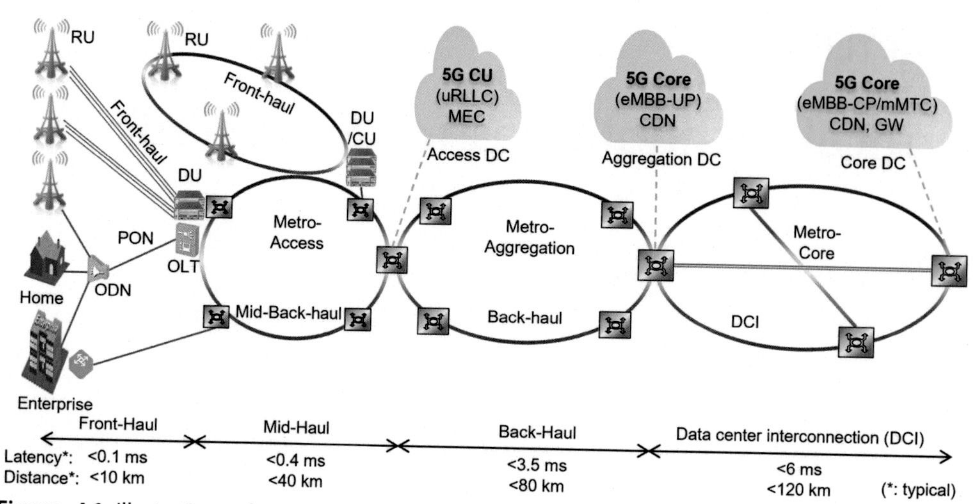

Figure 4.1 Illustration of a metro-area 5G C-RAN with front-, mid-, and back-haul network segments.

protection, as illustrated in Fig. 4.1. The key 5G X-haul network features of the illustrative metro network are listed in Table 4.1.

As shown in Table 4.1, the transmission distance and capacity requirements in the front-, mid-, and back-haul, and 5GC network segments are different. Consequently, different optical technologies in terms of modulation, detection, and multiplexing are needed to optimally support the various segments of 5G C-RAN, which will be discussed in the following sections.

4.2 Optical technologies for 5G front-haul

For the front-haul network segment, the required eCPRI data rate per AAU per direction for 200-MHz RF bandwidth is 50 Gb/s, as discussed in the previous chapter. To connect a 5G cell site with three AAUs for both downlink (DL) and uplink (UL) directions, twelve 25-Gb/s connections are needed. For the illustrative metro network with 12,000 AAUs, a total of 48,000 25-Gb/s connections are needed. These connections may be supported by the following implementation approaches [3–10].

- Direct fiber connection with one wavelength channel per fiber—This approach requires 48,000 fibers to connect all the 12,000 AAUs with 48,000 fixed-wavelength 25-Gb/s optical transceivers, which would be too expensive to deploy.
- Bi-directional (BiDi) direct fiber connection with two wavelength channels for DL and UL—The BiDi transmission halves the required number of fibers to 24,000, which would still be too expensive for deployment in most cases. Fig. 4.2 shows the schematic of a common 25-Gb/s BiDi optical transceiver that transmits and

Table 4.1 5G X-haul network features in an exemplary metro network.

Segment	Feature	Typical value
Front-haul	Number of AAUs[a]	12,000
	Bit rate to/from DU per AAU[b]	50 Gb/s
	Transmission distance	< 20 km (typically <10 km)
	Typical optical transceiver types	25 Gb/s NRZ, 100-Gb/s DMT or PAM4, BiDi, WDM, L-WDM, M-WDM
	Total bit rate to/from DUs	600 Tb/s
Mid-haul	Number of DUs[c]	400
	Bit rate to/from CU per DU[d]	300 Gb/s
	Transmission distance	< 80 km (typically <40 km)
	Typical optical transceiver types	50-Gb/s PAM4, O-band DWDM 100-Gb/s coherent, DWDM
	Total bit rate to/from CUs	120 Tb/s
Back-haul	Number of CUs[e]	20
	Bit rate to/from 5GC per CU[f]	3 Tb/s
	Transmission distance	< 80 km (typically)
	Typical optical transceiver types	200G/400G coherent, DWDM
	Total bit rate to/from 5GCs	60 Tb/s (per direction)
Metro-core	Number of 5GC nodes[g]	4
	Bit rate from one 5GC to another[h]	7.5 Tb/s
	Transmission distance	< 120 km (typically)
	Typical optical transceiver types	400G/800G coherent, DWDM
	Total bit rate to/from other 5GCs[i]	45 Tb/s

AAU, Active antenna unit; BiDi, bi-directional; *CU*, centralized unit; *DMT*, discrete multitone; *DU*, distributed unit; *DWDM*, dense WDM; *L-WDM*, 12-channel LAN-WDM; *M-WDM*, 12-channel medium-WDM; *NRZ*, non-return-to-zero; *PAM4*, 4-level pulse-amplitude modulation; *WDM*, wavelength-division multiplexing.

[a]Based on recent 5G base station deployment for large cities each having millions of people [2].

[b]Assuming eCPRI bit rate for a typical 5G AAU with 200-MHz RF bandwidth.

[c]Assuming that each DU is connected to 10 cell sites each having 3 AAUs.

[d]Assuming a fivefold reduction in interface bit rate after the DU processing in the eCPRI case.

[e]Assuming that each CU is connected to 20 DUs.

[f]Assuming a twofold reduction in outer interface bit rate after the CU processing.

[g]Assuming a typical metro network configuration where 5GC nodes are hosted in 4 core DCs.

[h]Assuming a twofold reduction in outer interface bit rate after the 5GC processing.

[i]Assuming 6 mesh connections for the 4 core DCs.

receives at two coarse wavelength-division multiplexing (CWDM) bands centered at 1271 and 1331 nm.

- C-band WDM with over 40 wavelength channels per fiber—This approach dramatically reduces the required number of fibers by over 40 times to save the costs associated with fiber cables and fiber deployments [7]. On the other hand, tunable optical transceivers, which are more expensive than fixed-wavelength transceivers, are desired. To further reduce the overall cost, 100-Gb/s optical transceivers based on spectrally efficient

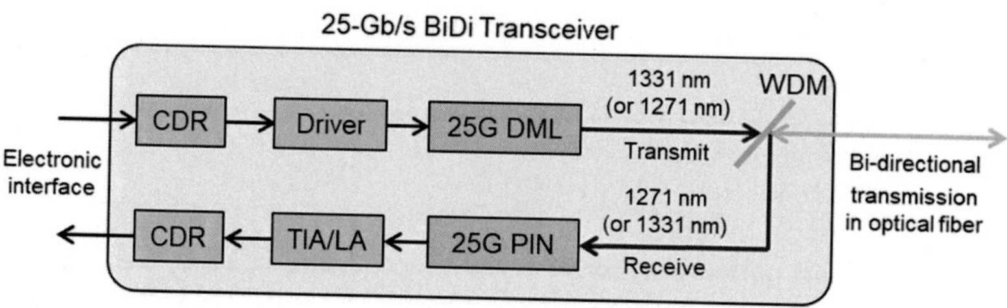

Figure 4.2 The schematic of a common 25-Gb/s BiDi optical transceiver that transmits and receives in two CWDM bands centered at 1271 and 1331 nm. *CDR*, Clock-data recovery; *DML*, directly modulated laser; *LA*, limiting amplifier; *PIN*, PIN photodetector; *TIA*, transimpedance amplifier; *WDM*, wavelength-division multiplexer.

intensity-modulation and direct-detection (IM/DD) modulation formats such as discrete multitone (DMT) [6] and 4-level pulse-amplitude modulation (PAM4) may be used.

- O-band WDM with 12 wavelength channels per fiber—This approach leverages low-cost optical transceiver ecosystem for local area network wavelength-division multiplexing (LAN-WDM) and CWDM, as well as the common 5G front-haul deployment scenario of 12 25-Gb/s connections per cell site, with the aim to provide an adequate balance between the fiber-associated costs and the transceiver-associated costs. The 12-channel LAN-WDM-based approach is referred to as L-WDM [9], while the 12-channel CWDM-based approach is referred to as medium wavelength-division multiplexing (M-WDM) [8].

For the above WDM-based approaches, they can be implemented under three categories:

- Passive WDM—The term "passive" here means that no active equipment requiring additional power supplier is needed at both the AAU sites and the DU sites (or central offices [COs]) to realize WDM transmission in the front-haul network. Tunable optical transceivers, typically in small format factor (SFP) pluggable modules, are inserted directly into AAUs and DUs without requiring additional power suppliers at both the AAUs and the DUs. WDM multiplexer/de-multiplexer (MUX/DMUX) is used to aggregate/de-aggregate the WDM channels that share one fiber to save fiber resources. BiDi WDM transmission to support both the DL and UL simultaneously in one fiber can be readily realized by assigning different wavelengths for the DL and UL. Note that front-haul link loss budget is generally supported without any optical amplification in the link (e.g., by erbium-doped fiber amplifier) so BiDi WDM transmission in the front-haul link is straightforward.
- Active WDM—The term "active" here means that active equipment requiring additional power supplier is needed at both the AAU sites and the DU sites (or COs) to realize WDM transmission in the front-haul network. Active optical transmission equipment, such as optical transport network (OTN) equipment, can be used at both the AAU sites and the DU sites to perform functions such as channel aggregation and

de-aggregation, frame encapsulation, synchronization, management, and protection. This solution supports various network topologies such as point-to-point (P2P), chain, and ring. In addition, it can provide advanced network functions for intelligent operations, administration and maintenance (OAM), as well as fault protection and location functions. However, the additional power supplies needed at the AAU sites may complicate the cell site deployment and increase the power consumption.

- Semiactive WDM—The term "semiactive" here means that active equipment requiring additional power supplier is needed at the DU sites (or COs) but not at the AAU sites to realize WDM transmission in the front-haul network. At the DU sites, active optical transmission equipment, such as OTN equipment, can be used to perform advanced network functions, as in the case of Active WDM. At the AAU sites, pluggable optical transceiver modules are inserted directly into AAUs without requiring additional power suppliers at the AAUs, as in the case of Passive WDM.

Fig. 4.3 illustrates 5G front-haul network configurations based on (A) direct fiber connection, (B) passive WDM, (C) active WDM, and (D) semiactive WDM. In the direct fiber connection configuration, the DUs in a CO are connected with N cell sites via 12N fibers each carrying a 25-Gb/s optical signal, using the aforementioned scenario of 12 25-Gb/s connections per 5G cell site. Each fiber connection is terminated by two pluggable optical transceivers at the two ends of the fiber. A large number of fibers and fiber distribution boxes are needed. In the passive WDM configuration, each fiber carries multiple (e.g., over 40) WDM channels, and optical add/drop multiplexers (OADMs) are deployed at the cell sites to distribute WDM channels to each AAU. Each WDM channel is terminated by two pluggable optical transceivers at the two ends of the fiber connection. In the active WDM configuration, active OTN equipment is added in the CO that contains the DUs, as well as at each cell site. Spectrally efficient modulation formats such as 100-Gb/s DMT can be used to further increase the capacity per fiber, thereby further reducing the number of fibers needed. Moreover, the channel aggregation and de-aggregation capability of OTN equipment readily supports various eCPRI and CPRI data rates, which is particularly useful for a cell site with multiple generations of radio technologies. In the semiactive WDM configuration, the active equipment at the CO can be an OTN equipment. No active optical equipment and additional power suppliers are needed at the cell sites, as the wavelength channels are transmitted and received by pluggable optical transceivers inside the AAUs.

Fig. 4.4 illustrates two additional implementation options of the semiactive WDM approach for 5G front-haul based on (A) one WDM MUX/DMUX shared by multiple neighboring cell sites and (B) one single fiber to each cell site via the 12-channel L-WDM or M-WDM. The active equipment at the CO can be a WDM-PON optical line terminal (OLT). Evidently, the semiWDM approach possesses the key advantage of the passive WDM approach, that is, not requiring additional power suppliers at the cell sites and the key advantage of the active WDM approach, that is,

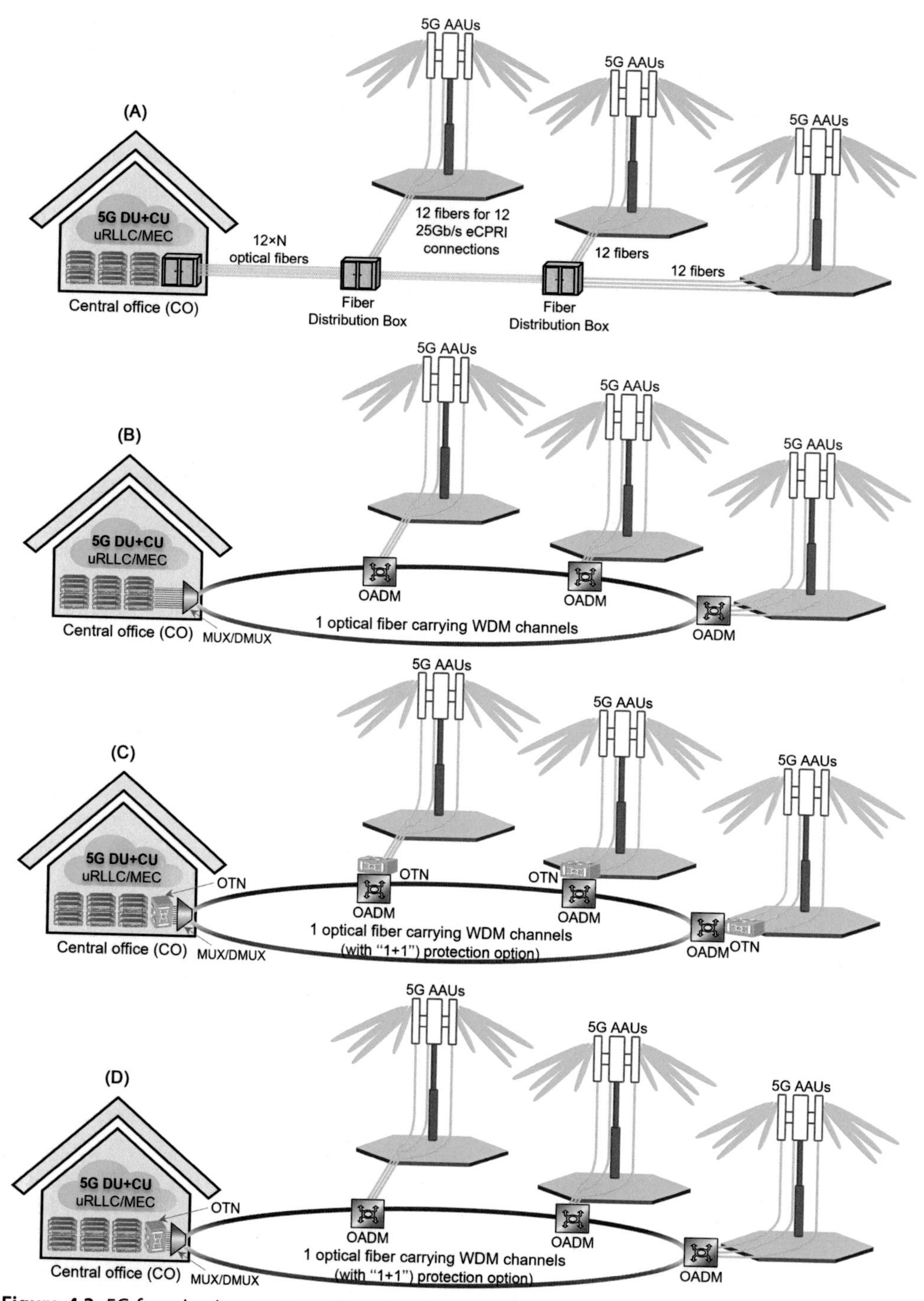

Figure 4.3 5G front-haul network configurations based on (A) direct fiber connection, (B) passive WDM, (C) active WDM (with OTN), and (D) semiactive WDM. Note that pluggable optical transceivers inside AAUs are not shown.

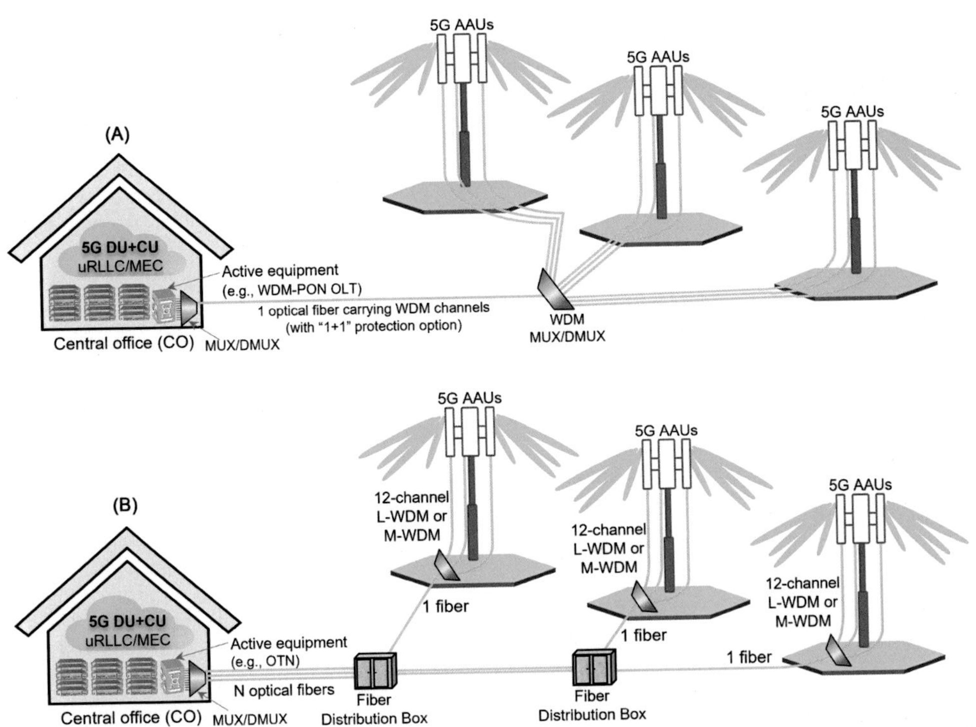

Figure 4.4 Semiactive WDM configurations for 5G front-haul based on (A) one WDM MUX/DMUX shared by multiple neighboring cell sites, and (B) one single fiber to each cell site via L-WDM or M-WDM.

providing advanced network OAM functions and enhanced network reliability and flexibility.

Table 4.2 compares four common 5G front-haul network solutions in terms of their advantages and disadvantages, as well as suitable deployment scenarios. Evidently, each solution has its most suitable deployment scenario. Altogether, the semiactive WDM solution is well suited for future large-scale cost-effective deployments with high front-haul capacity, OAM and reliability requirements after the L(M)-WDM ecosystem is fully developed.

4.3 The impact of fiber loss and dispersion on X-haul technology choices

As discussed in the previous sections, WDM is needed to meet the transmission capacity requirements in 5G X-haul networks. Since the transmission distance and capacity requirements in the front-, mid-, and back-haul, and 5GC network segments are

Table 4.2 Comparison of four 5G front-haul network solutions.

Solution	Advantages	Disadvantages	Suitable deployment scenarios
Direct fiber connection	• No extra power supplies required at AAU and DU sites • Only one type of fixed-wavelength optical transceivers required	• A massive number of fibers required • Lack of OAM and protection capabilities	Deployments in areas with abundant fiber resources and/or modest front-haul capacity requirements.
Passive WDM	• No extra power supplies required at AAU and DU sites • Much reduced number of fibers required	• Expensive tunable WDM transceivers required • Limited OAM and protection capabilities	Deployments in areas with high front-haul capacity requirements.
Active WDM	• Much reduced number of fibers required • Full-scale OAM and protection capabilities	• Extra power supplies required at AAU and DU sites • Expensive tunable WDM transceivers required	Deployments in areas with high front-haul capacity, OAM, and reliability requirements, where extra power supplies are allowed at AAU sites
Semiactive WDM	• Much reduced number of fibers required • No extra power supplies required at AAU sites • Sufficient OAM and protection capabilities • Potentially low optical transceiver cost when the L (M)-WDM ecosystem is fully developed	• Extra power supplies required at DU sites	Cost-effective deployments in areas with high front-haul capacity, OAM, and reliability requirements, especially after the L (M)-WDM ecosystem is fully developed

different, suitable WDM transmission systems need to be chosen for a given network segment to meet the transmission requirements with the lowest total cost of ownership (TCO). Optical signal transmission over an optical fiber is mainly impacted by two physical effects, namely, loss and chromatic dispersion. Fiber loss reduces optical signal power and decreases the received signal-to-noise ratio (SNR). Fiber chromatic dispersion causes different frequency components of an optical signal to travel at different speeds in the fiber, thus results in intersymbol-interference (ISI) at the receiver.

Although the dispersion-induced ISI may be compensated in a modern digital coherent receiver, it imposes a certain dispersion tolerance limit in a conventional direction detection optical receiver.

Fig. 4.5 shows the loss coefficient (in dB/km) and the dispersion coefficient (in ps/nm/km) of the most common standard single-mode fiber, the G.652 fiber [11], over the telecommunication window ranging from 1260 to 1625 nm. Owing to fiber manufacturing imperfections, the dispersion coefficient may vary between a maximum dispersion value (D_{max}) and minimum dispersion value (D_{min}). Conventionally, The telecommunication window is divided into five wavelength bands as follows.

- The original band (O-band) ranging from 1260 to 1360 nm, which offers the lowest dispersion (between −6.4 and 5.2 ps/nm/km) but relatively high loss of up to 0.48 dB/km.

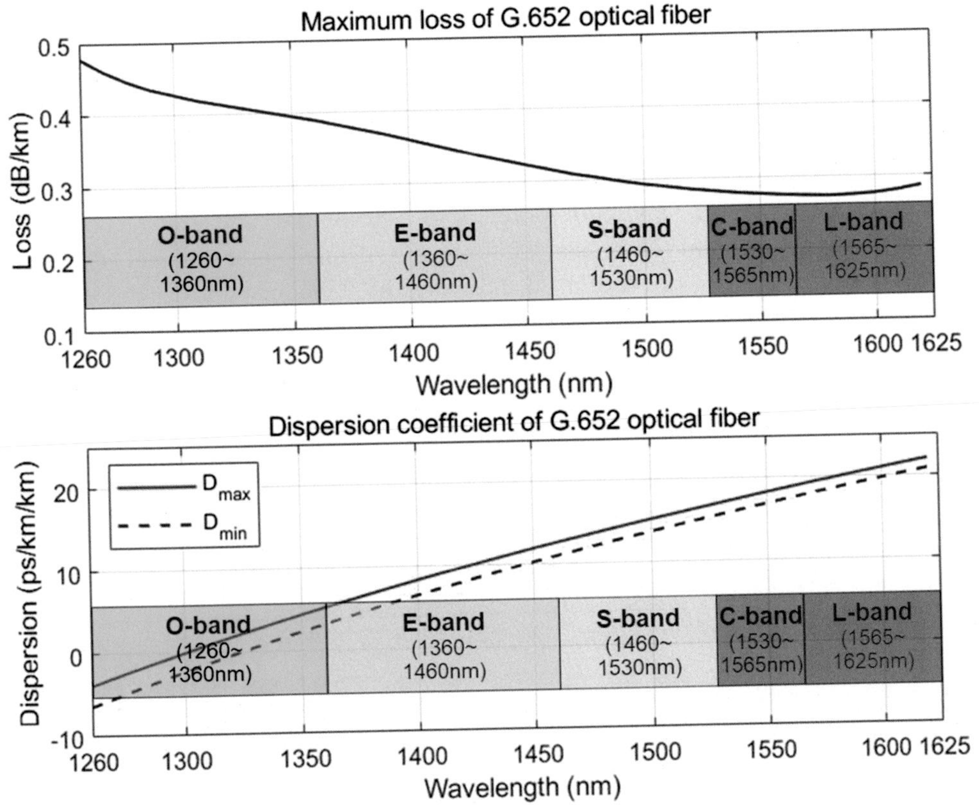

Figure 4.5 The loss and dispersion coefficients of the common G.652.C(D) standard single-mode fiber over the telecommunication window from 1260 to 1625 nm.

- The extended band (E-band) ranging from 1360 to 1460 nm, which offers a slightly lower loss than the O-band for G.652.C(D)-compliant fibers where the OH-absorption (or "water-absorption") peak at around 1383 nm is removed.
- The short-wavelength band (S-band) ranging from 1460 to 1530 nm, which offers a further reduced loss than the E-band.
- The conventional band (C-band) ranging from 1530 to 1565 nm, which offers the lowest loss of <0.3 dB/km but relatively high dispersion of up to 19 ps/nm/km.
- The long-wavelength band (L-band) ranging from 1565 to 1625 nm, which also offers the lowest loss of <0.3 dB/km but even higher dispersion of up to 22 ps/nm/km.

International Telecommunication Union (ITU) has specified CWDM over the entire telecommunication window, as shown in Fig. 4.6. Eighteen wavelength channels are spaced by 20 nm starting from 1271 nm and ending at 1611 nm. The useful bandwidth of each wavelength channel (B_{CH}) and the guard band (GB) between adjacent wavelength channels ($\Delta\lambda_{GB}$) are specified to be about 13 and 7 nm, respectively. Owing to the lack of a broadband optical amplifier to amplify all the CWDM channels together, CWDM transmission is usually limited to short distance (e.g., <40 km), and is well suited for front-haul connectivity.

ITU has also specified dense WDM (DWDM) over the C-band and the L-band [12], as shown in Fig. 4.7. Conventionally, the C-band contains 44 100-GHz-spaced DWDM channels between 195.9 THz (corresponding to 1530.33 nm) and 191.6 THz (corresponding to 1564.68 nm). The L-band contains 71 100-GHz-spaced DWDM channels between 191.5 THz (corresponding to 1565.50 nm) and 184.5 THz (corresponding to 1624.89 nm). Note that ITU recently defined finer channel spacings of 50, 25, and 12.5 GHz in the C and L bands with the frequency grids anchored to the common 193.1 THz (corresponding to 1552.52 nm) reference [12]. Owing to the availability of mature C-band optical components such as tunable C-band optical transceivers and WDM multiplexers/demultiplexers, C-band WDM transmission is also a viable option for high-capacity front-haul connectivity.

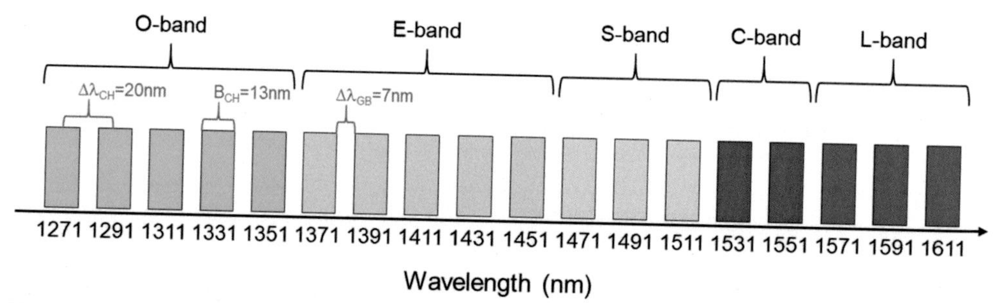

Figure 4.6 The ITU CWDM channel plan.

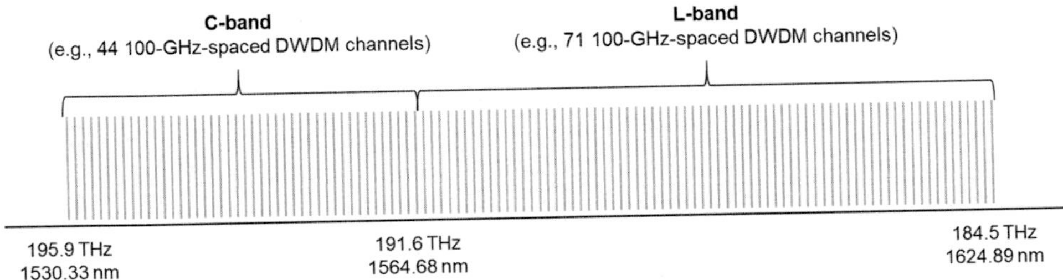

Figure 4.7 The ITU DWDM channel plans in the C-band and the L-band.

For short-distance front-haul links, low-cost optical transceivers based on simple IM/DD, instead of sophisticated digital coherent optical transceivers, are usually used. With DD, fiber dispersion effect cannot be fully compensated as in the case of digital coherent optical detection, thus the dispersion tolerance of a given IM/DD transceiver needs to be carefully examined. Without any loss of generality, the dispersion tolerance (D_{tol}) of a given IM/DD scheme can estimated using Eq. (4.1)

$$D_{tol} = C_{IM,DD}/R_{sym}{}^2 \tag{4.1}$$

where R_{sym} is the optical signal modulation symbol rate, and $C_{IM,DD}$ is a constant that depends on several transceiver settings such as modulation format, transmitter chirp, receiver-side equalization, the reference bit error ratio (BER) at which the dispersion tolerance is measured (as a higher reference BER usually leads to a larger dispersion tolerance), and the performance penalty allocated for the dispersion effect. The commonly used modulation format for front-haul transmission is nonreturn-to-zero (NRZ) on-off-keying, which is usually generated with the following transmitter types.

- Directly modulated laser (DML), which offers low cost and high output power but has a relatively high chirp that leads to a reduced dispersion tolerance.
- Electro-absorption modulated laser (EML), which offers a reduced chirp as compared to DML but incurs an extra loss of transmitter output power (of ~2 dB) due to the electro-absorption.
- Laser followed by an external Mach–Zehnder modulator (MZM), which can achieve chirp-free modulation but incurs an extra loss of transmitter output power (of over 4 dB typically) due to the loss of the external MZM.

The dispersion tolerances of 25-Gb/s NRZ based on a chirp-free MZM transmitter and an EML transmitter were measured to be about 150 and 80 ps/nm, respectively, for a 1-dB penalty at a reference BER of 10^{-3} [13]. With a typical DML transmitter whose chirp is larger than the EML, the dispersion tolerance of 25-Gb/s NRZ would be further reduced to about 50 ps/nm. Note that the use of a receiver-side equalizer, such as a feed-forward equalizer, can increase the dispersion tolerance to some extent [14].

Empirically, the dispersion-induced penalty (in dB), denoted as ΔQ_D, can be estimated from Eq. (4.2)

$$\Delta Q_D = \left(\left|D - D_{opt}\right| / D_{tol,1dB}\right)^{\beta} + \Delta Q_{opt} \quad (4.2)$$

where D_{opt} is the optimum dispersion value at which the dispersion penalty is minimized to ΔQ_{opt}, $D_{tol,1dB}$ is the dispersion tolerance at 1-dB penalty, and β is a transceiver-dependent parameter, which is usually around 2. For a chirp-free transmitter, D_{opt} and ΔQ_{opt} are both zero. For positively chirped DML and EML transmitters, D_{opt} and ΔQ_{opt} are both negative.

Fig. 4.8 shows the dispersion in the O-band and the estimated dispersion penalty for 25-Gb/s NRZ with DML transmitter after 10-km transmission over a standard G.652 fiber. The wavelength window between 1260 and 1320 nm exhibits a superior performance with the dispersion penalty limited to less than 1 dB. If up to 5 dB of dispersion penalty can be accommodated, the allowed maximum wavelength may

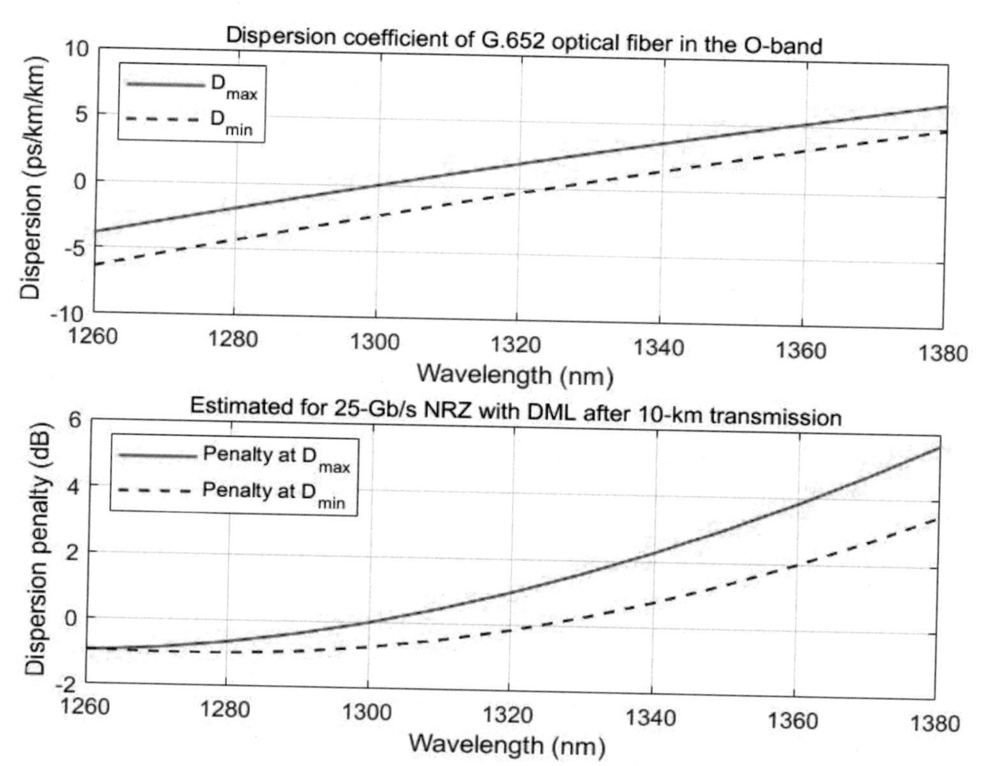

Figure 4.8 The dispersion coefficient (upper) and the estimated dispersion penalty of 25-Gb/s NRZ with DML after 10-km SSMF transmission (lower) in the O-band.

be increased to about 1375 nm. This shows the feasibility of using low-cost DML for WDM front-haul transmission at 25 Gb/s per wavelength channel in the O-band.

Fig. 4.9 shows the estimated dispersion penalty for 25-Gb/s NRZ with chirp-free MZM-based transmitter after 10-km transmission over a standard G.652 fiber in the C and L bands. The C-band (between 1530 and 1565 nm) shows superior performance with the dispersion penalty limited to below 1.5 dB, while the L-band (between 1565 and 1620 nm) shows a lightly higher dispersion penalty of up to about 2 dB. This shows the feasibility of using MZM, together with tunable laser, to support high-capacity DWDM front-haul transmission at 25 Gb/s per wavelength channel in the C and L bands.

For the mid-haul network segment, the transmission distance is typically up to 40 km, much longer than that for the front-haul. In the example shown in Table 4.1, the bit rate to/from a CU per DU is 300 Gb/s per direction, and the aggregated capacity to/from each CU reaches 6 Tb/s. It is thus desirable to use DWDM

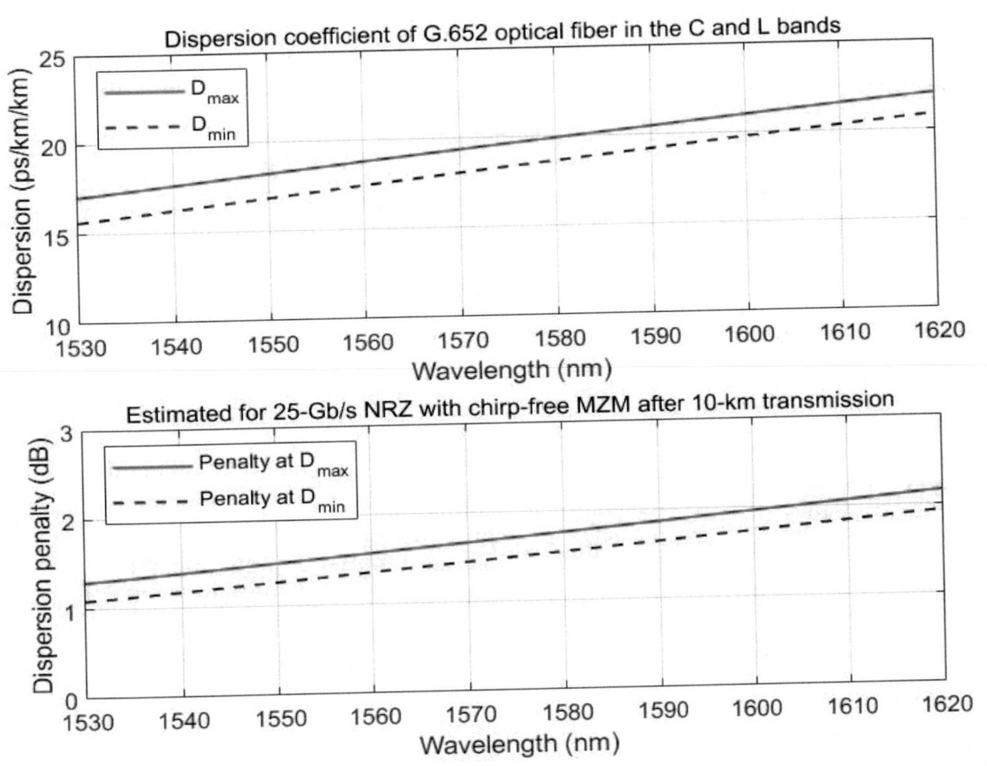

Figure 4.9 The dispersion coefficient (upper) and the estimated dispersion penalty of 25-Gb/s NRZ with chirp-free MZM after 10-km SSMF transmission (lower) in the C and L bands.

transceivers that are capable of higher interface bit rates and 40-km transmission distance. One option is to use 50-Gb/s PAM4 transceivers operating in the O-band. Fig. 4.10 shows the estimated dispersion penalty for 50-Gb/s PAM4 with an EML-based transmitter that has a moderate chirp (with $D_{\text{opt}} = -40$ ps/nm, $D_{\text{tol,1dB}} = 85$ ps/nm, $\beta = 2.4$, and $\Delta Q_{\text{opt}} = 1$ dB) after 40-km transmission over a standard G.652 fiber in the O-band. The dispersion penalty can be limited to less than 2 dB when the signal wavelength is between 1280 and 1322 nm. For backward-compatible and cost-effective operation in the O-band, the DWDM-over-CWDM approach is often desired. In accordance with the dispersion penalty consideration, the third CWDM channel window centered at 1311 nm is deemed suitable for supporting DWDM transmission with 50-Gb/s PAM4 signals. Assuming DWDM channel spacing to be 200 GHz, 16 DWDM channels can be supported in a single CWDM channel slot [15], which leads to a total per-fiber capacity of 800 Gb/s.

To further increase the interface bit rate and per-fiber capacity, spectrally efficient coherent modulation formats, such as polarization division–multiplexed quadrature-phase-shift keying (PDM-QPSK) carrying 4 bits per symbol, can be used. Remarkably, such a coherent signal is commonly received by a digital coherent receiver (DCR) that is capable of compensating linear transmission effects such as dispersion in the digital domain. Thus the combination of coherent modulation and DCR not only increases spectral efficiency but also allows DWDM transmission in any wavelength band without concerning about the dispersion penalty. A 100-Gb/s (net bit rate) PDM-QPSK signal has a spectral bandwidth of about 30 GHz (due to overheads for forward error correction [FEC] and other factors) and can be readily transmitted over the 50-GHz DWDM

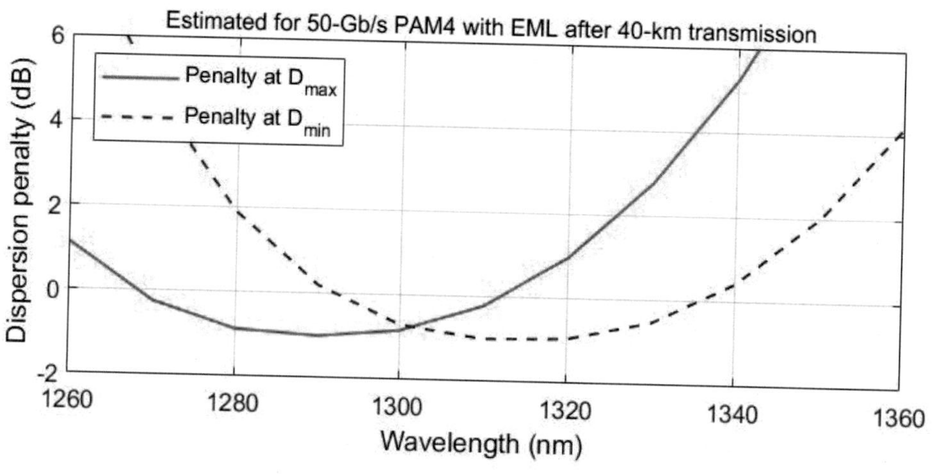

Figure 4.10 The estimated dispersion penalty of 50-Gb/s PAM4 with an EML-based transmitter having a moderate chirp after 40-km SSMF transmission in the O-band.

grid. Using the conventional ITU 50-GHz DWDM grid in the C-band, 80 wavelength channels can be transmitted in a single optical fiber, leading to a per-fiber capacity of 8 Tb/s. This per-fiber capacity is sufficient to carry all the 5G mid-haul traffic to (or from) a CU in the example presented in Table 4.1. Thus DWDM transmission with 100-Gb/s coherent signals is a viable approach to support mid-haul with suitable interface bit rate and high fiber utilization efficiency.

For the back-haul network segment, the transmission distance is typically up to 80 km, even longer than that for the mid-haul. In the example presented in Table 4.1, the bit rate to/from a 5GC per CU is 3 Tb/s per direction, and the aggregated capacity to/from each 5GC reaches 15 Tb/s. It is thus desirable to use DWDM transceivers that are capable of even higher interface bit rates and 80-km transmission distance. More spectrally efficient modulation formats, such as polarization division–multiplexed 16-ary quadrature amplitude modulation (PDM-16QAM) carrying 8 bits per symbol, can be used. To further increase the interface bit rate, a higher modulation symbol rate can be used. A 400-Gb/s (net bit rate) PDM-16QAM optical signal has a spectral bandwidth of about 60 GHz (due to overheads for FEC and other factors) and can be readily transmitted over the 100-GHz DWDM grid. Using the conventional ITU 100-GHz DWDM grid in the C-band, 40 wavelength channels can be transmitted in a single optical fiber, leading to a per-fiber capacity of 16 Tb/s. This per-fiber capacity is sufficient to carry all the 5G back-haul traffic to (or from) a 5GC in the example presented in Table 4.1. Consequently, DWDM transmission with 400-Gb/s coherent signals is a viable approach to support back-haul with suitable interface bit rate and high fiber utilization efficiency.

For the 5GC network segment, the 5GC nodes are usually hosted in metro core data centers. The DCI distance between two metro core data centers is typically up to 80 km, similar to that in the back-haul network segment. In the example presented in Table 4.1, the bit rate from one 5GC to another 5GC is 7.5 Tb/s per direction. Considering that the DCI also needs to carry other communication traffics for fixed broadband services, enterprise private line services, etc., the overall DCI capacity demand for each fiber connection can be similar or even larger than that in the back-haul network segment. Therefore the 5GC network segment can use a similar DWDM approach as in the back-haul case, for example, 40-channel DWDM in the C-band with 400-Gb/s PDM-16QAM optical signals, for an aggregated per-fiber capacity of 16 Tb/s. To achieve even higher capacity per fiber, the superC band with 6-THz of optical spectral bandwidth may be used to increase the per-fiber capacity to 24 Tb/s.

We will describe passive WDM, active WDM, semiactive WDM-PON, and L(M)-WDM for front-haul in Sections 4.4, 4.5, 4.6 and 4.7, respectively. Finally, Section 4.8 will summarize all the key technologies for the front-, mid-, and back-haul, and 5GC network segments.

4.4 Passive WDM–based front-haul

In the passive WDM configuration for 5G front-haul, each fiber carries multiple WDM channels, and OADMs can be deployed at the cell sites to distribute WDM channels to each AAU, as shown in Fig. 4.3B. Due to its technical maturity, C-band DWDM with 100-GHz channel spacing is a main approach for passive WDM–based front-haul. As illustrated in Fig. 4.7, 40 100-GHz-spaced DWDM channels can be readily supported in the C-band. Although fixed-wavelength optical transmitters may be used to deploy the 40-channel DWDM system, they cause great difficulty for wavelength planning, transceiver deployment, and WDM system maintenance. It is thus desirable to use tunable DWDM transmitters, each of which can be tuned to any one of the DWDM wavelengths. To save fiber resource, it is also desirable to realize BiDi transmission to support DL and UL simultaneously by using a single feeder fiber between the DWDM MUX/DMUXs. In the following, we will describe three examples of BiDi passive DWDM transmission systems.

Fig. 4.11 shows the first example of BiDi passive DWDM transmission system that supports BiDi passive DWDM transmission with a single feeder fiber and multiple tunable duplex transceivers (TRXs). Each duplex TRX consists of a tunable transmitter (TX) and a receiver (RX). The tunable TX can consist of a C-band tunable laser followed by an external MZM. The RX can be a PIN photodetector that is also operational in the entire C-band. Thus these tunable TRXs have a unified design for all the wavelength channels and become "colorless," which greatly eases the WDM system deployment and management. The TX and RX are connected to two adjacent ports of a DWDM MUX/DMUX via two drop fibers. Since the drop fibers are usually much shorter than the feeder fiber in real-world applications, it is acceptable to have two drop fibers per optical transceiver. Tunable 25-Gb/s duplex TRXs are now commercially available in compact SFP28 form factor [16], as shown in the inset of Fig. 4.11. Clearly, there are two fiber connection ports to the duplex SFP.

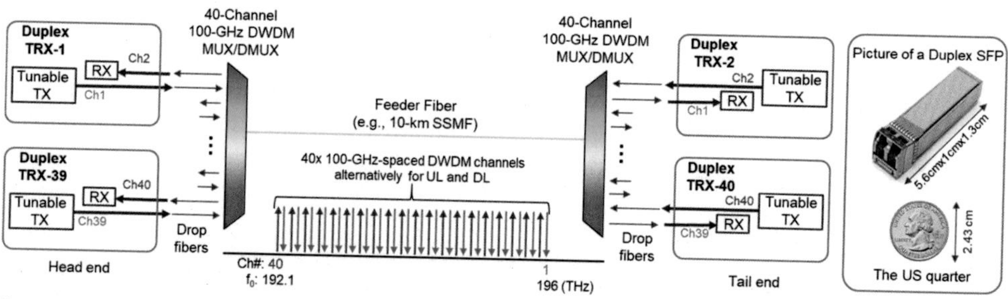

Figure 4.11 Illustration of the first exemplary BiDi passive DWDM transmission system with a single feeder fiber and tunable duplex TRXs.

To further save fiber resource, it is desirable to use only one drop fiber per TRX. In addition, using a single fiber for end-to-end BiDi transmission between two transceivers can minimize the propagation delay difference between the DL and UL channels and facilitate high-accuracy synchronization using precision time protocols such as the IEEE 1588 [17]. Fig. 4.12 shows the second exemplary BiDi passive DWDM transmission system that supports BiDi passive DWDM transmission with a single feeder fiber and multiple tunable BiDi TRXs, each of which is connected to one DWDM MUX/DMUX port via only one drop fiber. In this example, the DWDM MUX/DMUX is based on a cyclic arrayed waveguide grating (AWG) with a passband periodicity property, by which multiple spectral orders are routed to the same output port from an input port [18]. The frequency spacing between the adjacent spectral orders is referred to as the free spectral range (FSR) of the AWG, or Δf_{FSR}. For a 20-channel 100-GHz channelspacing AWG, we can set Δf_{FSR} to 2.6 THz to guarantee a guard band of at least 600 GHz between the DL and UL channels so that a single design of band separation filter (e.g., by a low-cost dielectric thin-film filter) can be used for the separation of the DL and UL wavelength channels in all the BiDi TRXs. Thus these tunable BiDi TRXs also have a unified design for "colorless" operation, which greatly eases the WDM system deployment and management.

The frequency pairing between the 20 DL and 20 UL channels shown in Fig. 4.12 follows the ITU G.698.4 recommendation for 100-GHz-spaced DWDM signals [19]. More specifically, the 20 DL channels are in the frequency range from 194.1 to 196 THz, which pair with the 20 UL channels that are in the frequency range from 191.5 to 193.4 THz. The frequency pairing between the DL and UL channels helps a UL transceiver know which UL wavelength to transmit once knowing which DL wavelength it receives (e.g., via DL in-band signaling). Tunable 25-Gb/s BiDi TRXs are also commercially available in compact SFP28 form factor [20], as shown in the inset of Fig. 4.12. Noticeably, there is only one fiber connection port to the BiDi SFP.

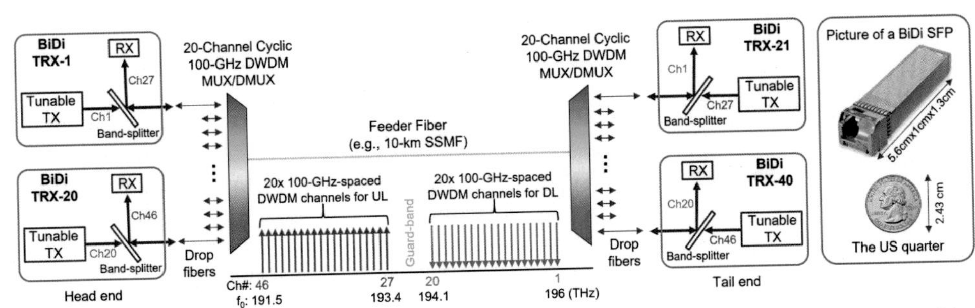

Figure 4.12 Illustration of the second exemplary BiDi passive DWDM transmission system with a single feeder fiber and tunable BiDi TRXs in which UL and DL channels are well separated in frequency.

Fig. 4.13 illustrates the third example of BiDi passive DWDM transmission system that supports BiDi passive DWDM transmission with a single feeder fiber and multiple tunable BiDi TRXs, in which each pair of UL and DL channels are in one 100-GHz frequency slot and are connected to one DWDM MUX/DMUX port via only one drop fiber. In this example, the separation of each pair of UL and DL channels is realized by a three-way circulator, which is functional over the entire C-band. Thus these circulator-based tunable BiDi TRXs also have a unified design for "colorless" operation to ease the WDM system deployment and management. In comparison to the first two approaches, the third approach allows for twice the number of DL and UL channels for the same 100-GHz MUX/DMUX. On the other hand, the effective bandwidth for each DWDM signal is halved. This approach is therefore more suitable for systems where the signal bandwidth is sufficiently small and the signal center wavelength is accurately controlled.

To facilitate the operation of the passive WDM systems, a control and management (C&M) communication link can be established between the head-end (or near-end) transceivers in the DUs and the tail-end (or far-end) transceivers in the AAUs to enable monitoring and control of far-end transceivers, such as automatic wavelength setting. To allow this important C&M communication to be interoperable among transceivers from different suppliers, it is desirable to specify and standardize the C&M interface. In March 2018, ITU-T published the first version of the G.698.4 recommendation (formerly known as "G.metro"), entitled "Multichannel bidirectional DWDM applications with port agnostic single-channel optical interfaces" [19].

The purpose of the G.686.4 recommendation is "to provide optical interface specifications toward the realization of transversely compatible BiDi DWDM systems, primarily intended for metro applications" [19]. Aided by a head-to-tail message channel (HTMC), each tail-end equipment transmitter has the capability to automatically adapt

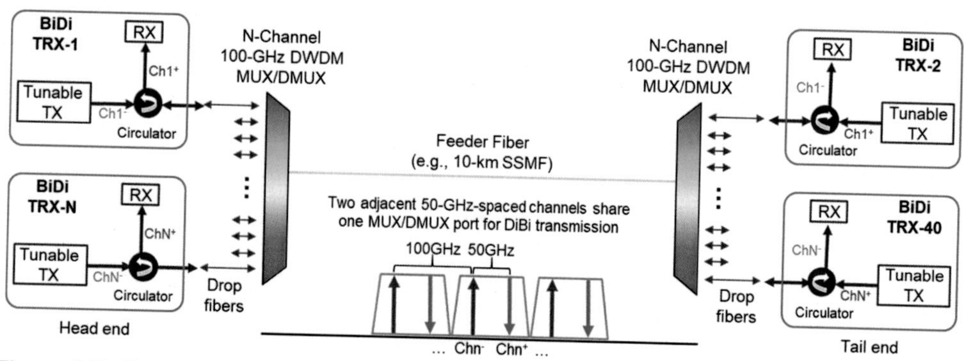

Figure 4.13 Illustration of the third example of BiDi passive DWDM transmission system with a single feeder fiber and tunable BiDi TRXs in which each pair of UL and DL channels are in one 100-GHz frequency slot.

its transmitted signal center frequency to match the connecting DWDM MUX/DMUX port or OADM port, thereby achieving port-agnostic DWDM transmission. The HTMC is carried by pilot tone modulation (with a modulation depth of 5%–8%), which is also specified in the recommendation. The first version of the G.698.4 recommendation describes BiDi DWDM systems with the following features [19]:

- Channel frequency spacing: 100 or 5 GHz.
- Bit rate of each signal channel: Up to 10 Gb/s.
- Transmission distance: Up to 20 km (without inline optical amplifiers).
- Per-fiber capacity: Up to 20 bidirectional channel pairs (i.e., 20 DL channels and 20 UL channels) for the case of 100-GHz channel spacing, and up to 40 bidirectional channels for the case of 50-GHz channel spacing.

For the case of 100-GHz channel spacing, the aggregated BiDi transmission capacity per fiber reaches up to 400 Gb/s. As discussed previously, 5G front-haul demands higher aggregated BiDi transmission capacity, and 25-Gb/s channel data rate is desired. It is thus reasonable for the ITU G.698.4 recommendation to be extended to include 25-Gb/s channel data rate, at least for transmission distance of up to 10 km in the C-band, to increase the aggregated BiDi transmission capacity per fiber to 1 Tb/s. It is worth noting that the ITU G.698.4 recommendation is also applicable to active and semiactive WDM configurations for the front-haul network segment.

4.5 Active WDM–based front-haul

In the active WDM configuration, active WDM equipment, such as OTN DWDM equipment, is added in the CO that contains the DUs, as well as at each cell site, as shown in Fig. 4.3C. With spectrally efficient modulation formats such as 100-Gb/s DMT [6,21], each DWDM channel can simultaneously support multiple eCPRI and CPRI tributaries, which is particularly useful for a cell site with multiple generations of radio technologies. For example, three 25-Gb/s eCPRI signals and three 10-Gb/s CPRI Option 7 signals (each of which has a net data rate of 8 Gb/s after the 8b/10b line coding is terminated) can share one 100-Gb/s (net data rate) and the aggregation of eCPRI/CPRI signals can be done by the OTN equipment, as illustrated in Fig. 4.14.

The use of OTN equipment also offers advanced network capabilities such as intelligent OAM, performance monitoring, and protection. Thus OTN equipment has been deployed to support the front-haul deployment with high bandwidth efficiency, high reliability, accurate synchronization among the connected nodes, and easy installation and configuration, as illustrated in Fig. 4.15 [6]. To achieve a high reliability, path protection can be applied. To achieve accurate synchronization among the connected nodes, precision time protocols such as that defined by the IEEE 1588 standard [17] can be applied. In order to maintain the accurate synchronization even in the event of path protection, latency compensation needs to be applied to ensure that the

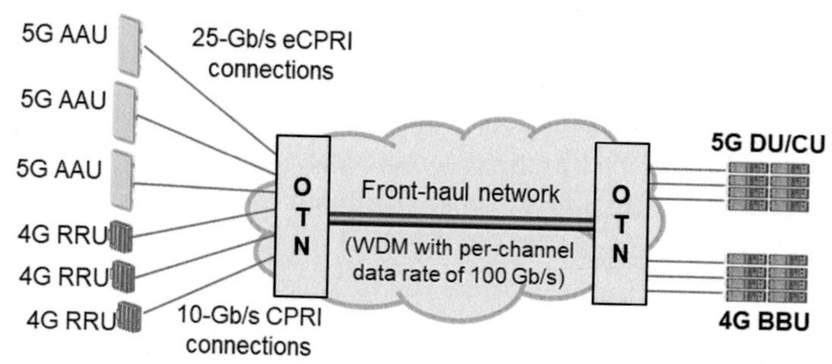

Figure 4.14 Schematic of an OTN-based active WDM system for front-haul with 100-Gb/s channels each aggregating 5G and 4G traffics.

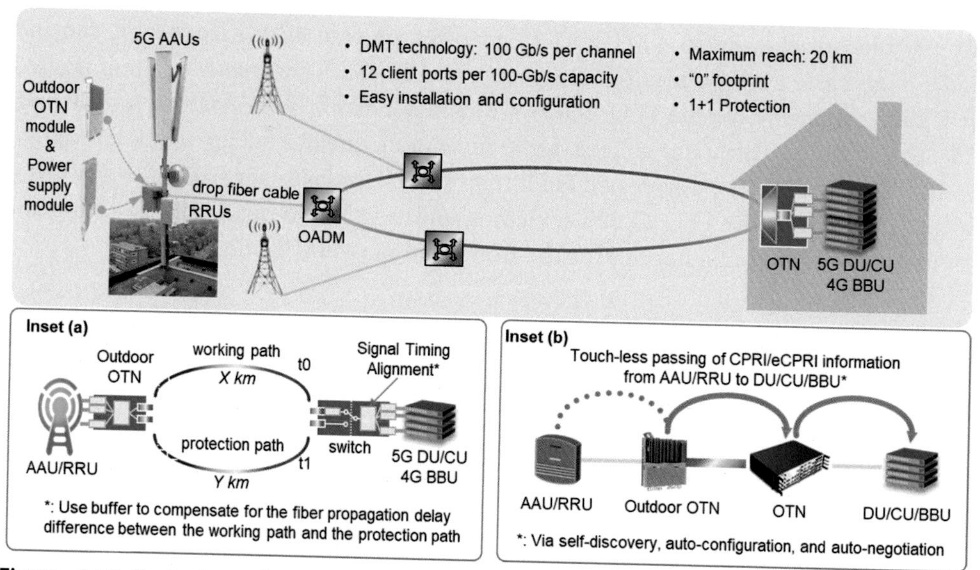

Figure 4.15 Illustration of an OTN-based front-haul network. Inset (a): latency compensation between the original and protection paths; inset (b): easy OAM via self-sensing of CPRI information and automatic negotiation and configuration.

protection path and the original path are equal in propagation delay, as illustrated in Inset (a). For an easy installation, it is desirable for OTN to be able to sense the CPRI information and conduct configuration and negotiation automatically, as illustrated in Inset (b).

To enable fast and cost-effective deployment of the OTN-based front-haul solution, innovations have been made to achieve the following features [22].

- Fully outdoor operation, by making the OTN equipment into a compact module that satisfies the outdoor environment requirements for temperature (i.e., −40°C to 55°C), humidity, etc.

- "0" footprint installation, by making the OTN equipment module, as well as the associated power supplier module, to have sufficiently small SWaP (size, weight, and power consumption) for mounting with other outdoor radio equipment modules (e.g., on a pole), thereby not requiring any equipment space on the ground, as shown in Fig. 4.15.
- Flexible aggregation of various interfaces, such as GE, 10GE, 25GE, CPRI and eCPRI, etc., with up to 12 client-side ports per 100-Gb/s line-side port, thus dramatically reducing the number of optical fibers needed for front-haul.
- Advanced OAM capabilities such as fault locating and visualized OAM management.

From the above, it can be seen that the ONT-based front-haul solution is an attractive option for the deployment of 5G front-haul links that connect cell sites having multiple generations of radio technologies with various interface bit rates and/or require advanced OAM capabilities.

4.6 Semiactive WDM-PON-based front-haul

In the semiactive WDM configuration, the active equipment at the CO can be a WDM-PON OLT, as shown in Fig. 4.3A. No active optical equipment and additional power suppliers are needed at the cell sites, as the wavelength channels are transmitted and received by pluggable optical network units (ONUs) inside the AAUs. Fig. 4.16 shows the schematic of a recently demonstrated WDM-PON-based semiactive WDM system for front-haul with 20 wavelength pairs for end-to-end BiDi transmission [7]. In this demonstration, the pairing of the downstream and upstream wavelengths follows the ITU G.698.4 recommendation for 100-GHz-spaced DWDM channels, as shown in Fig. 4.17. The 20 downstream channels are in the wavelength range from 1529.55 nm (corresponding to 196 THz) to 1544.53 nm (corresponding to 194.1 THz), which pair

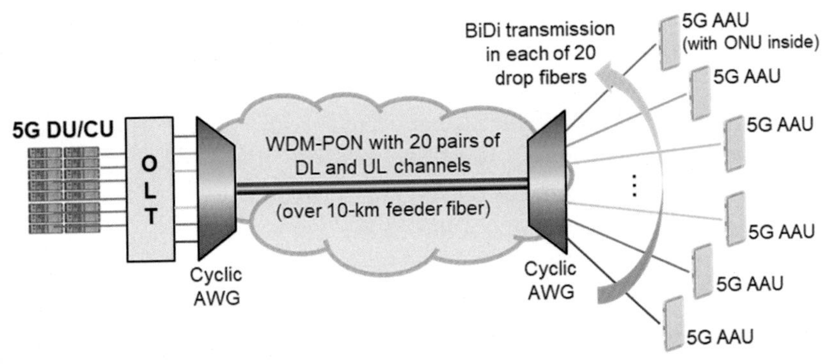

Figure 4.16 Schematic of a WDM-PON-based semiactive WDM system for front-haul with 20 wavelength pairs for end-to-end BiDi transmission.

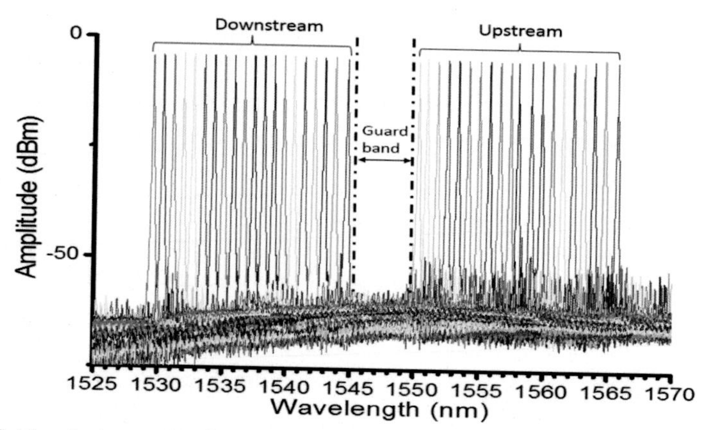

Figure 4.17 Pairing between 20 downstream and 20 upstream wavelengths in accordance with the ITU G.698.4 recommendation for 100-GHz-spaced DWDM channels [7].

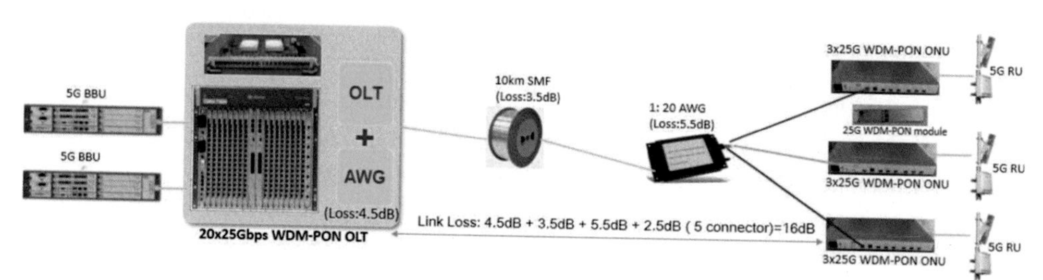

Figure 4.18 Experimental setup of a real-time WDM-PON system for 5G front-haul supporting 20 pairs of bi-directional 25-Gb/s signals [7].

with the 20 upstream channels that are in the wavelength range from 1550.12 nm (corresponding to 193.4 THz) to 1565.50 nm (corresponding to 191.5 THz), respectively. At the OLT side, a cyclic AWG with a FSR of 2.6 THz is used to multiplex all the downstream signals and demultiplex all the upstream signals. At the ONU side, another identical cyclic AWG is used to enable each pair of downstream and upstream signals to coexist in a drop fiber that is connected to the intended ONU for end-to-end BiDi transmission. The guard band between the downstream and the upstream channels is approximately 4.8 nm (or 600 GHz).

Fig. 4.18 shows the experimental setup of the WDM-PON-based front-haul supporting 20 pairs of BiDi 25-Gb/s signals [7]. From the figure, it is evident that the link loss for the WDM-PON system is 16 dB. If we allocate 2 dB for the optical path penalty and 3 dB for the system margin, the required total optical power budget becomes 21 dB. To reduce the cost of the optical transceiver modules, a unified 10-GHz-class commercial tunable transmitter optical subassembly, which consisted of a tunable laser and an external MZM, was used for both ONU and OLT optical transceiver modules, which were in the form of SFP28. To achieve high receiver sensitivity, an avalanche photodiode–based receiver was used. Fig. 4.17 shows the output spectrum of a representative WDM-PON transmitter capable of generating any of the 40 downstream and upstream signals with high SNR. The minimal optical output power of the transmitter is 0 dBm.

Fig. 4.19 shows the measured raw BER performance of the 25-Gb/s transceiver as a function of the received optical power in the back-to-back (B2B) case and after 10-km transmission in standard single-mode fiber (SSMF). The B2B eye diagram of the WDM-PON transmitter is shown as the inset. With electrical equalization, the extinction ratio of the transmitted signal reaches 8.2 dB. The dispersion penalty after 10-km transmission in SSMF is measured to be less than 1 dB, which is smaller than that estimated for chirp-free modulation (as shown in Fig. 4.9). This is because the MZM used in this demonstration was configured to generate a slight negative chirp to help mitigate the dispersion penalty [7]. At the input BER threshold of commonly used Reed–Solomon FEC (RS-FEC) (~5E-5), the receiver sensitivities for both B2B and 10-km transmission cases are better than −22 dBm, indicating a link loss budget of over 22 dB, which is sufficient to satisfy the loss budget requirement of 21 dB for the WDM-PON system.

To facilitate the control and management of the WDM-PON system, an auxiliary management and control channel (AMCC) was introduced into the system. Remodulation of the optical signal to carry the AMCC had been studied [23,24]. There is some sensitivity penalty caused by the remodulation method. Alternatively, the AMCC channel can be embedded in the codeword marker used for FEC synchronization [7]. This approach induces no power penalty and is compatible with the

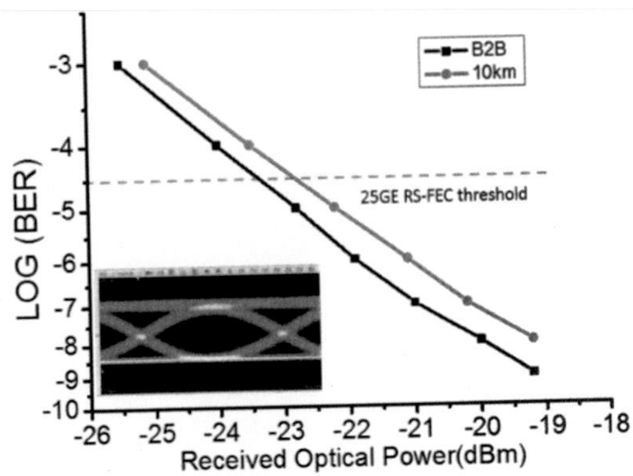

Figure 4.19 Experimentally measured raw BER performance of a 25-Gb/s signal as a function of the received optical power. Inset: typical measured 25-Gb/s NRZ eye diagram from the tunable transmitter [7].

standard 25GE interface. Fig. 4.20 presents the penalty-free AMCC method. When the RS-FEC is enabled, there is a codeword marker for every block of 1024 RS (528,514) codewords. The codeword marker is used for the reception side to synchronize the FEC codeword. The codeword marker has 257 bits with 1 bit as the synchronization header. For 25G Ethernet, only 64 bits of the codeword marker are used for FEC synchronization, and the rest 192 bits are filled with fixed patterns. The detailed information of the codeword marker can be found in IEEE 802.3 [25]. Thus the 192 bits of fixed pattern can be used for the AMCC channel. The OAM function can be realized through this AMCC channel. The data rate of the AMCC channel is up to 900 kb/s, which is much higher than that reported for the remodulation methods. In addition, this method of embedding the AMCC channel in the RS-FEC codeword marker neither induces any power penalty for the reception side nor requires any additional hardware. In the real-time WDM-PON demonstration, OAM functions such as ONU discovering, auto wavelength channel allocation, link BER reporting, optical module parameter reporting, and round-trip delay measurement have been realized [7].

To support 5G services with a high reliability, various PON protection schemes can be applied [26]. Commonly used PON protection schemes include

- Type A, where only the feeder fiber is protected
- Type B, where both the feeder fiber and the OLT equipment are protected
- Type C, where the OLT, optical distribution network, and ONUs are all protected.

In the demonstration of WDM-PON-based front-haul reported in Ref. [7], automatic Type-B protection was implemented with a short protection switch time of less than 20 ms. To ensure that the WDM-PON system also satisfies the latency and jitter requirements for 5G front-haul, real-time service performance test was conducted with the WDM-PON system connected to actual 5G equipment. The total system

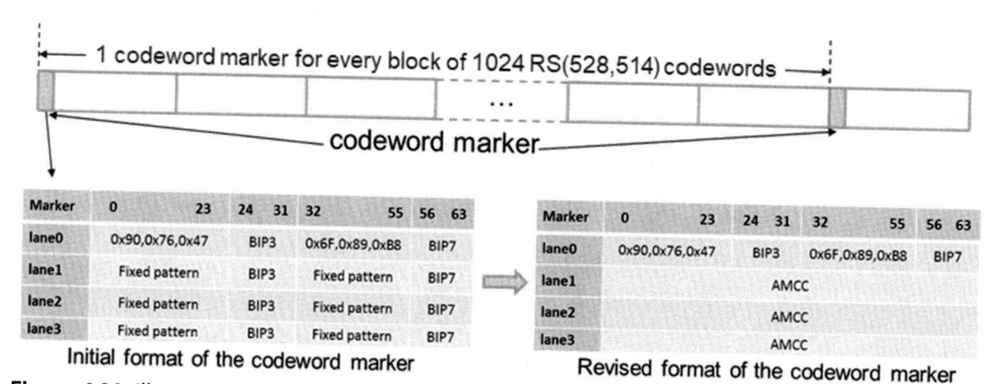

Figure 4.20 Illustration of the penalty-free method to carry the AMCC channel by embedding it in the FEC codeword marker [7].

capacity was 500 Gb/s for each direction. The maximal one-way latency after 10-km fiber transmission was measured to be 56 μs, which included a one-way fiber propagation latency of 10 μs and a one-way signal processing latency of 6 μs, well satisfying the one-way front-haul latency requirement of 100 μs. The jitter was measured to be less than 1 μs, also satisfying the corresponding requirement. Finally, the real-time service performance was measured for over 10 hours and there was no package loss observed. Owing to the quick protection witching time of less than 20 ms, there was no interruption of real-time voice service during Type-B protection events [7].

The above demonstration shows the promise of using WDM-PON to support 5G front-haul with a high capacity, low cost, easy management, and reliable protection. In 2020 ITU had started the standardization effort on WDM-PON for 5G front-haul applications [27], and more industry-wide developments on the WDM-PON technology can be expected.

4.7 Semiactive L- and M-WDM-based front-haul

As illustrated in Fig. 4.4B, there is another implementation option of semiactive WDM for 5G front-haul based on 12-channel L-WDM [9] or M-WDM [8] in the O-band. The 12-channel L-WDM leverages the ecosystem for LAN-WDM [28] and expands the number of wavelength channels from 8 (in LAN-WDM) to 12, as shown in Table 4.3.

Based on Fig. 4.8, the dispersion penalty of 25-Gb/s NRZ with DML after 10-km SSMF transmission in the L-WDM wavelength range (from ~1268 to ~1319 nm)

Table 4.3 Channel plan of the 12-channel L-WDM with respect to the 8-channel LAN-WDM.

L-WDM channel index	Corresponding LAN-WDM channel index	Center frequency (THz)	Center wavelength (nm)	Passband bandwidth (nm)
Lch1		236.2	1269.23	~2
Lch2	L0	235.4	1273.54	~2
Lch3	L1	234.6	1277.89	~2
Lch4	L2	233.8	1282.26	~2
Lch5	L3	233.0	1286.66	~2
Lch6		232.2	1291.10	~2
Lch7	L4	231.4	1295.56	~2
Lch8	L5	230.6	1300.05	~2
Lch9	L6	229.8	1304.58	~2
Lch10	L7	229.0	1309.14	~2
Lch11		228.2	1313.73	~2
Lch12		227.4	1318.35	~2

can be less than 1 dB, which shows the superior transmission performance of the L-WDM in carrying 12 25-Gb/s NRZ signals with an aggregated per-fiber capacity of 300 Gb/s. To allow for end-to-end BiDi transmission, two 1:6 cyclic AWGs can be used at the head end and the tail end such that each pair of BiDi channels can share the same fiber throughout the front-haul link (including the feeder fiber and drop fiber sections). The FSR of each of the 1:6 cyclic AWGs can be set to 4.8 THz, in which case the first six L-WDM channels, Lch1–Lch6, are paired with the last six channels, Lch7–Lch12, respectively.

To separate each pair of L-WDM BiDi channels in an optical transceiver, an optical circulator can be used. As the optical circulator is usually a broadband device that can operate over the entire O-band, this circulator-based band separation works universally for all the six pairs of BiDi channels. Alternatively, a WDM band separation filter can also be used, although the filter design has certain dependence on the wavelengths of the pair of BiDi channels. If a guard band is introduced (as in ITU G.698.4), then a universal WDM band separation filter can be used for all the L-WDM transceivers. As an example, the 800-GHz channel slot between LAN-WDM channels L3 and L4 can be used as the guard band, and the first six L-WDM channels can be shifted up in frequency by 800 GHz to accommodate this guard band. The channel passband bandwidth in the 12-channel L-WDM is the same as that in LAN-WDM, that is, ~2 nm, so the transmitter components developed for LAN-WDM to meet the laser wavelength stability requirement may be reused for L-WDM.

The 12-channel M-WDM leverages the ecosystem for CWDM and doubles the number of wavelength channels to 12 in the wavelength range of the first six CWDM channels, by shifting each CWDM channel by ±3.5 nm, as shown in Table 4.4. The channel plans of the 12-channel L- and M-WDM are shown in Fig. 4.21.

Based on Fig. 4.8, the dispersion penalty of 25-Gb/s NRZ with DML after 10-km SSMF transmission in the M-WDM wavelength range (from ~1266 to ~1376 nm) can span from −1 to 5 dB, so efforts need to be taken to mitigate the impact of the dispersion penalty. Using twelve 25-Gb/s NRZ signals in the M-WDM system, an aggregated per-fiber capacity of 300 Gb/s can also be achieved. To allow for the end-to-end BiDi transmission, each pair of channels that share the same CWDM channel slot can be routed together throughout the front-haul link by the CWDM multiplexer and de-multiplexer. For WDM systems with low channel count and large channel spacing, such as CWDM, thin-film filters (TFFs) are usually used for channel multiplexing and de-multiplexing. Each TFF is designed to pass a given CWDM channel and reflect all CWDM channels, which is realized by multiple layers of dielectric coatings designed to pass the given channel. A set of concatenated TFFs work together to pass (or filter) the CWDM channels one by one. As each TFF causes a loss to the reflected channels (e.g., ~0.5 dB for each reflection), TFF-based MUX/DMUX for M-WDM can be designed such that the longer the wavelengths of

Table 4.4 Channel plan of the 12-channel M-WDM with respect to the first six CWDM channels.

CWDM channel index	Center wavelength (nm)	M-WDM channel index	Center wavelength (nm)	Passband bandwidth (nm)
CWDM1	1271	C1 −	1267.5	~5
		C1 +	1274.5	~5
CWDM2	1291	C2 −	1287.5	~5
		C2 +	1294.5	~5
CWDM3	1311	C3 −	1307.5	~5
		C3 +	1314.5	~5
CWDM4	1331	C4 −	1327.5	~5
		C4 +	1334.5	~5
CWDM5	1351	C5 −	1347.5	~5
		C5 +	1354.5	~5
CWDM6	1371	C6 −	1367.5	~5
		C6 +	1374.5	~5

CWDM, Coarse wavelength-division multiplexing; *M-WDM*, medium wavelength-division multiplexing.

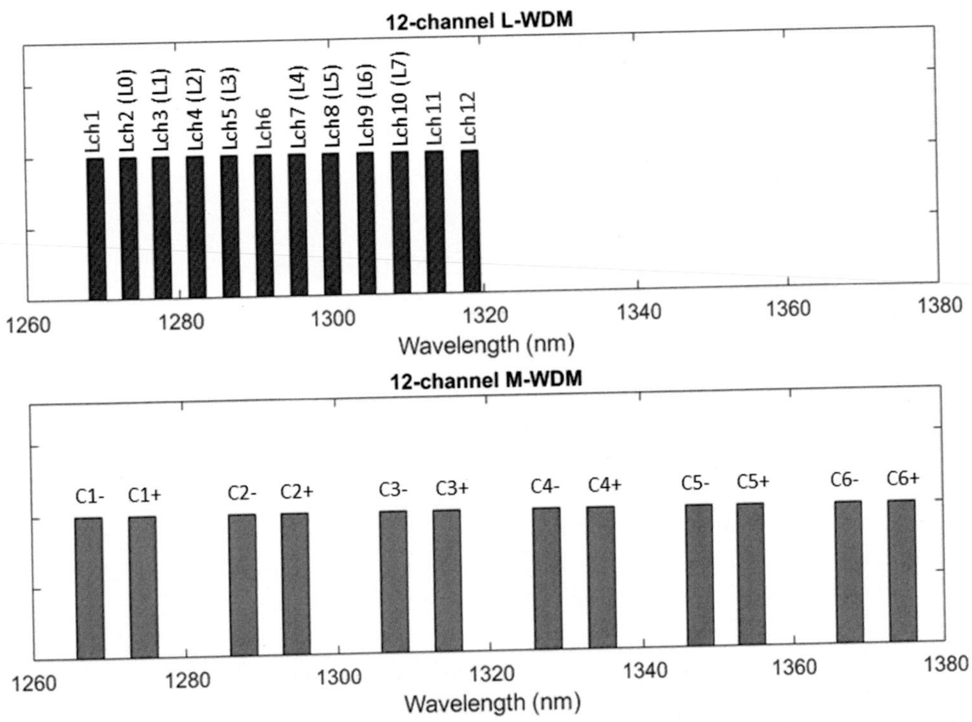

Figure 4.21 Channel plans of the 12-channel L-WDM (upper) and the 12-channel M-WDM (lower).

a pair of M-WDM channels, the sooner the pair gets passed, thereby reducing the MUX/DMUX loss for longer wavelength pairs that suffer more from the dispersion penalty [8] By doing so, the overall link budgets for all the M-WDM channels can become more even, and the sum of the dispersion penalty and MUX and DMUX losses can be limited to about 7 dB. Assuming a 4-dB fiber loss for 10-km SSMF, a 2-dB connector loss for all the needed connectors, and a 2-dB loss margin, the total link power budget needs to be around 15 dB, which can be readily achieved by using cost-effective 25-Gb/s DML with over 0 dBm output power and 25-Gb/s PIN-based receiver with receiver sensitivity better than −15 dBm.

To separate each pair of M-WDM BiDi channels in an optical transceiver, an optical circulator can be used universally for all the six pairs of BiDi channels. The channel passband bandwidth in the 12-channel M-WDM is about 5 nm, which is larger than that for the 12-channel L-WDM. Thus the laser wavelength stability requirement in M-WDM is more relaxed as compared to that in L-WDM. The thermal electronic cooler (TEC) temperature control technique can be applied to both L- and M-WDM lasers to satisfy the needed laser wavelength stability requirements. Note that the TEC may introduce a slight increase in the power consumption of the optical transceiver module (e.g., by ~0.5 W). On the other hand, the TEC may be needed anyway for outdoor deployment where the operating temperature range is large (e.g., the industrial temperature range from −40°C to 85°C). It is also worth noting that other CWDM channels outside the O-band may be used to support legacy back-haul connection needs via lower speed transceivers (e.g., 10-Gb/s NRZ transceivers). Thus it can be expected that L-WDM, M-WDM, and legacy CWDM will work together to collectively support the various front- and back-haul needs.

4.8 Summary and outlook on X-haul technologies

Table 4.5 summarizes the key optical transmission technologies and configurations for front-, mid-, and back-haul, and 5GC network segments that have been presented in this chapter.

Regarding the wavelength multiplexing schemes, L-WDM, M-WDM, WDM, DWDM are expected to be used to collectively address the large variety of 5G X-haul demands. Regarding the transceiver technologies, both IM/DD and coherent modulation/detection are needed, to support short- and long-reach applications. It is worth noting that coherent transceivers are more complex and expensive than IM/DD transceivers, and industry-wide efforts and collaborations to reduce the cost and power consumption of coherent transceivers are essential. As an effort to address the important common need of metro-network and metro-DCI for 400-Gb/s coherent transceivers that are optimized for 80-km DWDM transmission, the Optical Internetworking Forum (OIF) published the first implementation agreement for 400G-ZR (where ZR stands for 80-km

Table 4.5 Key optical transmission technologies and configurations for 5G X-haul.

Network segment	Front-haul		Mid-haul		Back-haul/5GC
Typical distance	< 10 km		< 40 km		< 80 km
Distance grade [29]	LR (long reach)		ER (extended reach)		ZR (~80 km reach)
Temperature range	− 40°C to 85°C		0°C−70°C		0°C−70°C
Wavelength band	O-band[a]	C-band	O-band[a]	C-band	C band[b]
Typical WDM type	L(M)-WDM	DWDM	WDM	DWDM	DWDM
Number of channels	12	40	16	80	40[b]
Channel data rate	25 Gb/s	25 Gb/s	50 Gb/s	100 Gb/s	400 Gb/s[c]
Total capacity per fiber	300 Gb/s	1 Tb/s	800 Gb/s	8 Tb/s	16 Tb/s[b]
Modulation format	NRZ	NRZ	PAM4	PDM-QPSK	PDM-16QAM
Transmitter/receiver type	DML/PIN	MZM/PIN/APD	EML/APD	PDM-IQ-MZM/DCR	PDM-IQ-MZM/DCR
Typical optical equipment	SFP28, OTN	SFP28, OTN	SFP56, OTN	QSFP-DD, OTN	QSFP-DD, OTN
Related standards	CCSA, ETSI-F5G	ITU G.698.4, ITU G.hsp	ITU G.698.2, OIF 400G-ZR, IEEE 802.3cw		OIF 400G-ZR, IEEE 802.3cw

APD, Avalanche photodiode; *CCSA*, China Communications Standards Association; *DCR*, digital coherent receiver; *DML*, Directly modulated laser; *DWDM*, dense WDM; *ETSI*, European Telecommunications Standards Institute; *ITU*, International Telecommunication Union; *MZM*, Mach−Zehnder modulator; *NRZ*, nonreturn-to-zero; *OIF*, Optical Internetworking Forum; *OTN*, optical transport network; *PAM4*, 4-level pulse-amplitude modulation; *PDM-QPSK*, polarization division−multiplexed quadrature-phase-shift keying; *PIN*, PIN photodetector.
[a] O-band DWDM with 100-Gb/s DMT channels may also be used.
[b] Super-C and/or L-band may be used to support more DWDM channels and per-fiber capacity.
[c] Other interface rates such as 800 Gb/s may also be used.

reach) coherent optical interface on April 29, 2020 [30]. The OIF 400G-ZR project aims to create an interoperable, low-cost 400 Gigabit coherent interface that addresses two key applications [30]:

- Amplified, point-to-point DWDM links with reaches of 120 km or less.
- Unamplified, single wavelength links with a loss budget of 11 dB.

Remarkably, it has been demonstrated that such a 400G-ZR coherent transceiver can be in the form of quad small form factor pluggable double density with a power consumption of <15 W [31]. In comparison, the primary DCI transceiver option used between 2016 and 2021 is 100G PAM4 (using two optical carriers) in the form of QSFP28 with a power consumption of 4.5 W, which is over 20% more than that of 400G-ZR for the same bit rate. This shows that through industry-wide innovations and collaborations, coherent transceivers have been made to be more energy-efficient than IM/DD transceivers for high-speed DWDM transmission over 80 km. On December 8, 2020, OIF launched the *800G Coherent* project, with the aim to define interoperable 800 Gb/s coherent line specifications for campus and DCI applications with transmission distance up to 80−120 km [32].

With continued advances in this field, it can be expected that coherent transceivers will find even more applications in short-reach optical transmission systems. Furthermore, most of the key optical technologies needed for 5G X-haul are similar to those needed for DCI and intra-data-center communications, so a shared optical fiber communication ecosystem can be developed to benefit both 5G and Cloud. In the following chapters, we will continue to discuss short-reach optical communication technologies, which include radio-over-fiber technologies, point-to-multipoint architectures for resource-efficient network traffic aggregation, and data center optics including both DCI and intra-data-center optics, before presenting other topics such as long-haul optical transmission technologies and advanced optical network technologies in the remaining chapters.

References

[1] The 3rd Generation Partnership Project (3GPP), <http://www.3gpp.org/>.

[2] See for example, <https://www.rcrwireless.com/20200609/5g/china-end-2020-over-600000-5g-base-stations-report>.

[3] China Telecom. 5G-Ready OTN technical white paper [Online]. <http://www.ngof.net/en/download/5G-Ready_OTN_Technical_White_Paper.pdf>; September 2017.

[4] J. Li. Photonics for 5G in China. In: Asia Communications and Photonics Conference (ACP), paper S4E.1; 2017.

[5] Chanclou P, Neto LA, Grzybowski K, Tayq Z, Saliou F, Genay N. Mobile fronthaul architecture and technologies: a RAN equipment assessment [invited]. IEEE/OSA J Opt Commun Netw 2018;10(1):A1−7.

[6] X Liu, N Deng. Chapter 17: Emerging optical communication technologies for 5G. In: Willner A, editor. Optical fiber telecommunications VII; 2019.

[7] X. Wu, D. Zhang, Z. Ye, H. Lin, X. Liu. Real-time demonstration of 20×25Gb/s WDM-PON for 5G fronthaul with embedded OAM and type-B protection. In: Optoelectronics and communications conference (OECC), Paper TuA3-2; 2019.

[8] H. Li. Vision and trend analysis for transport networks in 5G era. In: Asia Communications and photonics conference (ACP), Plenary Talk; 2020.

[9] J. Li. Recent advances in next-generation optical transport networks. In: European conference on optical communication (ECOC), Invited Talk in Workshop 14; 2020.

[10] F. Saliou, L.A. Neto, G. Simon, F.N. Sampaio, A.E. Ankouri, M. Wang, et al. 5G & optics in 2020 – Where are we now? What did we learn? In: European conference on optical communication (ECOC), Invited Talk We2J-1; 2020.
[11] ITU-T Recommendation G.652: Characteristics of a single-mode optical fibre and cable; 2016.
[12] ITU-T Recommendation G.694.1: Spectral grids for WDM applications: DWDM frequency grid; 2020.
[13] Houtsma V, van Veen D. A study of options for high-speed TDM-PON beyond 10G. J Lightwave Technol 2017;35(4):1059–66.
[14] M. Tao, L. Zhou, H. Zeng, S. Li, X. Liu. 50-Gb/s/λ TDM-PON based on 10G DML and 10G APD supporting PR10 link loss budget after 20-km downstream transmission in the O-band. In: Optical fiber communications conference (OFC), Paper Tu3G.2; 2017.
[15] See for example, <https://www.prnewswire.com/news-releases/o-band-athermal-awg-dwdm-muxdmx-technology-for-5g-fronthaul-optical-data-transport-network-solution-300804639.html>.
[16] See for example, <https://ii-vi.com/product/25ge-lr-10km-sfp28-optical-transceiver/>.
[17] IEEE Standard 1588-2019: IEEE standard for a precision clock synchronization protocol for networked measurement and control systems; 2019.
[18] Banerjee A, Park Y, Clarke F, Song H, Yang S, Kramer G, et al. Wavelength-division-multiplexed passive optical network (WDM-PON) technologies for broadband access: a review [Invited]. J Opt Netw 2005;4(11):737–58.
[19] ITU-T Recommendation G.698.4: Multichannel bi-directional DWDM applications with port agnostic single-channel optical interfaces; 2018.
[20] See for example, <https://compoundsemiconductor.net/article/106551/Finisar_Unveils_latest_Tech_at_OFC_2019>.
[21] T Takahara, T Tanaka, M Nishihara, Y Kai, L Li, Z Tao, et al. Discrete multi-tone for 100 Gb/s optical access networks. In: Optical fiber communication conference (OFC), paper M2I.1; 2014.
[22] See for example, <https://telecoms.com/intelligence/huawei-launches-full-range-of-5g-end-to-end-product-solutions/> and <https://info.support.huawei.com/network/ptmngsys/Web/WDMkg/en/48_5gfo.html>.
[23] G Nakagawa, K Sone, S Oda., S Yoshida, Y Aoki, M Takizawa, et al. Experimental investigation of AMCC superimposition impact on CPRI signal transmission in DWDM-PON network. In: European conference on optical communication (ECOC), Paper Th.1.D.2; 2016.
[24] Wagner C, et al. Impairment analysis of WDM-PON based on low-cost tunable lasers. J Lightwave Technol 2016;34(22):5300–7.
[25] IEEE 802.3by: IEEE standard for ethernet – amendment 2: media access control parameters, physical layers, and management parameters for 25 Gb/s operation; 2016.
[26] ITU-T G-series Recommendations – supplement 51: passive optical network protection considerations; 2017.
[27] Zhang D, Liu D, Wu X, Nesset D. Progress of ITU-T higher speed passive optical network (50G-PON) standardization. J Opt Commun Netw 2020;12:D99–108.
[28] IEEE 802.3bs: IEEE standard for ethernet amendment 10: media access control parameters, physical layers, and management parameters for 200 Gb/s and 400 Gb/s operation; 2017.
[29] See for example, <https://en.wikipedia.org/wiki/10_Gigabit_Ethernet>.
[30] See for example, <https://www.oiforum.com/oif-publishes-implementation-agreement-for-400zr-coherent-optical-interface/>.
[31] M. Filer. The coherent 400 Gb/s revolution. In: Lightreading optical networking digital symposium; May 28, 2020.
[32] See for example, <https://www.oiforum.com/oif-launches-800g-coherent-and-co-packaging-framework-ia-projects-elects-new-board-members-positions-officers-and-working-group-chairs/>.

CHAPTER 5

Digital signal processing–assisted radio-over-fiber

5.1 Overview on radio-over-fiber

Radio-over-fiber (RoF) generally refers to the transmission of an optical signal that carries one or more radio signals over an optical fiber [1–3]. Because of the use of optical fiber as the transmission medium, RoF offers low transmission loss and immunity to electromagnetic interference. RoF has been used for carrying wireless signals [4–7], as well as cable television signals, for example, via video overlay over passive optical networks in the fiber-to-the-home scenario [8–10]. In this chapter, we focus on the use of RoF for mobile front-haul transmission.

RoF is traditionally based on analog modulation and detection. Due to the availability of high-speed digital-to-analog converters (DACs) and analog-to-digital converters (ADCs) in recent years, digital signal processing (DSP)-assisted RoF techniques have been introduced to improve both the bandwidth efficiency and the signal fidelity for RoF transmission [11–23]. There are four common frequency-domain analog RoF (A-RoF) transmission schemes as follows.

- A-RoF carrying one radio frequency (RF) signal, as shown in Fig. 5.1A, where the RF signal waveform is directly modulated onto an optical carrier via electrical-to-optical conversion (E-to-O) such that the frequency spacing between the optical carrier and the center frequency of the RoF signal equals the RF carrier frequency, f_{RF}. Upon direct detection of the RoF signal for optical-to-electrical conversion (O-to-E), the original RF signal centered at f_{RF} is obtained without additional frequency conversion. Bi-direction transmission of downstream and upstream RoF signals over a single optical fiber can be achieved by using wavelength-division multiplexing. The drawback of this RoF scheme is its bandwidth inefficiency as f_{RF} is usually much larger than the RF signal bandwidth.
- A-RoF carrying one intermediate frequency (IF) signal, as shown in Fig. 5.1B, where the RF signal is first downconverted to an IF before being directly modulated onto an optical carrier via E-to-O such that the frequency spacing between the optical carrier and the center frequency of the RoF signal equals the IF, f_{IF}. Upon direct detection of the RoF signal for O-to-E, the IF signal centered at f_{IF} is obtained. With a frequency upconversion by (f_{RF}-f_{IF}), the original RF signal can

Optical Communications in the 5G Era
DOI: https://doi.org/10.1016/B978-0-12-821627-9.00010-3

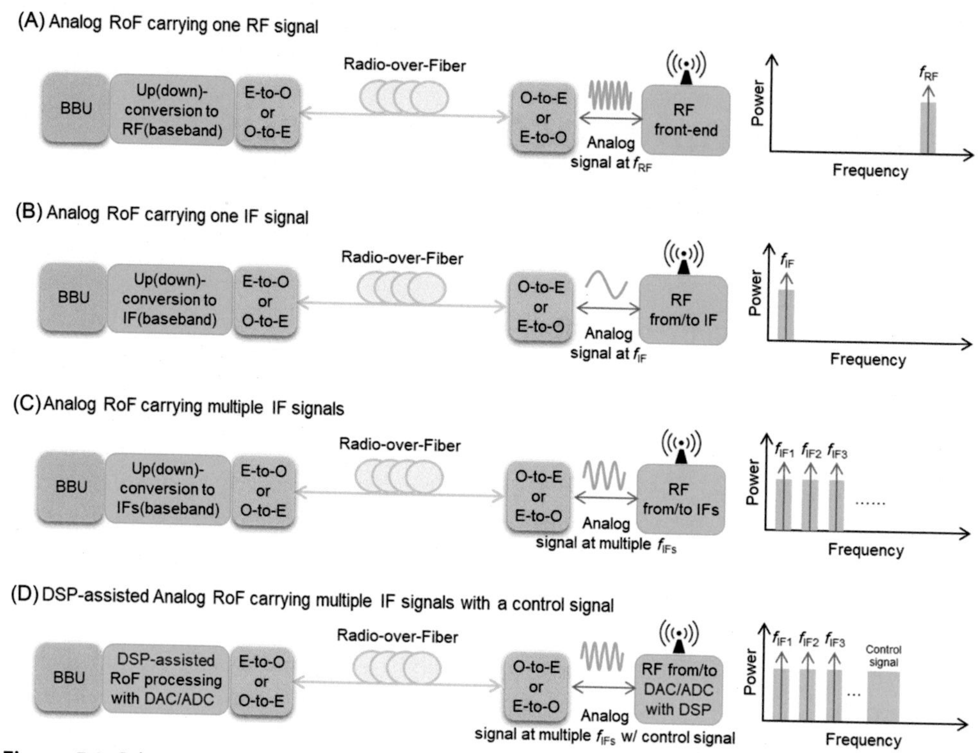

Figure 5.1 Schematics of four analog radio-over-fiber (RoF) transmission schemes: (A) analog RoF carrying one radio frequency signal, (B) analog RoF carrying one intermediate frequency (IF) signal, (C) analog RoF carrying multiple IF signals via frequency-division multiplexing (FDM), and (D) digital signal processing-assisted analog RoF carrying multiple IF signals with a control signal via FDM.

be obtained. This IF-based RoF scheme provides much higher bandwidth efficiency than the previous RF-based RoF scheme.

- A-RoF carrying multiple IF signals, as shown in Fig. 5.1C, where multiple IF signals are downconverted from RF signals to different IFs, which are then aggregated via frequency-division multiplexing (FDM) and modulated onto an optical carrier via E-to-O. Upon direct detection of the RoF signal for O-to-E, the IF signals are obtained. With corresponding de-aggregation and frequency upconversion of the received IF signals, the original RF signals can be obtained. This multi-IF-based RoF scheme can benefit from the bandwidth efficiency of the IF-based RoF and the large bandwidth offered by O-to-E and E-to-O converters to achieve much larger transmission capacity than the previous schemes based on a single RF signal or IF signal.
- DSP-assisted A-RoF carrying multiple IF signals with a control signal, as shown in Fig. 5.1D, where multiple IF signals with different center frequencies and a control signal with digital modulation are aggregated via FDM and modulated onto an

optical carrier via E-to-O. Upon direct detection of the RoF signal for O-to-E, RF signals are obtained by corresponding de-aggregation and frequency upconversion of the received IF signals, while the control signal can be downconverted to baseband for demodulation. DSP, together with DAC and ADC, is used at both the remote equipment site and the baseband unit site to facilitate (or assist) the aggregation and de-aggregation of various IF signals and the digitally modulated control signal. At the transmitter side, the aggregated signal generated by the DSP is converted to the analog domain via a high-speed DAC before the E-to-O conversion. At the receiver side, the A-RoF signal after the O-to-E conversion is converted to the digital domain via a high-speed ADC for subsequent DSP. As in the case of common public radio interface (CPRI), control words are needed for the control and management of the remote equipment in real time. So, a control signal that carries such control words and transmits with the radio signals synchronously is beneficial for front-haul applications.

With the commercial availability of high-resolution ADC, it has become feasible to convert the analog radio signals, at baseband or IF, to the digital domain with high fidelity, for example, with a resolution of 15 bits and an effective number of bits (ENOB) of 12. Similarly, high-resolution DAC has become available to convert the digital-domain signals to analog radio signals, at baseband or IF, with high fidelity. This enables the use of digital optical modulation and demodulation. Furthermore, the digital samples for different wireless signals can be aggregated or de-aggregated via time-division multiplexing (TDM), as in the case of CPRI-based front-haul transmission. Moreover, forward error correction (FEC) can be readily introduced to achieve virtually "error-free" transmission with a bit error ratio (BER) of less than 10^{-12}, thereby avoiding fiber transmission—induced degradation in wireless signal quality. This type of radio signal transmission over fiber with digital optical modulation and demodulation is referred to as digital RoF (D-RoF).

D-RoF based on CPRI is a well-established technique to support mobile fronthaul Cloud Radio Access Network (C-RAN) [24—26]. However, the original D-RoF transmission using binary modulation is highly bandwidth-inefficient as compared to A-RoF techniques as each wireless waveform sample may require 30 in-phase and quadrature (I/Q) bits to represent [26]. On the other hand, the error vector magnitude (EVM) performance of A-RoF needs to be much improved, especially for high-speed RoF, to support 5G wireless signals with 256- and 1024-quadrature amplitude modulation (QAM) formats [27]. Recently, several research groups have endeavored to improve the EVM performance of A-RoF by trading its bandwidth efficiency, for example, via various phase modulation (PM)-induced bandwidth expansion techniques [28—31]. Digital signal-to-noise ratio (SNR) adaptation of A-RoF carrying up to 1048576-QAM signals was demonstrated with PM, achieving a scaling of 6 dB SNR gain for each doubling of bandwidth or halving of spectral efficiency (SE) [31]. In parallel, it was recently found that the theoretical SE of D-RoF can approach that of A-RoF when Shannon capacity-approaching techniques are applied [32]. Moreover, the recovered SNR in D-RoF increases exponentially with the link bandwidth, offering a scaling of doubling SNR in decibels for each halving of SE. It was experimentally demonstrated that at a halved SE, D-RoF achieved an SNR gain of ~8.2 dB over A-RoF and an output EVM of 3.13% for a CPRI-equivalent data rate of about 60 Gb/second [32].

A recently presented approach is the novel hybrid digital-analog radio-over-fiber (DA-RoF) technique based on cascaded digital probabilistic constellation shaped (PCS) n-QAM modulation and analog pulse code modulation (PCM), where the digital n-QAM signal provides a natural approximation of the RoF waveform and the PCM provides the analog representation of the approximation error with an appropriate "magnification" to effectively increase the SNR of the received RoF signal [33]. It was further experimentally demonstrated that this DA-RoF technique enabled an SNR gain of 12.8 dB over A-RoF and an output EVM of 1.38% for a CPRI-equivalent data rate of 160 Gb/second. Fig. 5.2 shows the front-haul architecture based on CPRI and the modulation schemes for the CPRI-based D-RoF, CPRI-compatible A-RoF, and CPRI-compatible DA-RoF.

In a CPRI-based D-RoF, digital modulation is used for the I/Q bits of all the antenna-carriers (AxCs) with all the bits treated with equal importance. CPRI control word (CW) bits are also digitally modulated. In a CPRI-compatible A-RoF [17], PCM is used to provide an analog representation of the wireless signal from each AxC with the I/Q bits sorted from the most significant bit (MSB) to the least significant bit (LSB), while the CW bits are digitally modulated. In a CPRI-compatible

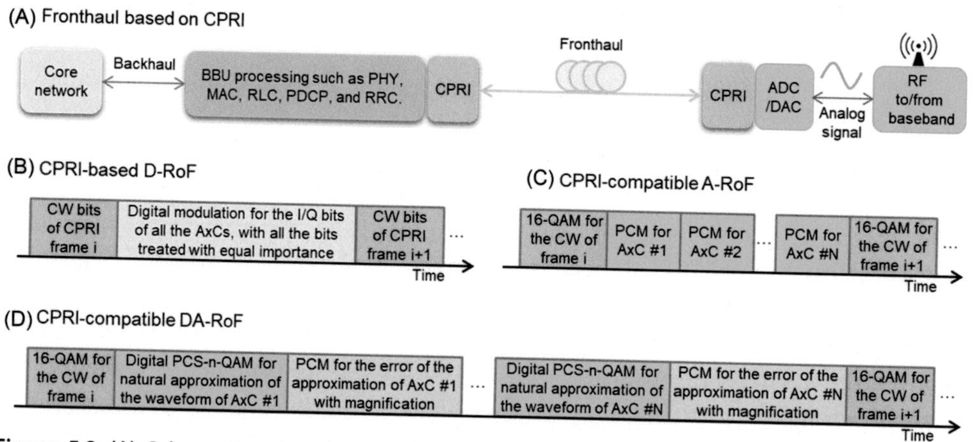

Figure 5.2 (A) Schematics of mobile front-haul based on common public radio interface (CPRI), (B) CPRI-based D-RoF modulation, (C) CPRI-compatible analog radio-over-fiber modulation, and (D) CPRI-compatible digital-analog radio-over-fiber modulation.

DA-RoF [33], a given analog RoF waveform, S, is represented as $S = W_1 + W_2$, where W_1 is a digital n-QAM signal for a natural approximation of S, and W_2 is an analog PCM signal representing the approximation error. The analog signal W_2 is magnified before being modulated via PCM to effectively increase its SNR. The CW bits are digitally modulated onto a 16-QAM control signal (CS) as in the case of the CPRI-compatible A-RoF [17].

In the following sections of this chapter, we will describe in more depth the DSP-assisted FDM-based A-RoF techniques in Section 5.2, CPRI-compatible TDM-based A-RoF techniques in Section 5.3, theoretical performance comparison between A-RoF and D-RoF in Section 5.4, and CPRI-compatible TDM-based DA-RoF in Section 5.5. Finally, concluding remarks will be outlined in Section 5.6.

5.2 Digital signal processing-assisted frequency-division multiplexing-based analog radio-over-fiber techniques

The principle of FDM-based RoF is to aggregate wireless signals to/from multiple antennas in the frequency domain and transmit them together over a fiber-based front-haul link using a single wavelength channel, as shown in Fig. 5.3. The wireless signals aggregated in the optical channel may have their spectral bandwidths unchanged, thereby leading to much improved bandwidth efficiency as compared to CPRI. In effect, the transmitted optical signal is an analog signal with SE much higher than the on-off keying signal used in CPRI. The channel aggregation and de-aggregation can be realized by using efficient DSP based on fast Fourier transform (FFT) and inverse FFT (IFFT) [18], as shown in Fig. 5.3. Dispersion-penalty-free transmission of six 100-MHz-bandwidth Long-Term Evolution (LTE)-A-like signals with 36.86-Gb/second CPRI-equivalent data rate over a 40-km standard single-mode fiber (SSMF) front-haul has been demonstrated by using a single 1550-nm directly modulated laser (DML) with a modulation bandwidth of only 2 GHz [12], as shown

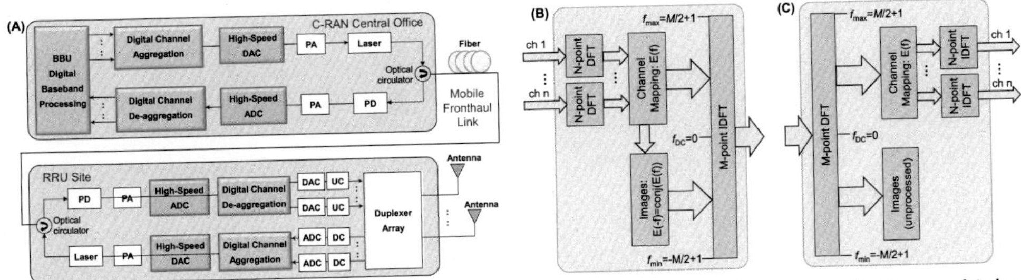

Figure 5.3 (A) Schematic of a digital signal processing (DSP)-assisted frequency-division multiplexing -based analog radio-over-fiber architecture with channel aggregation and de-aggregation, (B) Schematic of the DSP-based channel aggregator, and (C) Schematic of the DSP-based channel de-aggregator. (After [12]). *UC*, frequency upconverter; *DC*, downconverter.

in Fig. 5.4. The dispersion penalty due to the interplay between the fiber chromatic dispersion and the chirp of the DML has been avoided by a novel dispersion-penalty mitigation technique based on judicious mapping of the aggregated signals in the frequency domain.

The versatility of the DSP-based channel aggregation/de-aggregation approach has been shown in a front-haul experiment, where 36 evolved universal terrestrial radio access (E-UTRA) type wireless signals, six for each standardized bandwidth, 1.4, 3, 5, 10, 15, 20 MHz, are aggregated in a single wavelength channel [13], as shown in Fig. 5.5. The EVM performance of a representative recovered wireless signal is shown in Fig. 5.6. Reasonably small EVM of $<5\%$ was obtained when the received optical power was above -20 dBm, indicating a high link loss budget of over 25 dB when a DML with over 5 dBm output power is used [13].

To further improve the bandwidth efficiency (or SE) of the FDM-based A-RoF scheme, seamless aggregation of wireless channels in the frequency domain can be used. Fig. 5.7 shows the experimentally measured spectrum of 48 seamlessly aggregated 20-MHz LTE signals after 5-km SSMF transmission [15]. To achieve low latency for FFT/IFFT-based channel aggregation, new frequency-domain windowing (FDW) process was introduced after each N-point FFT, as shown in Fig. 5.8A. Similarly, an FDW process was added prior to each N-point IFFT in the channel de-aggregation stage, as shown in Fig. 5.8B. The FDW imposes a windowing function that attenuates the high-frequency components and suppresses the interchannel crosstalk, thereby reducing the FFT/IFFT size and thus the processing latency. With the use of FDW, the N and M values are reduced by 4 times, to 16 and 2048, respectively. The near-optimum windowing function for the 16-point FFT/IFFT case has been reported to be [0.44 0.66 0.94 1 1 1 1 1 1 1 1 1 1 0.94 0.66 0.44] [15]. The theoretical overall round-trip DSP-induced latency (including two channel aggregation stages and two de-aggregation stages) was calculated to be $\sim$1.6 μs. With a typical

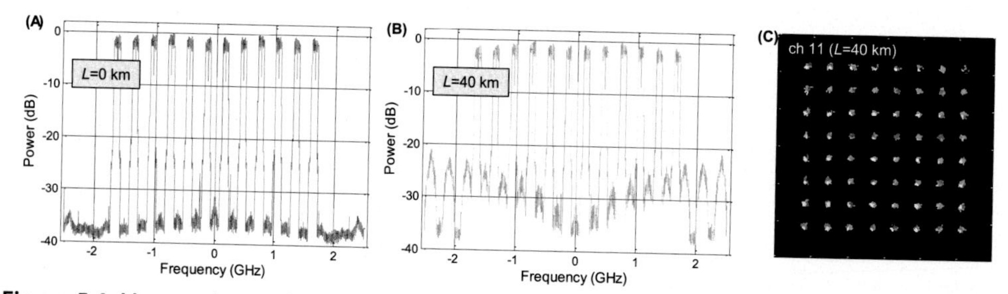

Figure 5.4 Measured optical spectra of the aggregated signals at $L = 0$ km (A) and 40 km (B), and a representative recovered constellation diagram of the highest-frequency channel at $P = -10$ dBm and $L = 40$ km (C), all under the odd-only channel mapping. (After [12]).

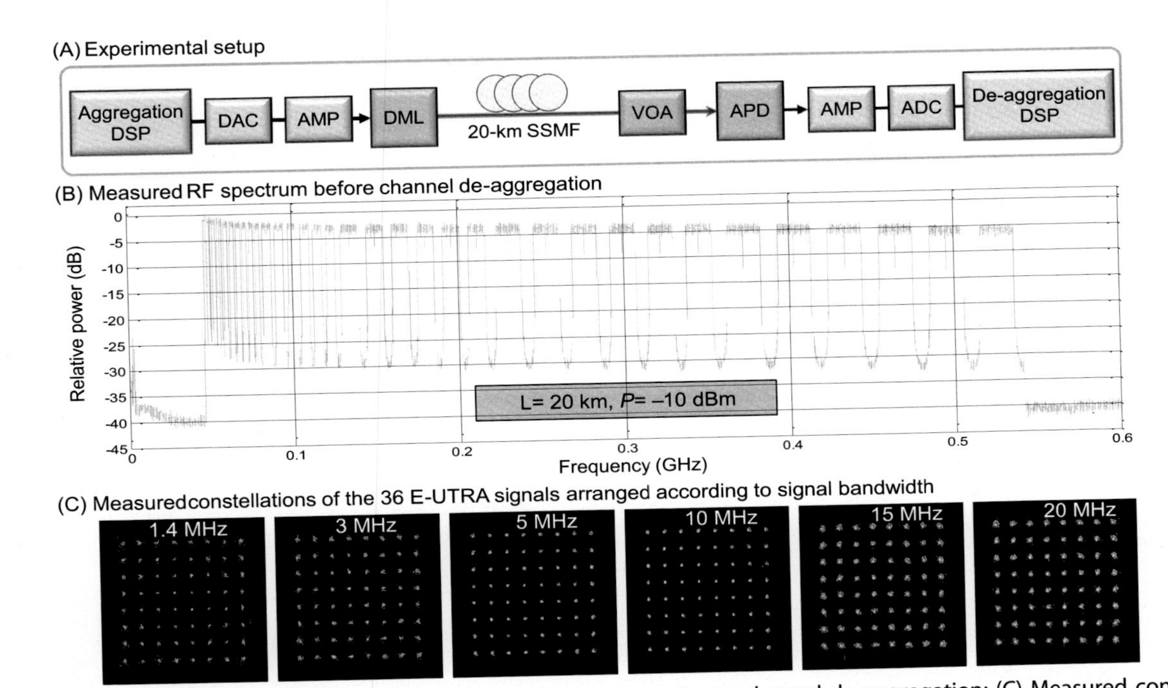

Figure 5.5 (A) Experimental setup; (B) Measured radio frequency spectrum prior to channel de-aggregation; (C) Measured constellation diagrams of the six groups of the universal terrestrial radio access-like signals. (After [13]) (*AMP*, RF amplifier; *VOA*, variable optical attenuator).

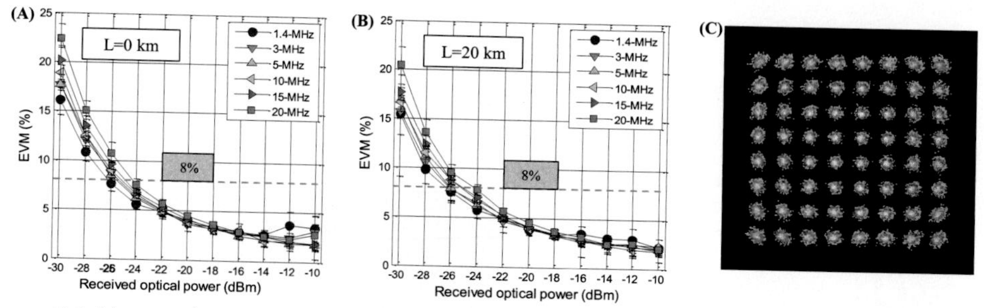

Figure 5.6 Measured error vector magnitude versus power for 64-quadrature amplitude modulation format at $L = 0$ km (A) and 20 km (B), and constellation diagram at $P = -20$ dBm and $L = 20$ km (C). (After [13]).

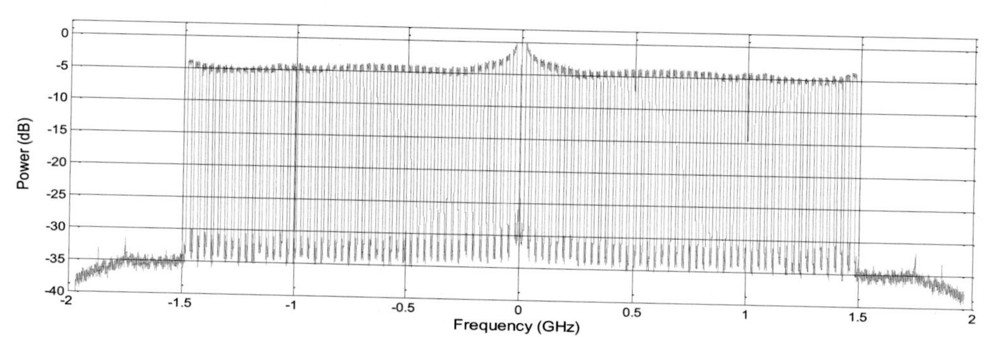

Figure 5.7 Experimentally measured spectrum of 48 seamlessly aggregated 20-MHz long-term evolution signals (and their images due to Hermitian symmetry) after 5-km standard single-mode fiber transmission. (After [15]).

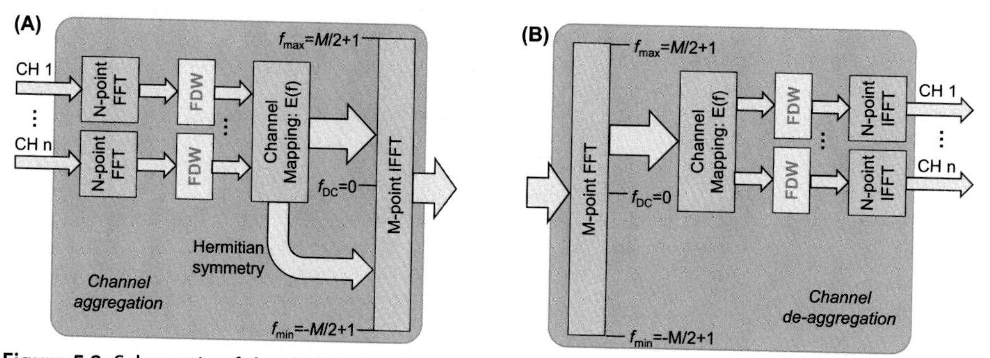

Figure 5.8 Schematic of the digital signal processing (DSP) blocks for fast Fourier transform/inverse FFT-based channel aggregation (A) and de-aggregation (B), both with the use of frequency-domain windowing to reduce the DSP processing latency. (After [15]).

implementation margin of <20%, the practically achievable round-trip DSP latency was estimated to be <2 μs [15].

Fig. 5.9 shows the EVMs of all the 48 LTE signals measured under three link conditions. It was found that all the 48 signals have similar EVM values, indicating a reasonable performance uniformity in the frequency domain. It was also found that the signal performances obtained after 5-km SSMF transmission are very similar to those obtained at $L = 0$ km. This indicates negligible fiber dispersion–induced penalty in this scenario. There was a small drop in performance at the 8th channel because of the nonuniform frequency response of the hardware used. The signal EVMs after 5-km SSMF transmission with $P_{RX} = -14$ dBm were about 3.5%, well below the typical 8% EVM threshold specified for 64-QAM [27].

To facilitate the control and management of the front-haul equipment and/or to be compliant with CPRI, it is necessary to transmit the CWs alongside the wireless signals. The CWs can be modulated by a signal-carrier QAM signal, which is then multiplexed with the wireless signals via FDM [16]. For CPRI Option 7, the control signal data rate is 491.52 Mb/second (=128 × 3.84 MHz). To achieve a BER of $<10^{-12}$ for the CWs, the constellation size of the control signal needs to be chosen based on the SNR of the fronthaul system. For a typical SNR of ~25 dB, 16-QAM can be used to achieve BER $<10^{-12}$ with ~1 dB margin. It is worth noting that FEC can be readily used for the 16-QAM control signal to relax its SNR requirement. When 16-QAM is used for CW modulation, the spectrum bandwidth needed for CWs is half of that needed for the wireless signals.

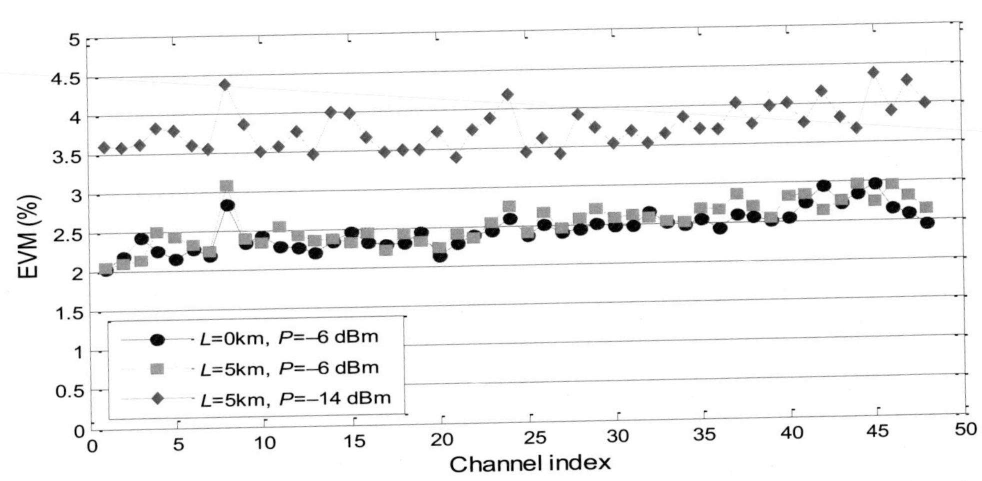

Figure 5.9 The error vector magnitudes of the 48 long-term evolution signals measured under three link conditions, (A) optical back-to-back ($L = 0$ km) at PRX = −6 dBm, (B) after 5-km transmission ($L = 5$ km) at −6 dBm received optical power, and (C) after 5-km transmission ($L = 5$ km) at −14 dBm received optical power. (After [15]).

Fig. 5.10A shows the aggregation process of a DSP-assisted FDM-based A-RoF with a control signal [16]. Without any loss of generality, we assume the channel aggregation is for n wireless signals with equal bandwidth as in the case of $n \times n$ Multiple-Input Multiple-Output (MIMO). The n digital baseband signals are first converted to the frequency domain by N-point FFTs. In parallel, the CW bits are modulated via 16-QAM. To facilitate the recovery of the CW bits, training symbols are added in the QAM signal to aid channel synchronization and equalization. The QAM signal is then converted to the frequency domain by a K-point FFT. The values of N and K are chosen on the basis of the bandwidths allocated to the wireless signals and the QAM signal. All the frequency-domain components are mapped onto the input of a single M-point IFFT. Hermitian symmetry is applied to obtain real-valued output from the IFFT to allow the use of low-cost intensity modulation and direct detection (IM/DD) for optical transmission. For channel de-aggregation, the reverse of the above process is applied, as shown in Fig. 5.10B. For the QAM control signal, channel estimation and equalization are performed. Finally, the CW bits are recovered by demodulating the equalized QAM control signal, optionally followed by FEC decoding to relax the SNR requirement.

Fig. 5.11 shows the RF spectrum of the generated FDM-based A-RoF signal consisting of 32 20-MHz LTE-like signals, each having a sampling rate of 30.72 MHz, and a 16-QAM control signal with a spectral bandwidth of 491.52 MHz and a total data rate of 1.96608 Gb/s [16]. The sizes of the N-FFT, K-FFT, and M-IFFT are 16, 256, and 2048, respectively. The sampling rate of the high-speed DAC and ADC is 3.9322 GSa/s. Additional pulse shaping is performed on the 16-QAM control signal.

Fig. 5.12 shows the experimental setup for evaluating the transmission performance of the above DSP-assisted FDM-based A-RoF with a control signal. We first generate

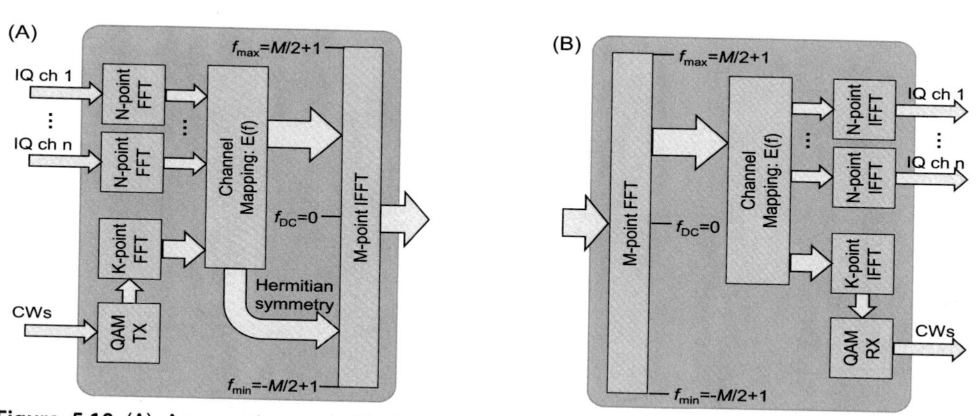

Figure 5.10 (A) Aggregation and (B) de-aggregation processes of a digital signal processing-assisted frequency-division multiplexing-based analog radio-over-fiber with a control signal. (After [16]).

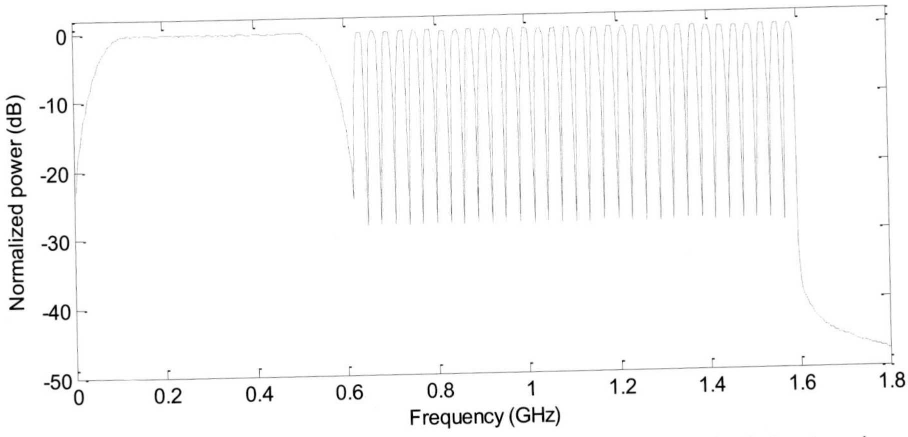

Figure 5.11 The radio frequency spectrum of an frequency-division multiplexing-based analog radio-over-fiber signal consisting of 32 20-MHz long-term evolution signals and a 491.52-Mbaud control signal. (After [16]).

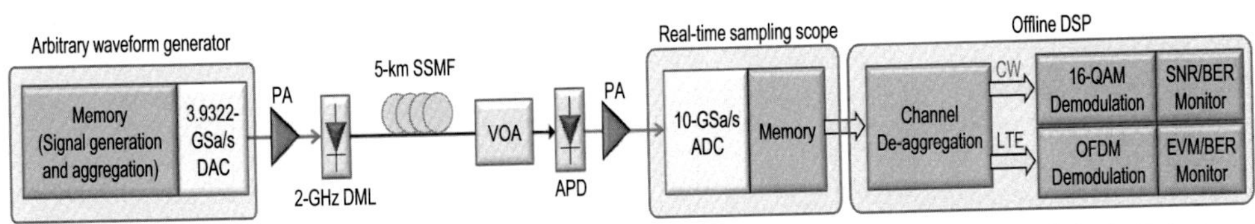

Figure 5.12 Experimental setup for evaluating the transmission performance of a digital signal processing-assisted frequency-division multiplexing-based analog radio-over-fiber with a control signal. (After [16]).

32 20-MHz LTE signals with orthogonal frequency division multiplexing (OFDM) with 64-QAM subcarrier modulation, and a 491.52-Mbaud 16-QAM signal to carry the CWs. These signals are then aggregated in the frequency domain using the DSP described above. The time-domain waveform of the aggregated signal is stored in an arbitrary waveform generator and outputted by a 3.9322-GSa/second DAC. This analog signal is then amplified by a power amplifier (PA) before driving a 1550-nm DML with a modulation bandwidth of about 2 GHz. The generated optical signal has a power of 9 dBm. It is launched into a 5-km SSMF. After fiber transmission, a variable optical attenuator (VOA) is used to vary the optical power received by an avalanche photodiode (APD). The detected signal is digitized by a 10-GSa/second ADC in a real-time sampling scope. The digitized samples are stored in the scope and later processed by offline DSP for downsampling and channel de-aggregation. The LTE signals are processed by OFDM channel equalization and demodulation, followed by the evaluation of the received signal EVM and BER. The 16-QAM signal is processed by QAM channel equalization and demodulation, followed by the evaluation of the received SNR and BER.

Fig. 5.13 shows the experimentally measured RF spectrum of the A-RoF signal after 5-km SSMF transmission at a received optical power of −8 dBm. All the signals are aggregated with the desired bandwidth and a small spectral overlap. There is a 5-dB drop in signal spectral power density from the low-frequency edge (~0.1 GHz) to the high-frequency edge (~1.6 GHz), primarily due to the bandwidth limitation of the DAC and ADC used. The signal spectral power density is 30–35 dB higher than the noise floor. Fig. 5.14 shows the experimentally measured constellations of the CW-carrying 16-QAM signal and the LTE OFDM-64QAM signals under the same

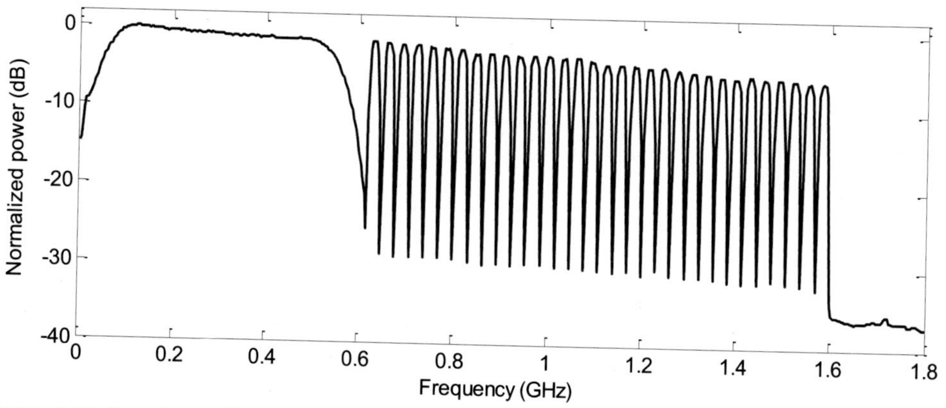

Figure 5.13 Experimentally measured radio frequency spectrum of the aggregated analog radio-over-fiber/control word signal after 5-km standard single-mode fiber transmission at a received optical power of −8 dBm. (After [16]).

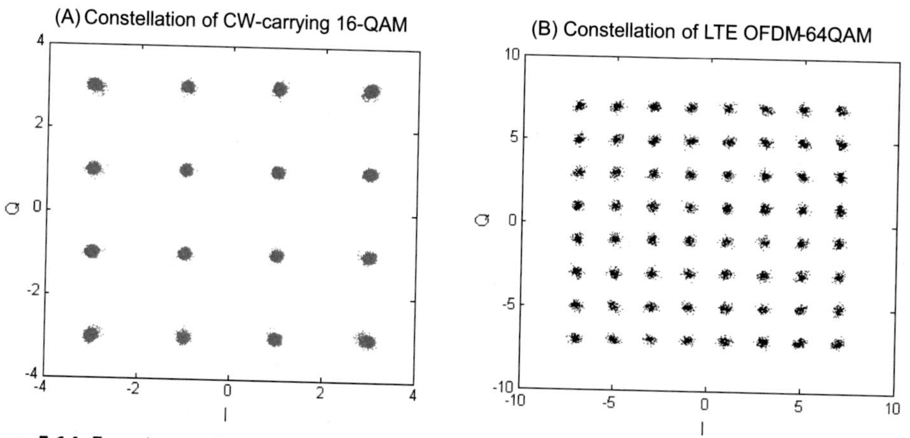

Figure 5.14 Experimentally measured constellations of the control word-carrying 16-quadrature amplitude modulation (QAM) control signal (A) and the long-term evolution orthogonal frequency division multiplexing-64QAM signals (B) after 5-km standard single-mode fiber transmission at a received optical power of −8 dBm. (After [16]).

condition. The recovered constellations are of high quality, and no errors are measured with over 10 million bits processed for the LTE signals and the CW carrying 16-QAM signals [16]. Fig. 5.15 shows the experimentally measured SNR performance of the control signal and the EVM performance of the LTE OFDM-64QAM signals after fiber transmission. The recovered constellations are of high quality, and no errors are measured with over 10 million bits processed for the LTE signals and the CW carrying 16-QAM signals [16]. When the received optical power is in the range between −12 and −4 dBm, the received EVMs of all the 32 LTE signals are below 4% and the SNR of the 16-QAM control signal is better than 30 dB, which leads to essentially error-free transmission of the CWs.

It is necessary to reduce the adjacent channel leakage ratio (ACLR) after front-haul transmission, in order to avoid interchannel crosstalk. It has been demonstrated that a sufficiently low ACLR can be obtained by DSP-based postfiltering during the channel de-aggregation stage [18]. The postfilter is implemented in the time domain with 72 taps. Fig. 5.16 shows a 20-MHz LTE signal after postfiltering. The ACLR is suppressed to below −73 dBc, well meeting the 3rd Generation Partnership Project's requirements on ACLR [27]. Moreover, the digital signal processors at the remote sites can also be configured to perform digital predistortion to compensate for the nonlinear distortion caused by the PAs before the antennas. This indicates the additional benefits of antenna-site DSP in improving the overall performance of mobile front-haul.

To increase the front-haul bandwidth efficiency, it is desirable to reduce the ratio between the number of the CW bits and the number of the I/Q bits from the typical ratio of 1:15. This is doable, especially for future wireless networks where the wireless

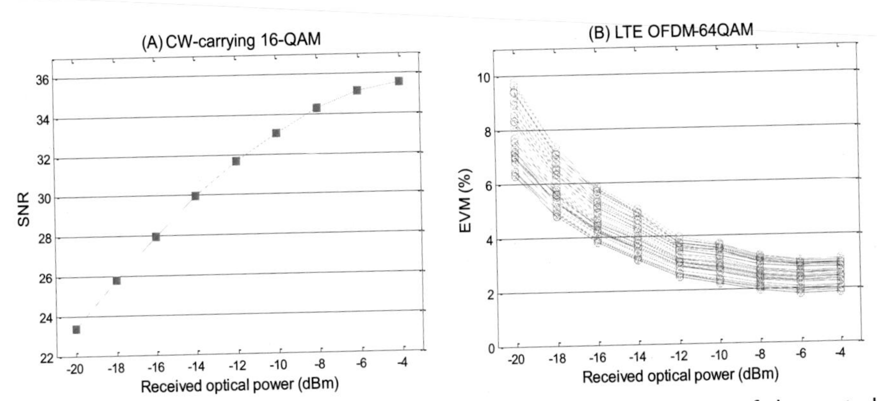

Figure 5.15 (A) Experimentally measured signal-to-noise ratio performance of the control signal, (B) Experimentally measured error vector magnitude performances of the 32 20-MHz long-term evolution orthogonal frequency division multiplexing-64 quadrature amplitude modulation signals. (After [16]).

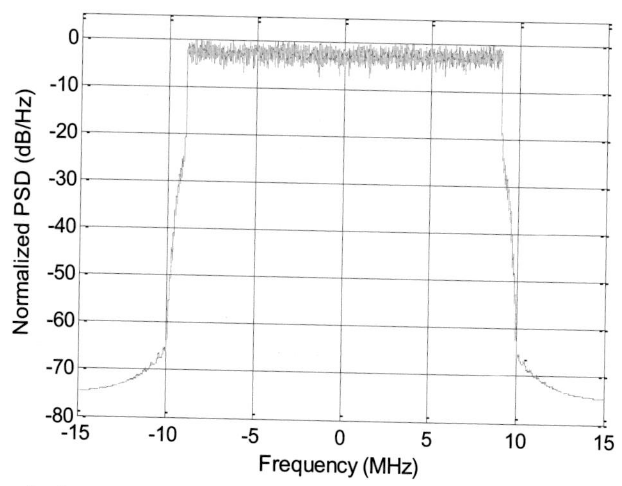

Figure 5.16 The radio frequency spectrum of a 20-MHz long-term evolution signal after adjacent channel leakage ratio suppression by a digital signal processing-based postfilter at the antenna site. (After [18]).

signal bandwidth is expected to be much increased while the data rates needed for CWs remain relatively constant. In fact, Release 7 of the CPRI specification now caps the number of CW bits per CPRI frame at 128 bits [26], corresponding to a control signal data rate of 491.52 Mb/second, as in the case of CPRI Option 7 with 16 CPRI containers. For CPRI Option 10 at 24.33024 Gb/second, there are 48 CPRI containers, and the ratio between the number of the CW bits and the number of the I/Q bits is reduced to 1:47. Scaling further for a typical 5G case with 100-MHz RF signal bandwidth and 64 × 64 MIMO, the CPRI-equivalent data rate would be about 40 times higher than Option 7, so the ratio between the number of the CW bits and the number of the I/Q bits could be further reduced to 1:639.

With the use of the DSP-assisted FDM-based A-RoF, the transmission of 32 200-MHz 5G-like wireless signals using 15-GHz-class DML and APD, as well as 40-GSa/second DAC and ADC, has been demonstrated [18]. Fig. 5.17 shows the experimental results. The EVMs before and after 1-km SSMF transmission in the C-band are measured to be ~3.5% and ~6%, respectively. When operating in the O-band, the dispersion-induced signal quality degradation can be much reduced and longer transmission distance (e.g., 10 km) can be supported. However, the EVM performance of the A-RoF approach needs to be further improved in order to support advanced 5G modulation formats such as OFDM-256QAM and OFDM-1024QAM, which desire the EVMs caused by the front-haul transmission to be much less than the end-to-end EVM allocations of 3.5% and 2.5%, respectively. We will describe how to improve the front-haul transmission performance via the DA-RoF approach in Section 5.5.

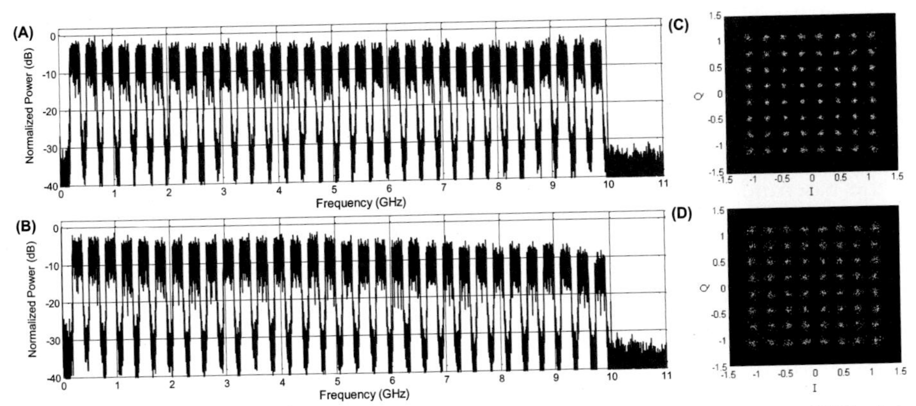

Figure 5.17 Experimentally measured radio frequency spectra (A,B) and constellations (C,D) of the aggregated 32 200-MHz orthogonal frequency division multiplexing-64quadrature amplitude modulation signals before fiber transmission (A,C) and after 1-km standard single-mode fiber transmission in the C-band (B,D). (After [18]).

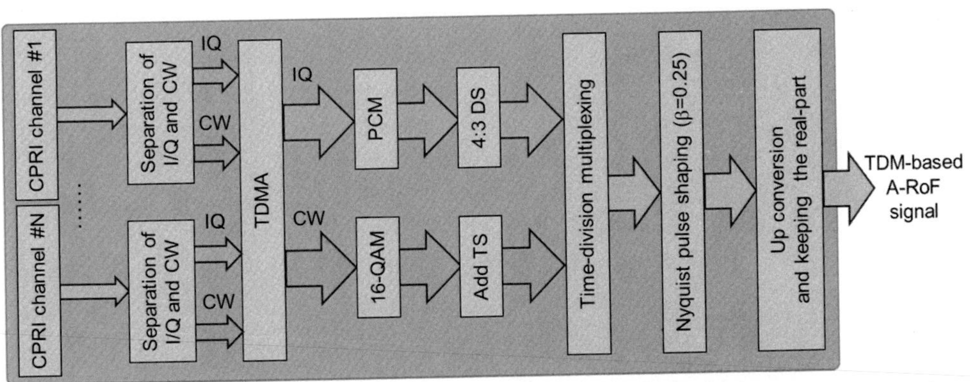

Figure 5.18 Transmitter digital signal processing diagram of the common public radio interface-compatible time-division multiplexing-based analog radio-over-fiber. (After [17]).

5.3 Common public radio interface-compatible time-division multiplexing-based analog radio-over-fiber techniques

As discussed earlier in this Chapter, CPRI-compatible A-RoF can also be implemented via TDM (instead of FDM), where the I/Q bits are modulated via PCM to represent the time-domain analog waveform from each AxC with high SE, while the CWs are modulated via a low-level QAM (e.g., 16-QAM) to achieve essentially error-free performance [17]. Fig. 5.18 shows the schematic diagram of the transmitter DSP. Multiple CPRI input channels are first processed so that I/Q bits are separated from the CW bits. TDM is then applied to aggregate the I/Q bits and CW bits separately, followed by

PCM for IQ bits and 16-QAM for CW bits. The PCM signal is downsampled by a factor of ¾, and the QAM signal is inserted with periodic training symbols (TSs) to facilitate synchronization. The two signal streams are then multiplexed in the time domain, Nyquist pulse shaped, and frequency upconverted to generate a real-valued A-RoF signal for IM/DD. Fig. 5.19 shows the receiver DSP. The input signal is downconverted to the baseband, and TS-based synchronization and time-division de-multiplexing are performed to separate the PCM signal stream from the QAM signal stream. The QAM signals are used to train a finite impulse response–based equalizer, which then performs channel equalization for both the PCM and QAM signals. With the use of the equalizer, the SNR performance of the front-haul can be accurately estimated. After equalization, time-division de-multiplexing is applied to de-aggregate the I/Q and CW bits for their corresponding CPRI channels. Finally, the I/Q and CW bits of each CPRI channel are reconstructed and outputted.

In the CPRI-compatible TDM-based A-RoF scheme, CWs can be modulated by 16-QAM, which leads to a BER of well below 10^{-12} for a typical SNR of about 24 dB (or below 18 dB when a typical Reed–Solomon FEC is applied). Assuming 4:3 downsampling for the PCM signal, the spectral bandwidth needed to transmit the I/Q and CW bits in CPRI Option 7 is only 0.3072 GHz (=491.52 MHz/ $4 + 8 \times 30.72$ MHz $\times 3/4$), indicating that the A-RoF approach has a bandwidth efficiency that is 32 (=9.8304/0.3072) times higher than that of the typical CPRI approach. Fig. 5.20 shows the experimental setup for demonstrating the CPRI-compatible TDM-based A-RoF scheme. An 8-Gbaud A-RoF signal, corresponding to a CPRI-equivalent data rate of 256 Gb/second (=32×8 Gb/second), is generated by the transmitter DSP. The signal is stored in an arbitrary waveform generator and outputted by a 64-GS/second DAC. The generated analog signal is then amplified before

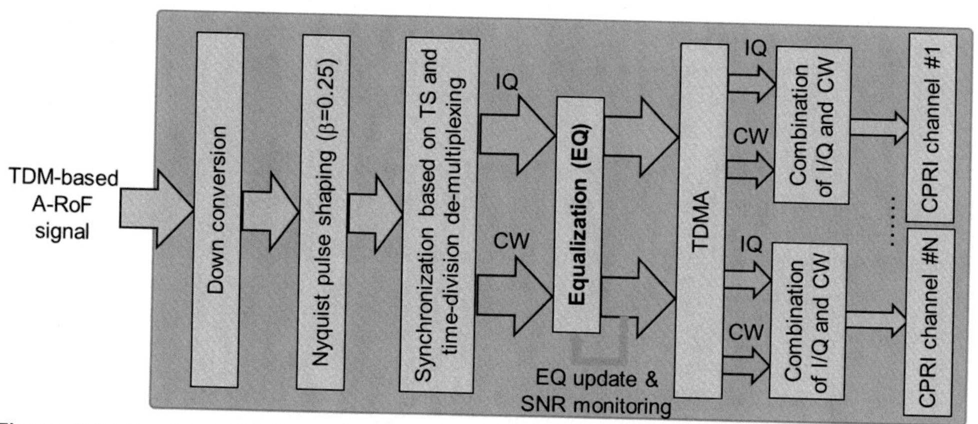

Figure 5.19 Receiver digital signal processing diagram of the common public radio interface-compatible time-division multiplexing-based analog radio-over-fiber. (After [17]).

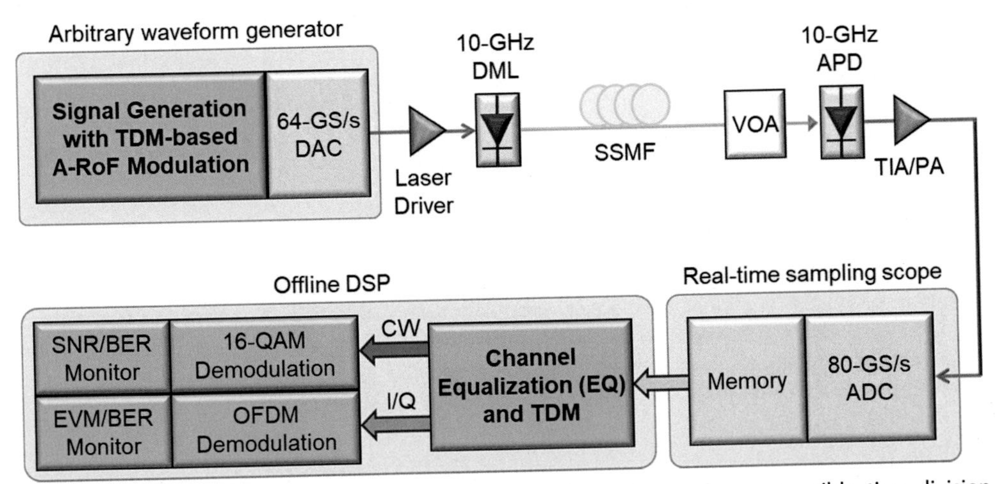

Figure 5.20 Experimental setup for the common public radio interface-compatible time-division multiplexing-based analog radio-over-fiber.

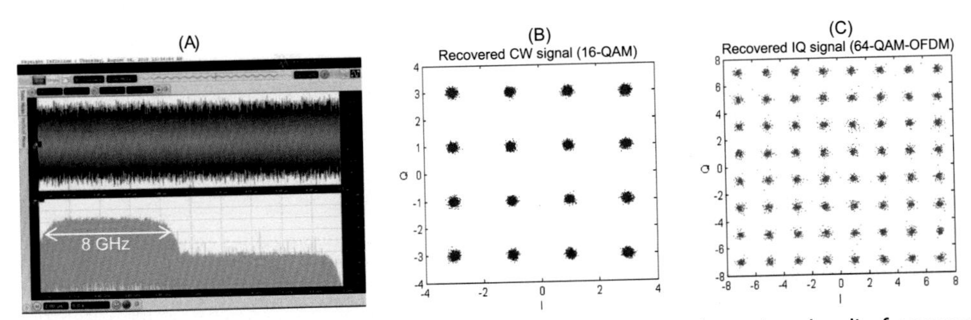

Figure 5.21 (A) Screen capture showing the time-domain waveform (upper) and radio frequency spectrum (lower) of the 8-Gbaud time-division multiple access (TDMA) radio-over-fiber/control word (CW) signal, (B) recovered constellation of the CW signal, and (C) recovered constellation of the 64-quadrature amplitude modulation-orthogonal frequency division multiplexing wireless signal under the electronic back-to-back configuration. (After [17]).

driving a 1550-nm 10-GHz-bandwidth DML. The generated optical signal has a power of 8 dBm and is launched into a 1-km SSMF. After fiber transmission, a VOA is used to vary the optical power received by an APD. The detected signal is digitized by an 80-GS/second ADC in a real-time sampling scope. The digitized samples are stored in the scope, and later processed by the receiver DSP.

Fig. 5.21 shows the measured time-domain waveform and RF spectrum of the 8-Gbaud TDM A-RoF signal, as well as the constellations of the recovered CW signal and wireless signal (64-QAM-OFDM) under the electronic back-to-back configuration. Both signals are well recovered. Fig. 5.22A shows the spectra of the recovered 8-Gbaud TDM A-RoF signal after optical modulation and detection in the optical

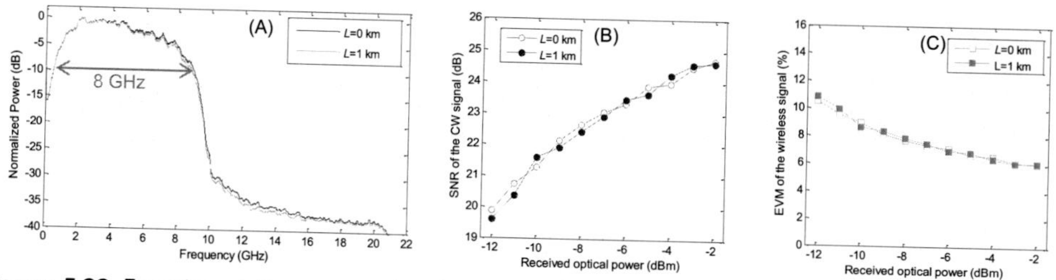

Figure 5.22 Experimentally measured radio frequency spectra of the 8-Gbaud TDMA radio-over-fiber/control word (CW) signal after optical modulation and detection (A), signal-to-noise ratio of the CW signal versus received optical power (B), and error vector magnitude performance of the 64-quadrature amplitude modulation-orthogonal frequency division multiplexing signal versus received power (C). (After [17]).

back-to-back case ($L = 0$ km) and after 1-km SSMF transmission ($L = 1$ km). There is a roll-off of ~8 dB at the high-frequency edge of the signal spectrum (due to the bandwidth limitations of the optical transmitter and receiver), while the dispersion-induced power fading is negligible. Fig. 5.22B and C respectively show the SNR of the CW signal and the EVM of the wireless 64-QAM-OFDM signal versus the received optical power (P_{RX}). For the CW signal, BER $< 10^{-12}$ is obtained without FEC (or with a received SNR of larger than 23.9 dB) at $P_{RX} \geq -4$ dBm. For the wireless 64-QAM-OFDM signal, EVM $< 8\%$ is achieved at $P_{RX} \geq -8$ dBm. Moreover, negligible fiber transmission penalty is observed by comparing the cases of $L = 0$ km and $L = 1$ km. When operating in the O-band, 10-km transmission distance can be readily supported.

The above experimental demonstration was based on offline DSP [17]. It is desirable to implement the TDM A-RoF scheme using real-time DSP to quantify its implementation complexity, processing latency, and long-term operational stability. Real-time implementation of a CPRI-compatible TDM A-RoF transceiver using a Xilinx Virtex-7 field-programmable gate array, together with a pair of 5-GSa/second DAC and ADC, has been demonstrated [19,20]. It is capable of aggregating 43 20-MHz LTE signals with their corresponding CWs as specified by CPRI, leading to a CPRI-equivalent aggregated throughput of about 53 Gb/second. The implementation complexity is shown to be reasonably low, and the one-way processing latency is measured to be as low as 1.8 μs [19,20]. Long-term measurements over 12 hours have shown the stable operation of the real-time EMF transceiver. This indicates the practicality of the TDM A-RoF technique. Higher CPRI-equivalent data rate can be achieved by implementing the DSP in application-specific integrated circuits that offer more DSP resources.

As in the case of the FDM-based A-RoF approach discussed in the previous section, the EVM performance of the TDM-based A-RoF approach is insufficient to support advanced 5G modulation formats such as OFDM-256QAM and

OFDM-1024QAM in high front-haul link capacity scenarios. We will describe how to improve the front-haul transmission performance via the DA-RoF approach in Section 5.5.

5.4 Theoretical performance of digital signal processing-assisted radio-over-fiber

In the experimental demonstrations of the high-speed FDM-based and TDM-based A-RoF approaches described in Sections 5.3 and 5.4, the best EVM performance was about 6%, corresponding to an effective SNR of 24.4 dB. This is primarily limited by the ENOBs of the high-speed DAC and ADC used, as well as the noise and signal distortion introduced in fiber optical transmission link. To support advanced 5G modulation formats such as OFDM-1024QAM, the EVM of the received wireless signal needs to be less than 2.5%, corresponding to a SNR of higher than 32 dB. To reserve margins for signal distortions occurring outside the fronthaul link, it is desirable for the front-haul link segment to have an even higher SNR, for example, 36 dB, corresponding to an EVM of 1.6%. Therefore it is of value to explore techniques that can substantially improve the SNR, for example, by over 10 dB.

PM has been demonstrated to possess the ability to trade the bandwidth efficiency (or SE) of A-RoF for improved SNR performance [29]. The E-field of the phase-modulated RoF signal can be expressed as [29].

$$E_{PM}(t) = e^{j[\omega t + K_{PM}E(t)]}, \tag{5.1}$$

Where $E(t)$ is the E-field of the original RoF signal, ω is the carrier frequency, and KPM is the phase modulation index (PMI). A larger PMI leads to a larger bandwidth expansion and a higher SNR according to the following relationship [29].

$$\mathrm{SNR} \propto (K_{\mathrm{PM}})^2, \tag{5.2}$$

which means that the SNR increases quadratically with the PMI, that is, the SNR increases by 6 dB for each doubling of the signal bandwidth or halving of the SE. Such SNR adaptation has been experimentally demonstrated by flexibly adjusting the PM index in the transmitter DSP, achieving a wide SNR range of between 29 and 62 dB [31].

To estimate the maximum SNR gain that is theoretically possible by trading the SE, we can resort to the Shannon capacity theorem. In the high-SE region, we can relate the achievable SE with the link SNR, $\mathrm{SNR_{link}}$, as follows

$$\mathrm{SNR_{link}} \approx 2^{\mathrm{SE}}, \tag{5.3}$$

which means that the required SNR_{link} decreases exponentially with the decrease of the SE. Thus for each doubling of the signal bandwidth or halving of the SE, the required SNR_{link} in dB is reduced by one-half. This also implies that the received SNR doubles for each doubling of bandwidth or halving of SE. For example, an RoF signal that requires an SNR of 36 dB can be transmitted over a fiber link with an SNR_{link} of only 18 dB if twice the RoF signal bandwidth is used for transmission, indicating a dramatic SNR gain of 18 dB to support high-fidelity RoF transmission with halved transmission SE. This can be readily understood by looking at QAM based transmission. For a D-RoF signal with n-point QAM modulation, its theoretical SE is $\log_2(n)$, so we have

$$SNR_{link} \approx n, \tag{5.4}$$

which means that the n-QAM signal can be supported theoretically in a link whose SNR_{link} is n. For the fiber link with an SNR_{link} of 18 dB, the maximum QAM constellation would be 64-QAM carrying 6 bits per symbol. Thus 12 bits can be transmitted over two 64-QAM symbols. If these 12 bits are transmitted over one symbol at doubled SE, we need to use 4096-QAM, which requires an SNR_{link} of 4096 or 36 dB.

The above analysis indicates a very effective way to trade the transmission SE for the achievable signal fidelity. In fact, this scheme is applied in the CPRI-based D-RoF transmission where a high-fidelity wireless signal can be digitalized to 30 I/Q bits per sample, corresponding to a SNR of 90 dB. These 30 I/Q bits are transmitted over an optical link by multiple optically modulated symbols, to effectively lower the SNR requirement on the optical link (SNR_{link}). Originally, CPRI uses binary optical transmission, for example, 10-Gb/second nonreturn-to-zero for CPRI Option 7 carrying one bit per optically modulated symbol. To increase the SE of D-RoF, advanced modulation and coding techniques can be used. It has been found that the theoretical SE of D-RoF can approach that of A-RoF when Shannon capacity-approaching techniques are applied [32]. In a proof-of-concept experiment, D-RoF was shown to achieve an SNR gain over A-RoF of ~8.2 dB at halved SE and an output EVM of 3.13% for a CPRI-equivalent data rate of about 60 Gb/second [32]. The SNR gain is still a few decibels lower than the theoretical gain due to the implementation limitations.

The same concept of trading the transmission SE for the achievable signal fidelity can be applied to DSP-assisted A-RoF. Recently, a novel hybrid DA-RoF technique based on cascaded PCS-n-QAM modulation and PCM, where the digital PCS-n-QAM signal provides a natural approximation of the RoF waveform and the PCM provides the analog representation of the approximation error with a suitably chosen "magnification" to effectively increase the SNR of the received RoF signal [33]. This DA-RoF technique has been demonstrated in an 8-Gbaud single-carrier transmission experiment using a cost-effective 10-GHz-bandwidth IM/DD system, achieving an SNR gain over A-RoF of 12.8 dB at halved SE and an output EVM of 1.38% for a CPRI-equivalent data rate of 160 Gb/second. Remarkably, this SNR gain of 12.8 dB was obtained without resorting to FEC for the digital n-QAM modulation portion, so energy-efficient and low-latency RoF transmission can be supported. In addition more cascaded modulations can be used to flexibly increase the bandwidth expansion factor, thereby further improving the RoF signal fidelity and/or relaxing the link SNR requirement. More details on the DA-RoF scheme will be provided in the next section.

5.5 Common public radio interface-compatible time-division multiplexing-based digital-analog radio-over-fiber

As discussed in the previous section, one can trade transmission SE for signal fidelity. As A-RoF possesses superior SE, it can be beneficial to trade some of its SE for improved signal fidelity, especially in links with insufficient SNR_{link}. This can be realized by the hybrid DA-RoF technique where cascaded digital modulation and analog modulation are applied [33]. For the digital modulation part, digitization is needed to recover the original signal. For the analog modulation part, no digitization is needed in the signal recovery process. Fig. 5.23 shows the modulation scheme of one DA-RoF implementation. A given analog RoF signal, S, can be represented as

$$S = W_1 + W_2, \tag{5.5}$$

where W_1 is a digital n-QAM signal for a natural approximation of S and W_2 is a PCM signal representing the approximation error. The digital part of the signal, W_1, can be expressed as

$$W_1 = \text{round}\left(\frac{(\sqrt{n}-1)S}{2E_{\max}}\right)\cdot\frac{2E_{\max}}{(\sqrt{n}-1)}, \tag{5.6}$$

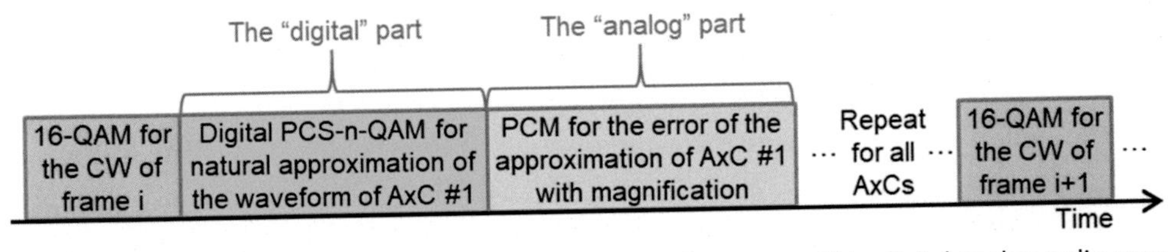

Figure 5.23 Illustration of the common public radio interface-compatible digital-analog radio-over-fiber modulation concept based on cascaded digital probabilistic constellation shaped-n-quadrature amplitude modulation modulation and analog pulse code modulation modulation. (After [33]).

where round( ) is a rounding function that rounds a complex number to its nearest Gaussian integer and E_{max} is the maximum amplitude of the I/Q components of S (set by a suitable clipping). For a typical wireless baseband signal with OFDM modulation, its time domain samples are complex Gaussian—distributed, so its n-QAM approximation is naturally a PCS constellation [34]. The achievable SE of a PCS-n-QAM signal is linked to its entropy (in bits per symbol), which is generally defined as

$$H(X) = -\sum_{i=1}^{n} P(x_i)\log_2 P(x_i), \tag{5.7}$$

where X denotes the set of the constellation points at $x_1, x_2, \ldots, x_n$, with probabilities $P(x_1), P(x_2), \ldots, P(x_n)$, respectively. The entropy of the PCS-n-QAM modulation can be flexibly adjusted by varying n and E_{max}. The analog part of signal, W_2, is magnified by a suitable scaling factor before being modulated via PCM to effectively increase the SNR. The CW bits are digitally modulated onto a 16-QAM CS, as also illustrated in Fig. 5.23.

Fig. 5.24A shows the DSP flow diagram for the DA-RoF modulation. Proper normalization of the incoming RoF waveform is performed to realize n-QAM for W_1. The analog signal W_2 is magnified by a factor of (c_2/c_1) with respect to W_1 such that the power of the magnified W_2 becomes similar to that of W_1. Fig. 5.24B shows the DSP flow diagram for the DA-RoF demodulation. Subsequent to the front-haul transmission, the W_1 portion of the received RoF waveform is demodulated back to the original n-QAM, while the magnified W_2 is shrunk by the same factor of (c_2/c_1) to its original amplitude. The analog signal S is then reconstructed by summing the recovered W_1 and W_2. The other DSP processes are similar to those described in Figs. 5.18 and 5.19.

Fig. 5.25 shows the experimental setup, which is similar to that shown in Fig. 5.20. The key difference is that the DA-RoF modulation and demodulation are now used in the transmitter and receiver signal processing. The digital part is based on PCS-121-QAM with an entropy of 5.12 bits/symbol. The constellation diagrams of original/recovered RoF waveform, its digital PCS-121-QAM part, and its analog PCM part are added as insets to illustrate the DA-RoF modulation and demodulation. The recovered DA-RoF signal spectra before and after fiber transmission (with 17 ps/nm dispersion) at −8 dBm received optical power (RoP) are shown in inset (*g*). Representative constellation diagrams of the recovered 16-QAM CW signal and OFDM-64-QAM wireless signal after fiber transmission are shown in insets (*h*) and (*i*), respectively. The EVM of the wireless signal is much reduced compared to that in A-RoF [17].

Fig. 5.26A shows the SNR of the recovered wireless signal as a function of the RoP. At an RoP between −8 and −2 dBm, DA-RoF provides an SNR gain

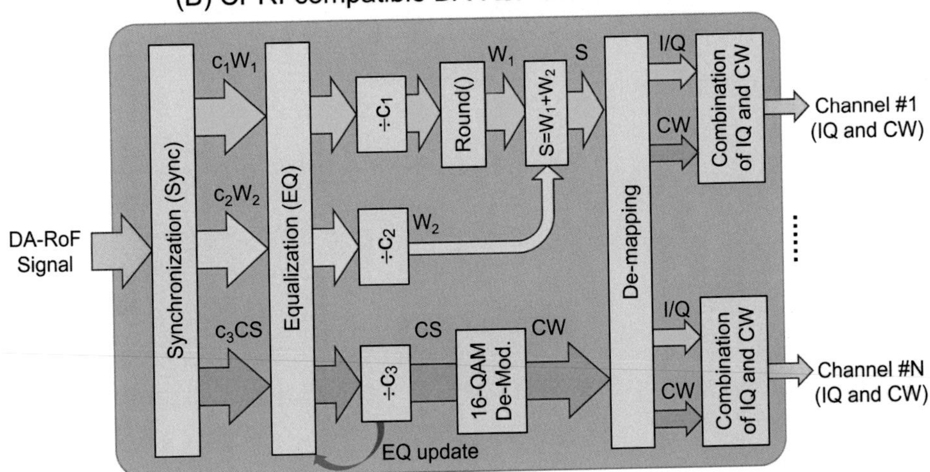

Figure 5.24 (A) The digital signal processing (DSP) diagram of the common public radio interface (CPRI)-compatible digital-analog radio-over-fiber (DA-RoF) modulation; (B) The DSP diagram of the CPRI-compatible DA-RoF demodulation. (After [33]).

between 10.5 and 12.8 dB, respectively. Fig. 5.26B shows the EVM performance of the recovered wireless signal. At an RoP between −8 and −2 dBm, the EVM for the DA-RoF case is reduced substantially to 2.28% and 1.38%, respectively. Representative constellation diagrams of the recovered wireless signals with 64- and 1024-QAM subcarrier modulations are shown as insets in Fig. 5.26 to indicate the dramatic performance improvements enabled by DA-RoF. The largest 5G wireless signal constellations applied so far are 256- and 1024-QAM, which require an EVM of

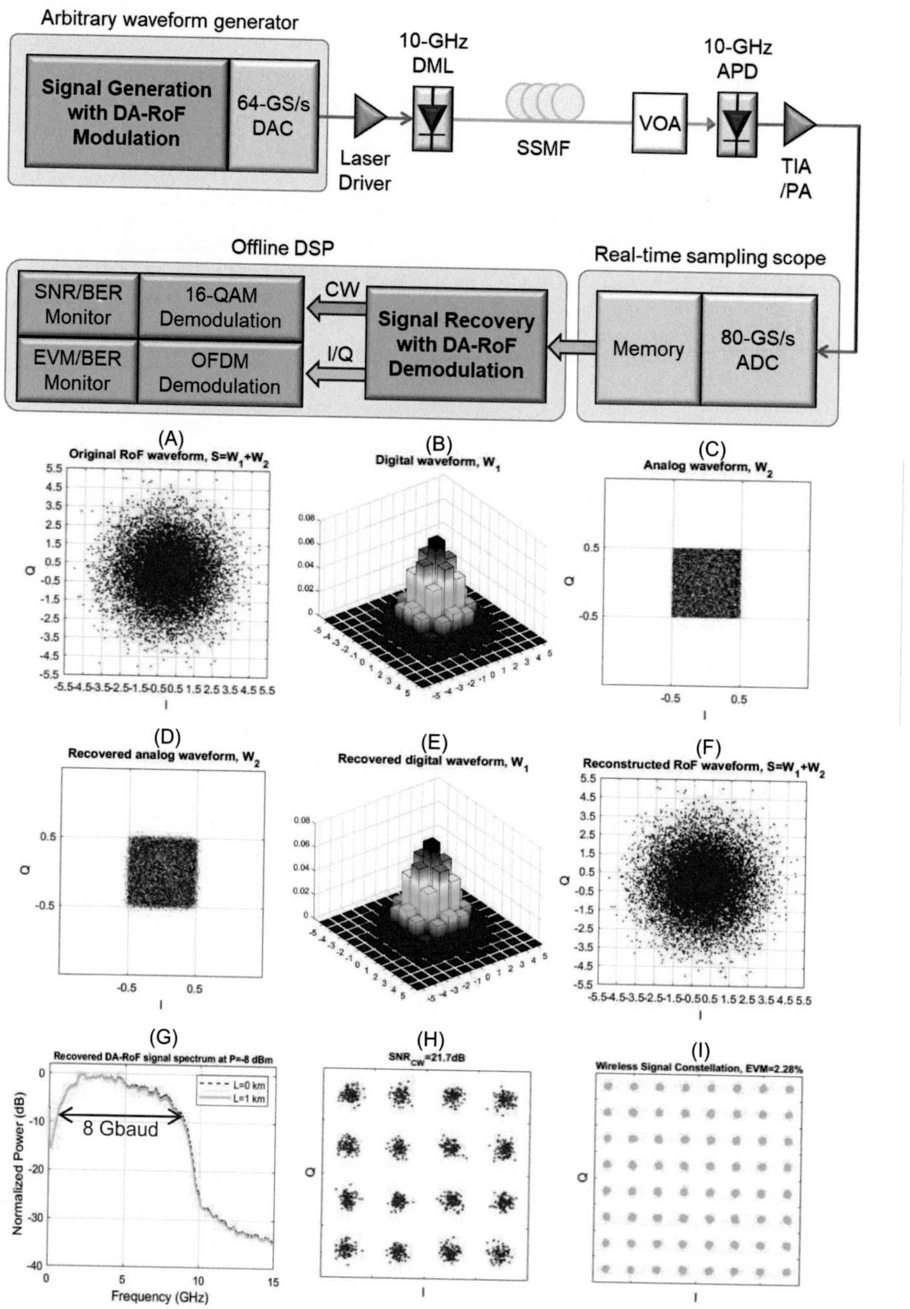

Figure 5.25 Schematic of the experimental setup for digital-analog radio-over-fiber (DA-RoF). Insets (A)/(F), (B)/(E), and (C)/(D) are the original/recovered constellation diagrams of the RoF waveform, its digital probabilistic constellation shaped-121-quadrature amplitude modulation (QAM) part, and its analog pulse code modulation part, respectively. Inset (*g*) is the recovered DA-RoF signal spectra before and after the fiber transmission. Insets (*h*) and (*i*) are representative recovered constellation diagrams of the 16-QAM control word signal and the orthogonal frequency division multiplexing-64-QAM wireless signal, respectively, at −8 dBm received optical power. (After [33]).

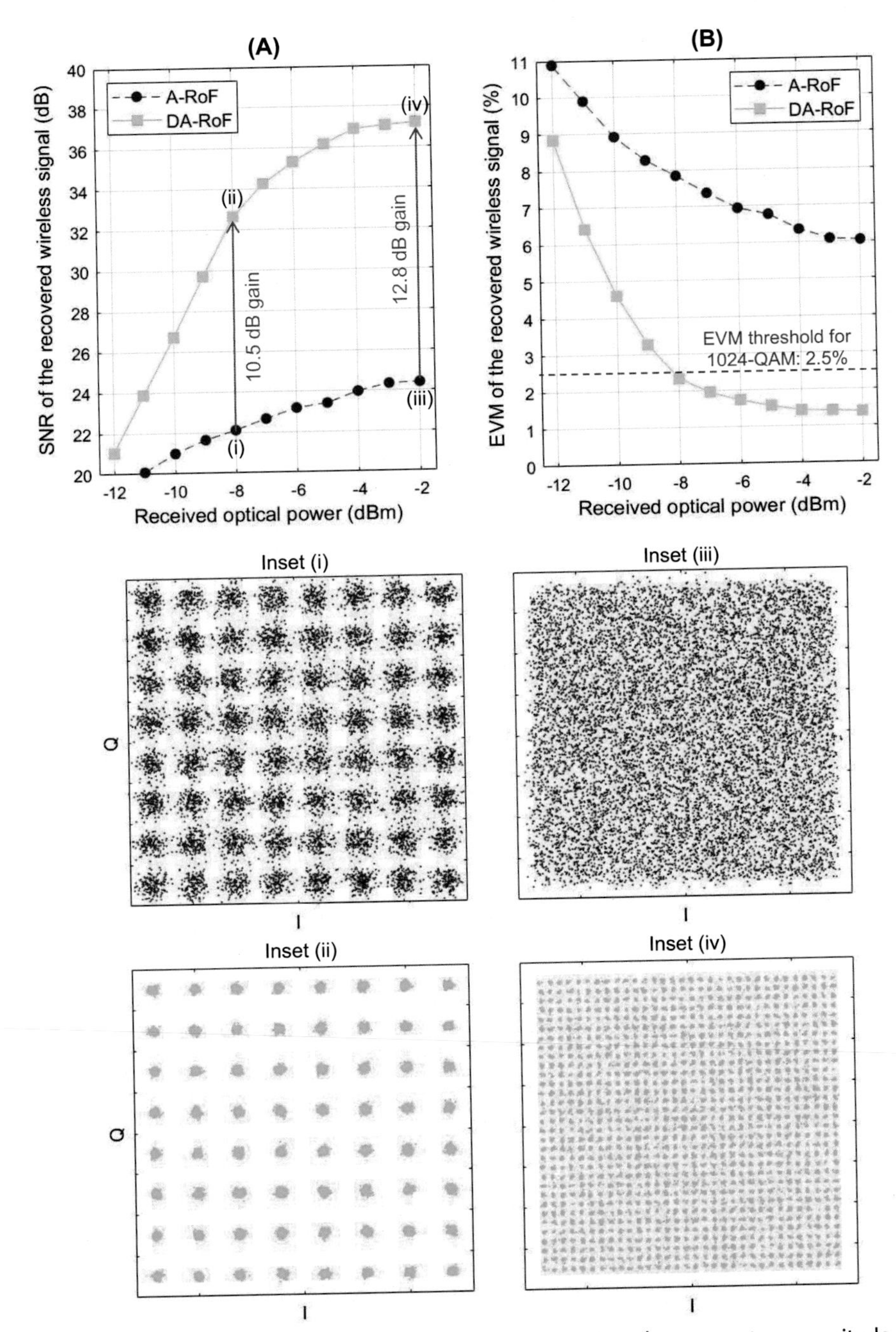

Figure 5.26 Experimentally measured signal-to-noise ratio (A) and error vector magnitude (B) of the recovered wireless signal as a function of the received optical power for both digital-analog radio-over-fiber (DA-RoF) and analog-RoF (A-RoF). Insets (i)/(iii) and (ii)/(iv) are representative constellation diagrams for the recovered wireless signals at the received optical powers indicated in (A) via A-RoF and DA-RoF, respectively. (After [33]).

below 3.5% and 2.5%, respectively [27,31]. Thus the DA-RoF technique enables the support of these large constellations with additional margin for signal distortions outside the front-haul segment.

It is worth noting that the SNR gain provided by DA-RoF is at the expense of reduced SE. In the case of CPRI-compatible DA-RoF, the CPRI-equivalent data rate for the 8-Gaud signal is 160 Gb/second, which is 62.5% of that of CPRI-compatible A-RoF (256 Gb/second) [17]. In the absence of the CS, the SE of DA-RoF is 50% of that of A-RoF, because each complex sample of an A-RoF waveform is represented by two samples in DA-RoF, one in W_1 and the other in W_2. Thus DA-RoF is capable of achieving a SNR gain over A-RoF of >10 dB at halved SE, which can enable the transmission of high-fidelity wireless signals over links with limited SNRs. As compared to the capacity-approaching D-RoF [32], the DA-RoF shows superior performance and does not require computation-intensive FEC, and thus energy-efficient and low-latency RoF transmission can be readily supported.

DA-RoF can also provide the SNR adaptation capability [31,32], for example, achieving higher SNR gain for links with higher SNR_{link} by using a larger entropy in the digital PCS-n-QAM part and a larger magnification in the analog PCM part. Fig. 5.27 shows the SNR of the received wireless signal as a function of the link SNR

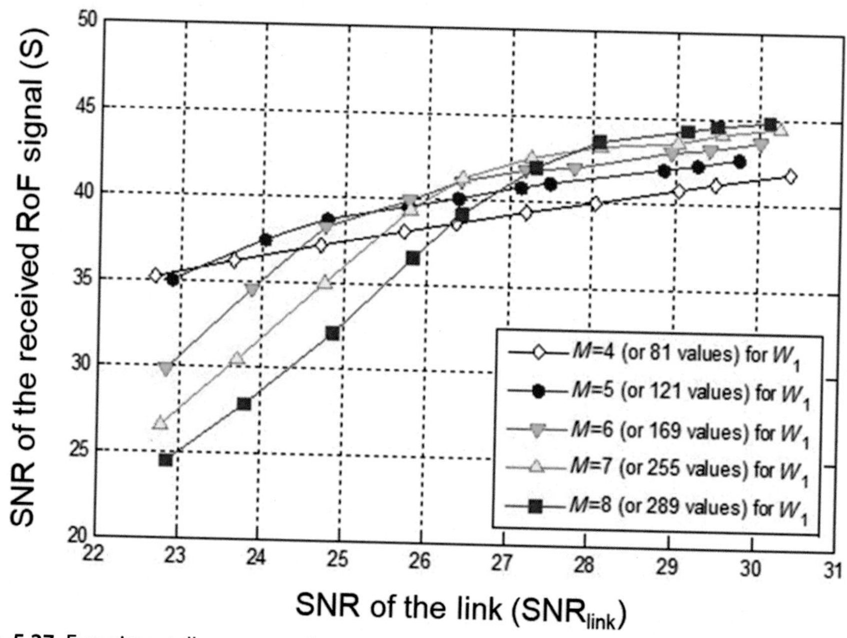

Figure 5.27 Experimentally measured signal signal-to-noise ratio (SNR) as a function of the link SNR for different *n* values in the digital probabilistic constellation shaped-n-quadrature amplitude modulation part of digital-analog radio-over-fiber.

measured in a 1.66-Gbaud transmission experiment [35]. The size of the n-QAM constellation (n) is represented as $(2M+1)^2$, where M is set to 4, 5, 6, 7, or 8, corresponding to 81, 121, 169, 255, or 289 complex values used in the approximation for obtaining W_1, respectively.

Evidently, the SNR gain increases with SNR_{link}. At a high SNR_{link} of 30 dB, the SNRs of the recovered RoF signal are improved to by approximately 11, 12.5, 14, 14.5, and 15 dB (to 41, 42.5, 44, 44.5, and 45 dB) when the M values are set to 4, 5, 6, 7, and 8, respectively. The best-performing n-QAM for $SNR_{link} = 30$ dB is 289-QAM (at $M = 8$). On the other hand, at a low SNR_{link} of 23 dB, the SNRs of the recovered RoF signal are improved to by approximately 12.5, 12, 7.5, 4, and 1 dB (to 35.5, 35, 30.5, 27, and 24 dB) when the M values are set to 4, 5, 6, 7, and 8, respectively. The best-performing n-QAM for $SNR_{link} = 23$ dB is 81-QAM (at $M = 4$). For a given SNR_{link}, there is an optimum value of M that provides the best SNR performance for the recovered RoF signal. As SNR_{link} increases from 23 to 30 dB, the optimum SNR gain obtained by DA-RoF (with halved SE) also increases from about 12.5 to 15 dB. This is in agreement with the theoretical results in Section 5.4 showing that the higher the SNR_{link}, the larger the SNR gain obtained by doubling the bandwidth. More bandwidth expansion in the DA-RoF scheme can be straightforwardly realized by cascading more digital modulations in the overall modulation process, and the requirement on SNR_{link} can be further relaxed. Thus SNR adaptation with a large dynamic range can be achieved.

It is worth noting that the SNR gains achieved in the DA-RoF demonstrations [33] are not yet close to the theoretical limit, for which capacity-approaching FEC codes need to be used. It is feasible to introduce FEC in the digital modulation part of the DA-RoF modulation. For example, the MSBs representing a wireless signal can be digitally transmitted with capacity-approaching modulation and coding schemes such as PCS and low-density parity check codes, while the remaining LSBs can be transmitted via PCM. As the PCM is uncoded, the overall power consumption and processing latency of the coded DA-RoF scheme can still be less than its fully digital RoF counterpart. Thus the DA-RoF technique offers a new option to attain the optimum balance between the signal transmission performance and the DSP complexity for a given RoF system.

5.6 Concluding remarks

On the basis of the studies presented in the previous sections, we can draw some concluding remarks on the three RoF techniques, A-RoF, D-RoF, and DA-RoF, enumerated as follows.

- A-RoF provides high transmission SE but requires relatively high link SNR, which has traditionally prevented it from achieving sufficient signal fidelity and dynamic

range. PM can be applied to allow A-RoF to trade SE for signal fidelity at a scaling of 6 dB SNR gain for each halving of SE.
- With high-resolution DAC/ADC and error-free digital optical transmission based on binary modulation, D-RoF supports sufficiently high signal fidelity and large dynamic range and has become the mainstream technology for mobile front-haul in 4G networks. On the other hand, D-RoF based on binary optical modulation is bandwidth-inefficient. To increase the bandwidth efficiency of D-RoF, high-level QAM modulation with capacity-approaching PCS and FEC can be used, at the expense of increased power consumption and processing latency. SNR adaption can be realized at a theoretical scaling of doubled SNR for each halving of SE.
- Combining the advantages of A-RoF and D-RoF, DA-RoF offers high signal fidelity, flexible SNR adaptation, low DSP complexity and power consumption, as well as low processing latency. DA-RoF can provide an SNR gain of >12 dB at halved SE when the link SNR is above 23 dB.

The advanced RoF techniques may find applications in future 5G and 6G mobile networks with high-bandwidth wireless signals and massive MIMO antennas. For example, to support a 256 × 256 MIMO with 200-MHz wireless signal bandwidth, the CPRI-equivalent data rate would be as high as 2.56 Tb/second (even with the assumption of the efficient 64b/66b line coding). On the other hand, the aggregated RF bandwidth is only 51.2 GHz. When a bandwidth expansion factor of 2 is used to improve the RoF signal fidelity (e.g., via DA-RoF), the achievable aggregation bandwidth of the RoF system needs to be 102.4 GHz, which can be supported by a 400G-ZR type coherent transceiver having two polarizations each modulated with an RF bandwidth of about 60 GHz [36,37].

By dint of the advances in the field of digital coherent transceivers, such 400G-ZR type coherent transceiver can fit into a quad small form-factor pluggable double density (QSFP-DD) format factor with a total power consumption of less than 15 W [36,37]. This example shows that a high-end 5G/6G front-haul connection that requires a CPRI-equivalent data rate of multi-Tb/second per wavelength could be supported by a compact power-efficient dual-polarization coherent transceiver when DSP-assisted DA-RoF is applied. This supports one of the visions of 5G to be "*green and soft*" [38] by providing high energy efficiency and software-defined adaptation in C-RAN front-haul links. Continued research in advanced DSP-assisted RoF technologies are anticipated to be fruitful and valuable.

References

[1] Cooper AJ. 'Fibre/radio' for the provision of cordless/mobile telephony services in the access network. Electron Lett 1990;26:2054–6.
[2] Fye D.M., Design of fiber optic antenna remoting links for cellular radio applications.In: Proceedings of IEEE Vehicular Technology Conference; 1990. p. 622–625.

[3] Ogawa H, Polifko D, Banba S. Millimeter-wave fiber optics systems for personal radio communications. IEEE Trans Micro Thy Tech 1992;40(12):2285–93 Dec.
[4] Wake D, Webster M, Wimpenny G, Beacham K, Crawford L. Radio over fiber for mobile communications. In: Proceeding of International Topical Meeting on Microwave Photonics; 2004. p. 157–160.
[5] Wake D, Nkansah A, Gomes NJ. Radio over fiber link design for next generation wireless systems. J Light Technol 2010;28(16):2456–64.
[6] Novak D, et al. Radio-over-fiber technologies for emerging wireless systems. IEEE J Quant Electron 2016;52(1):1–11.
[7] Lim C, Tian Y, Ranaweera C, Nirmalathas TA, Wong E, Lee K-L. Evolution of radio-over-fiber technology. J Light Technol 2019;37:1647–56.
[8] ITU-T Recommendation G. 983.3. A broadband optical access system with increased service capability by wavelength allocation; 2001 March.
[9] Phillips MR, Ott DM. Crosstalk due to optical fiber nonlinearities in WDM CATV lightwave systems. J Light Technol 1999;17(10):1782–92 Oct.
[10] Kim H, Jun SB, Chung YC. Raman crosstalk suppression in CATV overlay passive optical network. IEEE Photon Technol Lett 2007;19(9):695–7 May 1.
[11] Cho S-H, Park H, Chung HS, Doo KH, Lee S, Lee JH. Cost-effective next generation mobile fronthaul architecture with multi-IF carrier transmission scheme. In: Proceeding OFC 2014, paper Tu2B.6; 2014.
[12] Liu X, Effenberger F, Chand N, Zhou L, Lin H. Efficient mobile fronthaul transmission of multiple LTE-A signals with 36.86-Gb/s CPRI-equivalent data rate using a directly-modulated laser and fiber dispersion mitigation. In: ACP 2014, post-deadline paper AF4B.5; 2014.
[13] Liu X, Effenberger F, Chand N, Zhou L, Lin H. Demonstration of bandwidth-efficient mobile fronthaul enabling seamless aggregation of 36 E-UTRA-like wireless signals in a single 1.1-GHz wavelength channel. In: Proceeding OFC 2015, paper M2J.2; 2015.
[14] Zhu M, Liu X, Chand N, Effenberger F, Chang G-K. High-capacity mobile fronthaul supporting LTE-advanced carrier aggregation and 8 × 8 MIMO. In: Proceeding OFC 2015, paper M2J.3; 2015.
[15] Liu X, Zeng H, Chand N, Effenberger F. Experimental demonstration of high-throughput low-latency mobile fronthaul supporting 48 20-MHz LTE signals with 59-Gb/s CPRI-equivalent rate and 2-μs processing latency. In: Proceeding ECOC, invited paper We.4.4.3; 2015.
[16] Liu X, Zeng H, Effenberger F. Bandwidth-efficient synchronous transmission of I/Q waveforms and control words via frequency-division multiplexing for mobile fronthaul. In: Globecom 2015, paper SAC 21–3; 2015.
[17] Liu X, Zeng H, Chand N, Effenberger F. CPRI-compatible efficient mobile fronthaul transmission via equalized TDMA achieving 256 Gb/s CPRI-equivalent data rate in a single 10-GHz-bandwidth IM-DD Channel. In: Proceeding OFC'16, paper W1H.3; 2016.
[18] Liu X, Zeng H, Chand N, Effenberger F. Efficient mobile fronthaul via DSP-based channel aggregation. J Light Technol 2016;34(6):1556–64 March 15.
[19] Zeng H, Liu X, Megeed S, Chand N, Effenberger F. Demonstration of a real-time FPGA-based CPRI-compatible efficient mobile fronthaul transceiver supporting 53 Gb/s CPRI-equivalent data rate using 2.5-GHz-class optics. In: Proceeding ECOC 2016, paper W1E. l; 2016.
[20] Zeng H, Liu X, Megeed S, Chand N, Effenberger F. Real-time demonstration of CPRI-compatible efficient mobile fronthaul using FPGA. J Light Technol 2017;35(6):1241–7.
[21] Ishimura S, Bekkali A, Tanaka K, Nishimura K, Suzuki M. 1.032-Tb/s CPRI-equivalent rate IF-over-fiber transmission using a parallel IM/PM transmitter for high-capacity mobile fronthaul links. J Light Technol 2018;36(8):1478–84.
[22] Le ST, Schuh K, Chagnon M, Buchali F, Buelow H. 1.53-Tbps CPRI-equivalent data rate transmission with Kramers–Kronig receiver for mobile fronthaul links. In: Proceeding ECOC'2018, paper We4B.4; 2018.
[23] Alimi IA, Teixeira AL, Monteiro PP. Toward an efficient C-RAN optical fronthaul for the future networks: a tutorial on technologies, requirements, challenges, and solutions. IEEE Commun Surv Tutor 2018;20(1):708–69 First quarter.

[24] I C-L, Huang J, Duan R, Cui C, Jiang J,Li L. Recent progress on C-RAN centralization and cloudification. In: IEEE Access, vol. 2; 2014. p. 1030–1039.
[25] Pizzinat A., Chanclou P, Diallo T, Saliou F. Things you should know about fronthaul. In: Proceeding ECOC, Tu.4.2.1, Cannes; 2014.
[26] CPRI Specification V7.0. Common public radio interface (CPRI); interface specification; 2015 October.
[27] 3GPP Technical specification 36.104.
[28] Caballero A, Zibar D, Monroy IT. Performance evaluation of digital coherent receivers for phase-modulated radio-over-fiber links. J Light Technol 2011;29(21):3282–92.
[29] Che D, Yuan F, Shieh W. High-fidelity angle-modulated analog optical link. Opt Express 2016;24:16320–8.
[30] Ishimura S, Kao H, Tanaka K, Nishimura K, Suzuki M. SSBI-free 1024QAM single-sideband direct-detection transmission using phase modulation for high-quality analog mobile fronthaul. In: Proceeding ECOC'2019, post-deadline paper PD.1.2; 2019.
[31] Che D. Digital SNR adaptation of analog radio-over-fiber links carrying up to 1048576-QAM signals. In: Proceeding ECOC 2020, post-deadline paper PDP 2.1; 2020.
[32] Ji H, Sun C, Shieh W. Spectral efficiency comparison between analog and digital RoF for mobile fronthaul transmission link. J Light Technol 2020;38(20):5617–23.
[33] Liu X. Hybrid digital-analog radio-over-fiber (DA-RoF) modulation and demodulation achieving a SNR gain over analog RoF of >10 dB at halved spectral efficiency. submitted to OFC 2021; 2021.
[34] Buchali F, Steiner F, Böcherer G, Schmalen L, Schulte P, Idler W. Rate-adaptation and reach increase by probabilistically shaped 64-QAM: an experimental demonstration. J Light Technol 2016;34(7):1599–609.
[35] Liu X, Zeng H. Cascaded waveform modulation with an embedded control signal for high-performance mobile fronthaul, US patent 10,205,522 10; 2019.
[36] Optical Internetworking Forum (OIF). Implementation agreement 400ZR [Online]. https://www.oiforum.com/wp-content/uploads/OIF-400ZR-01.0_reduced2.pdf, 2020.
[37] Filer M. The coherent 400 Gb/s revolution. Lightreading optical networking digital symposium; 2020 May 28.
[38] I C-L, Rowell C, Han S, Xu Z, Li G, Pan Z. Towards green & soft: a 5G perspective. IEEE Commun Mag 2014;52(2):66–73.

CHAPTER 6

Point-to-multipoint transmission

6.1 TDMA-based P2MP

6.1.1 TDMA-based P2MP for connecting densely installed compact AAUs/RRUs

As the 5G deployment reaches deeper into our communities, there is a need for densely installed compact antenna units to complement the coverage of macro cells. Fig. 6.1 illustrates a deployment scenario of densely installed compact active antenna units (AAUs) and remote radio units (RRUs) for a highly populated residential area. Instead of massive multiple-input multiple-output (m-MIMO) antennas used in 5G macro cells, these compact AAUs and RRUs usually use 2×2 MIMO or 4×4 MIMO with less than 100 MHz of radio frequency (RF) bandwidth. The needed enhanced Common Public Radio Interface (eCPRI) or Common Public Radio Interface (CPRI) (with compression) bit rates for these compact AAUs/RRUs are of the order of 2.5 Gb/s per unit. To cost-effectively connect the massive number of such compact antenna units to their corresponding distributed units (DUs) and base-band units (BBUs), point-to-multipoint (P2MP) optical network architectures are naturally suitable. P2MP communication can be based on time-division multiple access (TDMA) or frequency-division multiple access (FDMA). In this section, we will present TDMA-based P2MP communication, while FDMA-based P2MP communication will be presented in the next section.

6.1.2 TDM-PON for CPRI connections

Time-division multiplexing passive optical network (TDM-PON) is a well-established technology for TDMA-based P2MP optical communication [1–10]. As shown in Fig. 6.1, the optical distribution network (ODN) in PON is fully passive, therefore no power suppliers and active components are needed in the ODN, which makes it easy and cost-effective to deploy. In addition, only one feeder fiber is needed to connect the equipment room with a passive optical splitter that is in close proximity to the compact AAUs/RRUs bidirectionally, thereby saving fiber resources. Moreover, PON has been the primary solution for fiber-to-the-home (FTTH) and fiber-to-the-building (FTTB) applications; therefore the ODNs deployed for FTTH and FFTB applications may be shared or rearchitected for fiber-to-the-antenna applications [11–15]. In 2021, 50-Gb/s PON has been standardized by the international

Optical Communications in the 5G Era
DOI: https://doi.org/10.1016/B978-0-12-821627-9.00008-5

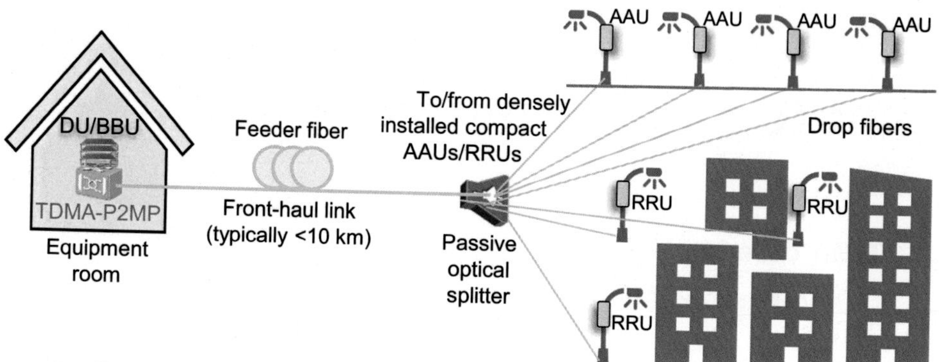

Figure 6.1 Illustration of a cost-effective time-division multiple access-based point-to-multipoint optical network that connects densely installed compact active antenna units/remote radio units in a highly populated residential area to the distributed unit/baseband unit in a nearby equipment room.

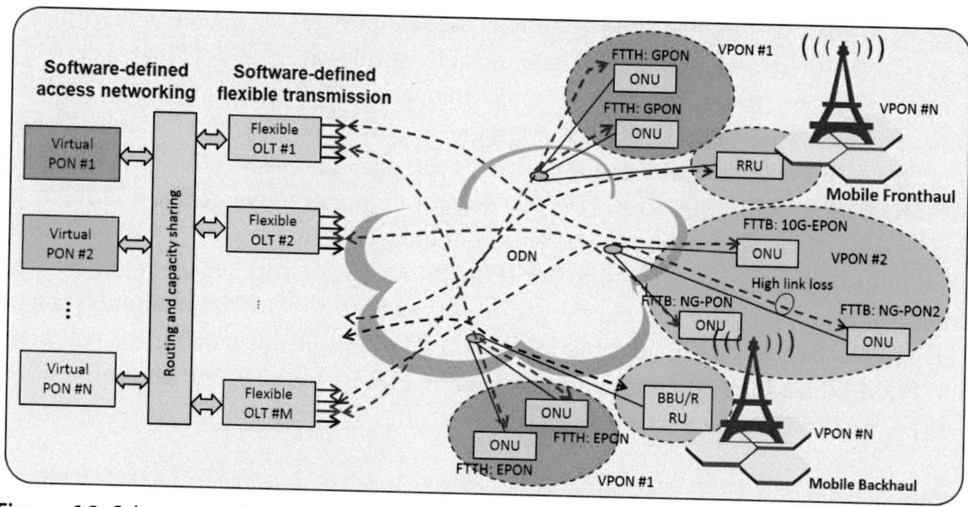

Figure 6.2 Schematic of a flexible-passive optical network that addresses the demands such as customer-specific data rates, network optimization and virtualization, and fiber/wireless convergence. *OLT*, Optical line terminal; *ODN*, optical distribution network; *ONU*, optical network unit. *VPON*, virtual PON; *FTTH(B)*, fiber-to-the-home (business); *RRU*, remote radio unit; *BBU*, baseband unit. *(After [5]).*

telecommunication union (ITU) [9,16]. The net payload data rate of 50G-PON is over 40 Gb/s per direction, thus a meaningful number (e.g., 16) of compact AAUs/RRUs can be supported by a single 50G-PON optical line terminal (OLT).

Fig. 6.2 illustrates a flexible-PON architecture with software-defined networking and network function virtualization implemented to optimize the network in terms of throughput, energy efficiency, and control and management [5]. Multiple virtual

PON units are interconnected to better address various applications, such as FTTH and FTTB, through software-defined routing and capacity sharing. Flexible OLTs are used to realize software-defined transmission with flexible data rates and link loss budgets, depending on network topology and dynamics.

There is a clear need for coordination between radio access network (RAN) and PON to simplify the overall access network, reduce the network latency, and improve the network cost- and power-efficiency. To meet the stringent latency requirements of 5G mobile front- and back-haul, joint media access control (MAC) scheduling between the RAN-MAC and PON-MAC is needed. For example, wireless scheduling can be shared with the PON scheduling in advance, such that the upstream traffic from wireless user equipment can be seamlessly carried over by the TDM-PON system without having to wait for a negotiation between the ONU and OLT, as illustrated in Fig. 6.3. This concept has been recently proposed and described in Refs [17–19]. With this coordination and advances in fast burst-mode channel tracking [e.g., via advanced burst-mode digital signal processing (DSP)], accelerated burst scheduling of PON can also be realized.

Fig. 6.4 illustrates an accelerated burst scheduling in TDM-PON upstream transmission for timing-critical mobile front-haul applications. Preferably, the cycle time period (T_{cycle}), during which the OLT scans through all the ONUs once, is a multiple of the CPRI basic frame period, 260.416667 ns [or 1/(3.84 MHz)], so that multiple CPRI frames can be

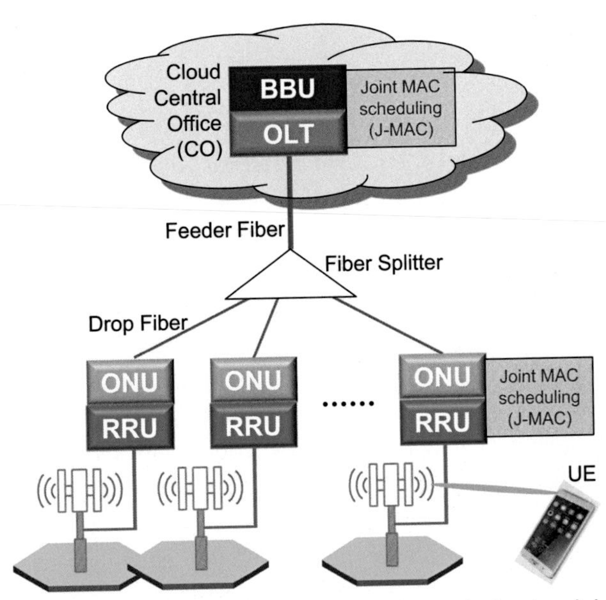

Figure 6.3 Illustration of a time-division multiplexing passive optical network based mobile front-/back-haul network with RAN-passive optical network coordination for joint media access control scheduling to reduce the overall network latency. *(After [5]).*

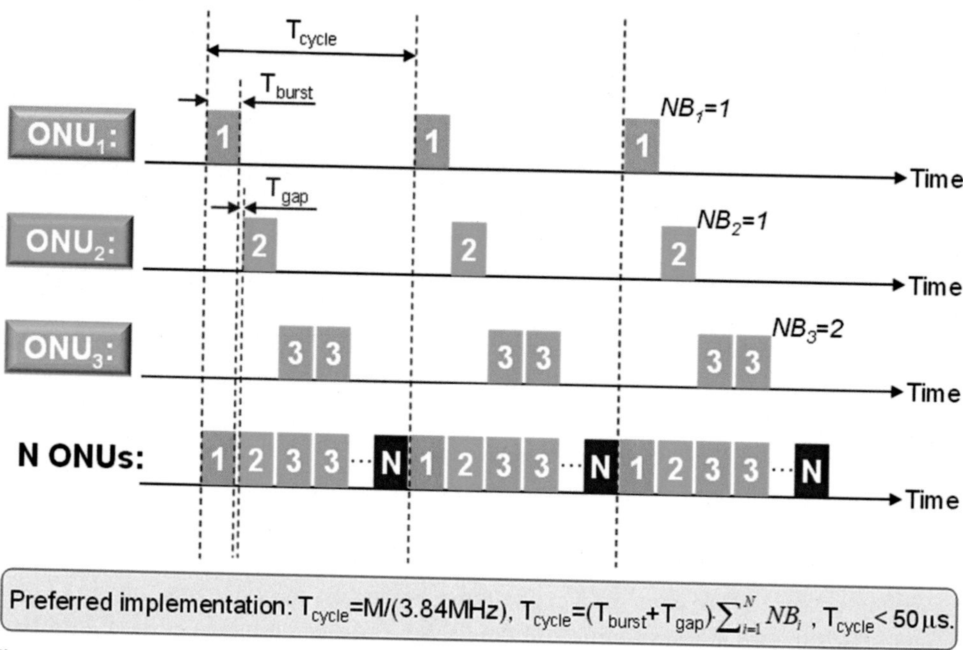

Figure 6.4 Illustration of a low-latency synchronous burst scheduling design in time-division multiplexing passive optical network upstream transmission for timing-critical mobile front-haul applications. *(After [5]).*

transported per cycle. T_{cycle} needs to be shorter than the processing latency specified. For example, T_{cycle} is preferred to be less than 50 μs for some front-haul applications. Each ONU transmits CPRI frames in a burst-by-burst basis with a predetermined burst period, T_{burst}. Flexible bandwidth allocation can be realized by assigning each ONU a given number of bursts per cycle (NB). A suitable gap period, T_{gap}, is allocated between successive bursts to avoid implementation imperfection—induced burst collision. Thus T_{cycle} can also be expressed as $T_{\text{cycle}} = (T_{\text{burst}} + T_{\text{gap}}) \cdot \sum_{i=1}^{N} \text{NB}_i$, where N is the total number of ONUs in the PON and NB_i is the number of bursts per cycle assigned to the i-th ONU.

Note that in this specific PON architecture, the OLT and the all the ONUs have a common clock frequency that is locked to the source clock of the BBU pool. In the case of CPRI, a fundamental frequency is the CPRI basic frame rate (or the basic universal mobile telecommunications system chip rate), 3.84 MHz. In essence, it is beneficial for PON and RAN to share both the MAC-layer scheduling and the physical-layer clock to achieve low-latency, synchronous forwarding of mobile signals through the low-cost PON system. By doing so, timing-critical mobile communication demands can be effectively supported. This new type of PON system may coexist with current generation of PON systems by using different wavelength plans.

6.1.3 TDM-PON for eCPRI connections

Passive optical network has a P2MP architecture, which is uniquely suited for carrying eCPRI traffic to obtain the statistical multiplexing gain and reduce fiber cost via the sharing of the common fiber infrastructure by multiple RRU sites [20]. Fig. 6.5 illustrates a low-latency eCPRI-over-PON architecture. To carry eCPRI packets over TDM-PON with high efficiency, we need to aggregate upstream eCPRI packets into

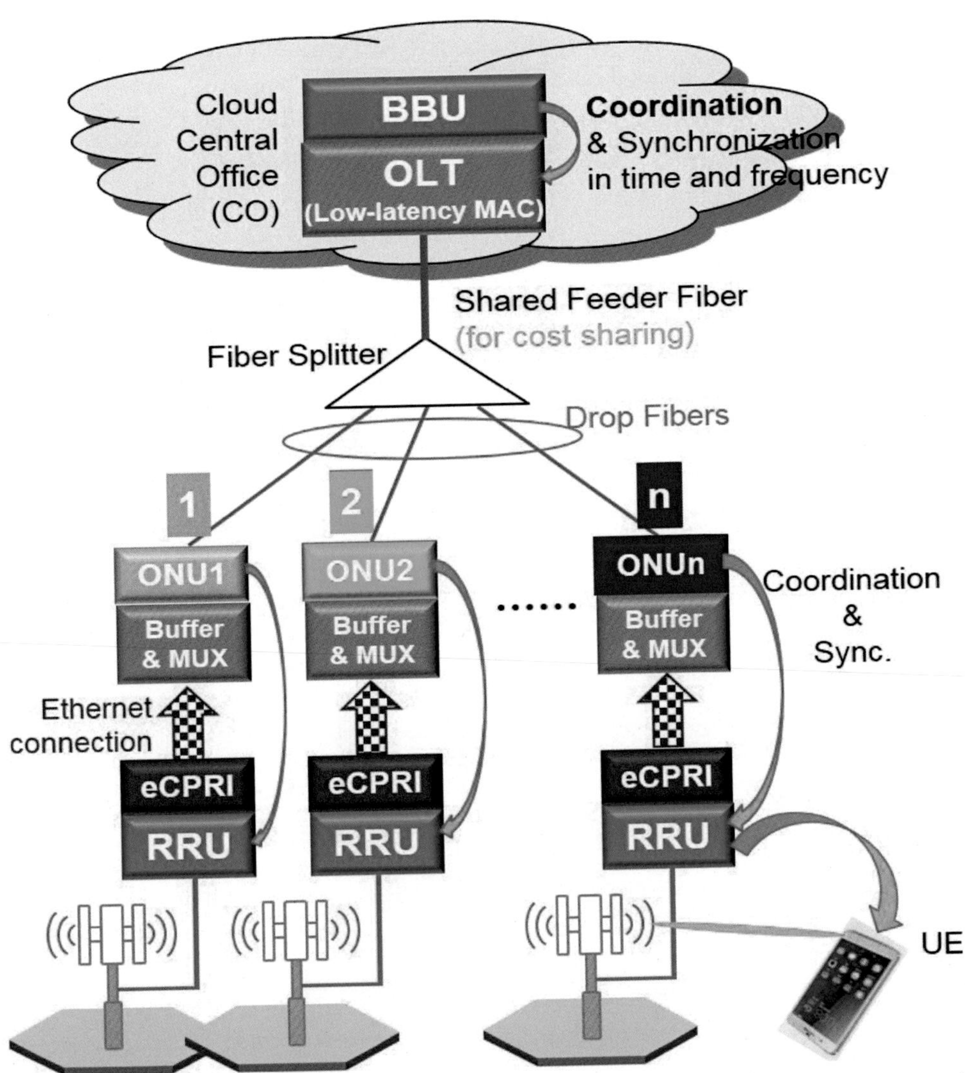

Figure 6.5 Illustration of a low-latency enhanced common public radio interface-over- passive optical network architecture. *(After [20]).*

fixed-duration TDM-PON bursts as there is an overhead to be paid for each burst for synchronization and channel tracking purpose. The fixed-duration TDM-PON bursts can then be transmitted in a similar way as the CPRI-over-PON case. To carry eCPRI packets over TDM-PON with low latency, we propose to transmit each TDM-PON burst right after the aggregation of multiple upstream eCPRI packets into this burst is completed, as illustrated in Fig. 6.6. This can be achieved by the coordination between the RAN-MAC and PON-MAC.

As an example, we can have the following arrangement.

1. PON's cycle time (T_{cycle}) equals the largest RRU symbol period (T_{sym});
2. Each PON cycle is divided into M ONU transmission bursts (e.g., $M = 16$);
3. Assuming a total upstream transmission rate from all the ONU's to the OLT of R_{US} (e.g., 40 Gb/s), each burst corresponds to a rate of $R_{burst} = R_{US}/M$ (e.g., 2.5 Gb/s);

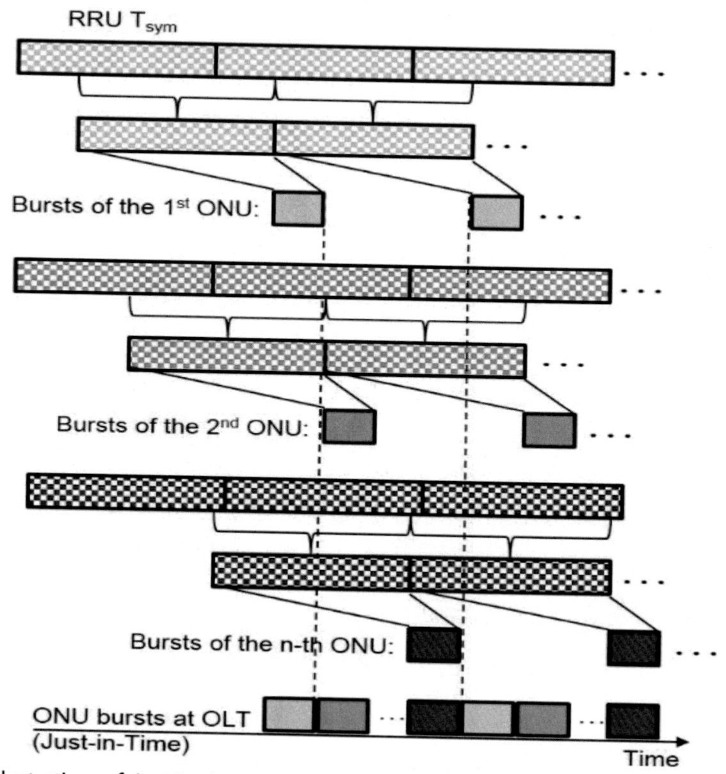

Figure 6.6 Illustration of just-in-time aggregation of enhanced common public radio interface (eCPRI) packets into time-division multiplexing passive optical network bursts in the low-latency eCPRI-over-passive optical network. *(After [20]).*

4. The peak transmission rate between an RRU and an ONU can be up to $4 \times R_{burst}$ (e.g., 10 Gb/s by using a 10GE connection);
5. The ONU buffer size is $2 \times N_{sym}$ (to avoid buffer overflow and packet drop).

To carry eCPRI packets over TDM-PON with high performance or signal fidelity, we propose to apply forward-error correction (FEC) to each TDM-PON burst by adding FEC parity check bits into the burst during the aggregation process. The downstream transmission arrangement can be made to be similar to the upstream transmission. Fig. 6.7 illustrates such a symmetric downstream and upstream eCPRI-PON frame structure.

As an implementation example, we have the following.

- $T_{cycle} = 40\ \mu s = 2$ Mb/frame at 50 Gb/s
- FEC RS(216, 248) codeword = 1984 $b = \sim 40$ ns. ~ 1000 codewords per frame
- Total overhead time = 2 codewords = 3968 bits
- $T_{gap} = 53.76$ ns (2688 bits), Preamble 102.4 ns (5120 bits), delimiter 2.56 ns (128 bits)
- Efficiency $= 216/248 \times (1000-2 \times 8)/1000 = 85.7\%$ (assuming 8 ONUs)
- Net payload rate = 50 G $\times$ 84.3% = 42 Gb/s.

Here, downstream signal consists of preamble and delimiter for each payload segment destined to a given ONU, similar to the upstream case, with the following features.

- The data recoveries in both upstream and downstream are similar, so the DSP implementation may be simplified. For example, same FEC and data recovery method can be used. Each ONU only needs to recover the data destined to it, thus reducing the power consumption at the each ONU.
- The downstream signal does not need the guard spaces, and the laser can operate in the continuous mode. More control and management bits may be added in those "guard spaces."
- The downstream transmission efficiency is the same as the upstream efficiency.

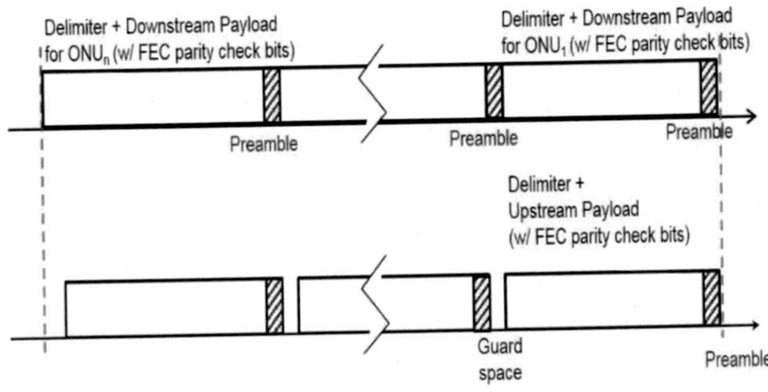

Figure 6.7 Illustration of a symmetric downstream and upstream enhanced common public radio interface-passive optical network frame structure. *(After [20])*.

The downstream frame structure can also be made to be different from the upstream. Fig. 6.8 shows such an asymmetric downstream and upstream eCPRI-PON frame structure. Here, the downstream frame header can be 1 codeword, or 1984 bits, which is sufficient to contain all the needed ONU information. The downstream transmission efficiency is $216/248 \times (1000-1)/1000 = 87\%$, which is slightly higher than that of the upstream.

The symmetric downstream and upstream eCPRI-PON frame structure has been implemented in a real-time platform based on FPGA [20]. The actual processing latency associated with the aggregation of the eCPRI packets and FEC encoding is only ~ 45 μs, which is considered acceptable for low-latency 5G applications where the end-to-end latency needs to be limited to 1 Ms.

6.1.4 Modulation formats for TDMA-P2MP

Low cost and power consumption per bit is of crucial importance in mobile fronthaul applications. This calls for low-cost low-power-consumption optical transceivers running at a high speed, for example, 50 Gb/s. One attractive approach is to use advanced modulation and detection techniques, enabled by DSP, to support high-speed transmission with low-cost narrow-bandwidth optics [5]. For example,

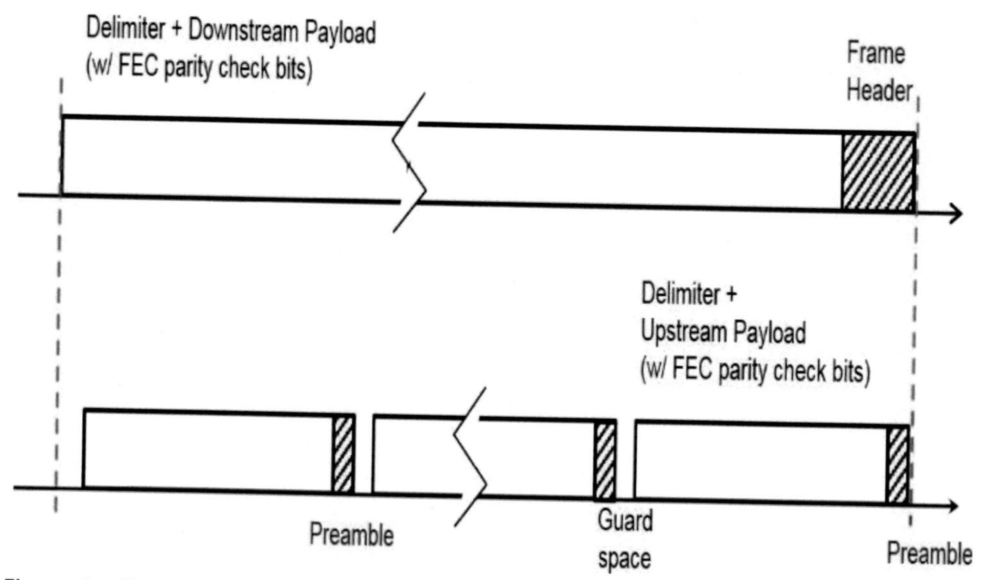

Figure 6.8 Illustration of an asymmetric downstream and upstream enhanced common public radio interface-passive optical network frame structure. *(After [20]).*

10-GHz-class lasers can be used to generate 40 and 50 Gb/s signals via digital equalization (EQ) [21–29]. Fig. 6.9 shows a downstream modulation and detection scheme based on DSP-enabled high-speed transmission for CPRI/eCPRI-PON. At the transmitter side, payload data are modulated via non-return-to-zero (NRZ), 3-level pulse-amplitude modulation (PAM3), or 4-level pulse-amplitude modulation (PAM4) modulation format, carrying 1, 1.5, or 2 bits per symbol, respectively. The modulation format information can be sent to the receiver in the frame header by using the most reliable NRZ format. The signal is then oversampled to 64 GSa/s and outputted by a high-speed digital-to-analog converter (DAC), to drive a directly modulated laser (DML) whose 3-dB bandwidth is about 10 GHz. After 20-km standard single-mode fiber (SSMF) transmission, the optical signal is converted to electronic domain by an avalanche photo-detector (APD). The converted electronic signal is then digitized by an 80-GSa/s analog-to-digital converter (ADC), followed by DSP-based clock recovery, EQ, and demodulation. The T/2-fractionally spaced equalizer is used in the experiment. In order to overcome the bandwidth limitation of 10G-class optics, 60 EQ taps are used. The recovered bit sequence is then compared with the original bit sequence for error counting.

Fig. 6.10 shows the gray-coded modulation mapping of these NRZ, PAM3, and PAM4 modulation formats. The encoded signal is modulated at 25 Gbaud, leading to 25, 37.5, and 50 Gb/s data rates for NRZ, PAM3, and PAM4, respectively. Fig. 6.11 shows the recovered signal symbols as a function of time for 25-Gb/s NRZ, 37.5-Gb/s PAM3, and 50-Gb/s PAM4 at a received optical power of −20 dBm. Fig. 6.12 shows experimentally measured bit error ratio (BER) performance as a function of received optical power for these three formats. At typical FEC raw BER threshold of 10^{-3}, 25-Gb/s NRZ, 37.5-Gb/s PAM3, and 50-Gb/s PAM4 require a received optical power (ROP) of approximately −26, −22, and −19 dBm, respectively. Assuming a transmitter power of 7 dBm, link budgets of 33,

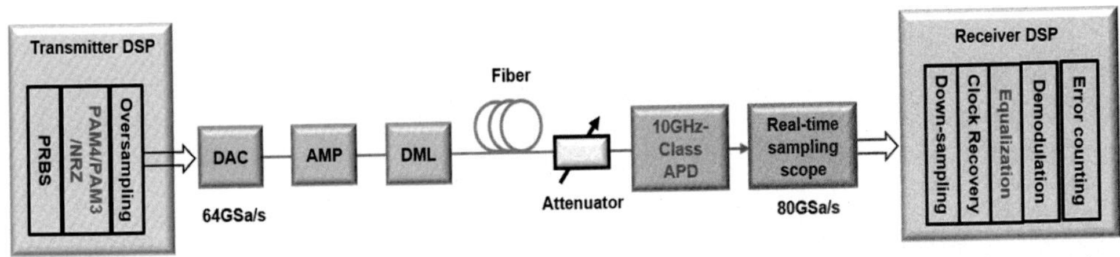

Figure 6.9 Downstream modulation and detection scheme based on digital signal processing-enabled high-speed transmission for common public radio interface/enhanced common public radio interface-passive optical network. *(After [20]).*

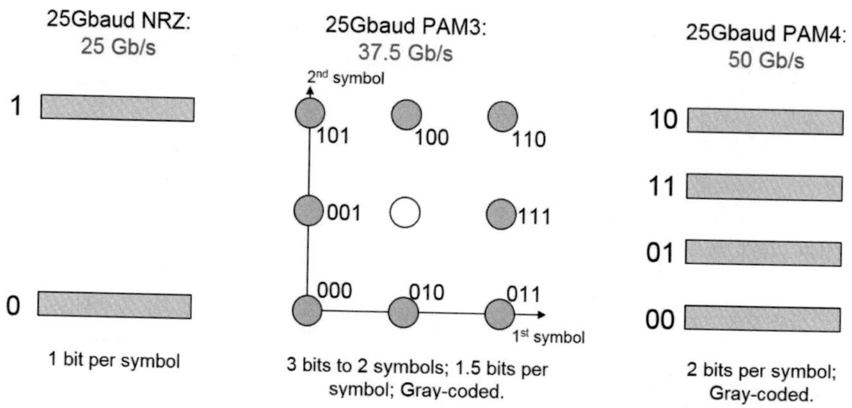

Figure 6.10 Illustrations of nonreturn-to-zero, PAM3, and PAM4 modulation formats, carrying 1, 1.5, and 2 bits per symbol, respectively. *(After [20]).*

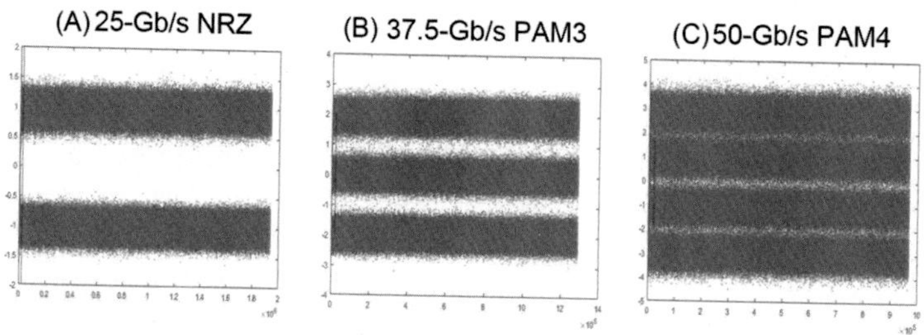

Figure 6.11 Experimentally recovered nonreturn-to-zero, PAM3, and PAM4 symbols at −20 dBm received optical power. *(After [20]).*

29, and 26 dB can be supported, respectively. When advanced FEC is used, raw BER threshold of 10^{-2} may be realized [9], and the link budgets can be further increased by about 2 dB.

At the Nyquist frequency of 12.5 GHz, the channel response is about 15 dB down from the low-frequency response, which clearly indicates severe bandwidth limitation in the system. By dint of DSP-enabled channel EQ, reasonably good BER performances have been achieved. It can be expected that when the bandwidths of DML and APD are doubled, 50-Gbaud transmission can be supported with similar implementation penalties. In this case, 50-Gb/s NRZ, 75-Gb/s PAM3, and 100-Gb/s PAM4 can be supported with receiver sensitivities of about −24, −20, and −17 dBm at an FEC input BER of 10^{-2} [20]. With a moderate transmitter power of 5 dBm, link budgets of 29, 25, and 22 dB, respectively can be achieved, sufficient to

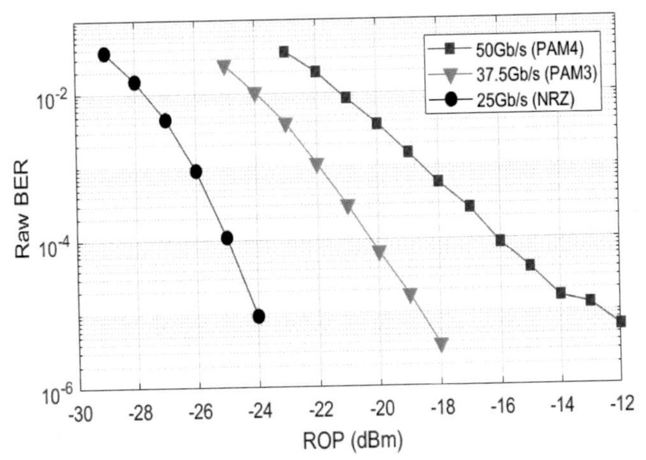

Figure 6.12 Experimentally measured bit error ratio performance as a function of received optical power. *(After [20]).*

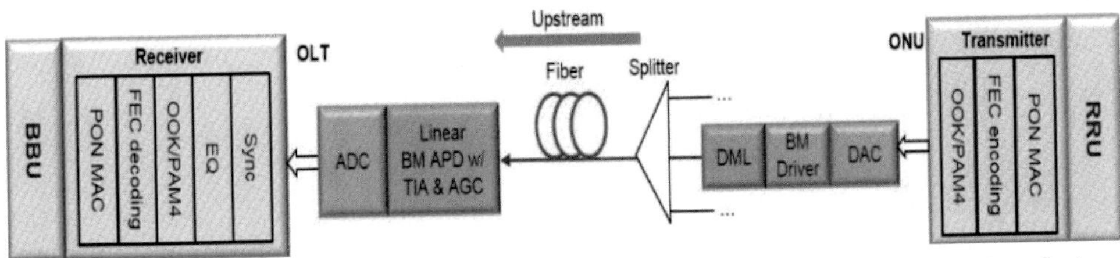

Figure 6.13 Schematic upstream modulation and detection scheme based on digital signal processing-assisted on-off-keying and PAM4 in a common public radio interface/enhanced common public radio interface-passive optical network. *(After [20]).*

support most of the CPRI-over-PON and eCPRI-over-PON needs for densely installed compact AAUs/RRUs.

6.1.5 Burst-mode DSP for TDMA-P2MP

Digital signal processing can also be used to support upstream transmission in CPRI/eCRPI-PON to achieve fast burst-mode tracking capability and thus low latency for future wireless applications. Fig. 6.13 shows P2MP CPRI/eCPRI-PON architecture, where a single feeder fiber is shared by multiple RRUs and statistical multiplexing gain may be obtained, potentially reducing the overall network cost. Compared to the downstream transmission, the upstream transmission additionally requires the APD,

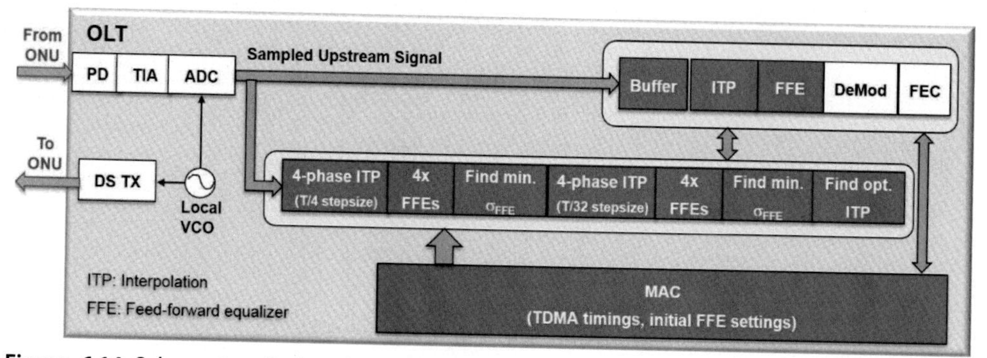

Figure 6.14 Schematic of digital signal processing-enabled burst-mode detection for fast channel synchronization and equalization. *(After [20]).*

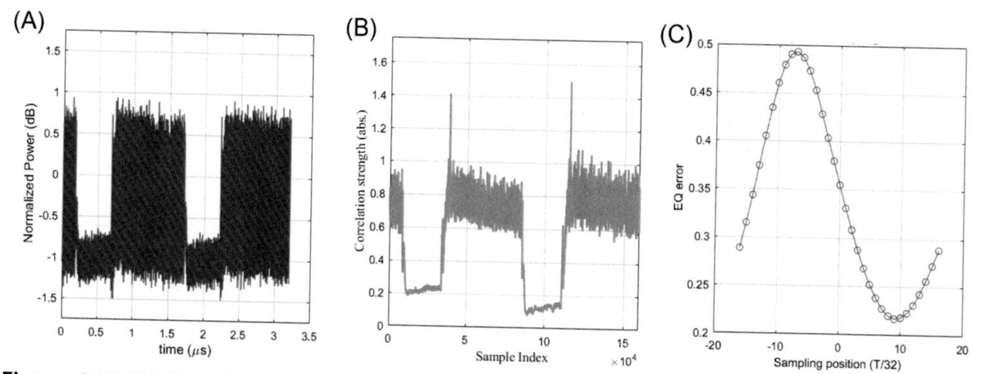

Figure 6.15 (A) Experimentally measured 25-Gb/s upstream signal bursts; (B) Correlation-based burst synchronization; (C) Measured equalization error as a function of the sampling position. *(After [20]).*

trans-impedance amplifier (TIA) and automatic gain control (AGC) to be able to operate in burst-mode.

The existing upstream receiver is based on the clock data recovery (CDR). Usually it requires 0.5−1 μs to converge. The proposed DSP-based scheme aims to achieve fast convergence. As a result, PON system efficiency can be improved. Fig. 6.14 shows the schematic of a DSP-assisted burst-mode detection scheme for fast channel synchronization and EQ. To achieve fast channel synchronization and EQ, we propose to take the advantage of the fact that each ONU clock is automatically synchronized with the OLT clock, and thus there is no need to perform clock recovery when receiving the upstream bursts from different ONUs. We only need to recover the clock phase or the optimal sampling position at the beginning of each burst from an

ONU, during which the past EQ coefficients for this ONU can be reused, that is, no EQ adaptation is needed. This helps to substantially increase channel tracking speed. Fig. 6.15 shows representative experimental results measured at 25 Gbaud based on NRZ modulation. Fig. 6.15A shows the experimentally measured time-domain waveform of the upstream signal bursts, each of which has a short duration of 1 μs. The gap between the successive bursts is 0.5 μs. Fig. 6.15B shows the correlation strength between the known delimiter pattern and the received signal waveform as a function of time. The correlation peaks at the delimiter positions of the upstream bursts are evident and indicate valid burst synchronization. Fig. 6.15C shows the EQ error as a function of the sampling position. The optimal sampling position can be found by minimizing the EQ error. As this process can be performed in parallel, the optimal sampling position can be found without time-consuming adaption and convergence. We have found that fast burst synchronization can be achieved within 100 ns. Using the example of $T_{cycle} = 40$ μs and $T_{burst} \sim 5$ μs, this 100-ns period needed for burst-mode channel tracking accounts for an overhead of only ~2%.

The BER performance of the burst-mode upstream transmission was experimentally appraised and found to be similar to that in the continuous-mode transmission, indicating negligible burst-mode penalty. This proof-of-concept experiment clearly indicates the benefits of DSP-enabled burst-mode detection in achieving both high receiver sensitivity and fast channel tracking speed. On the whole, TDM-PON offers a cost-effective solution to connect a large number of compact AAUs and RRUs to their corresponding DUs and BBUs in the 5G network desertification application.

6.2 FDMA-based P2MP

6.2.1 FDMA-based P2MP for connecting DUs and CUs

Frequency-division multiple access is another well-established technology for P2MP communication [30–39]. As the optical carrier frequency in the telecommunication window is on the order of 200 THz, controlling and stabilizing the optical carrier frequency with high accuracy requires sophisticated wavelength control and temperature stabilization. Digital coherent detection is capable of finding the frequency offset between the signal to be received and the optical local oscillator (OLO) and consequently locking the frequency of the OLO with respect to the signal. In addition, digital coherent detection can precisely extract a frequency subchannel from an FDMA channel. Thus digital coherent detection is necessary for high-performance FDMA-based P2MP transceivers. Since digital coherent receivers are more complex and expensive than direct-detection receivers, it makes economic sense to use the coherent-detection FDMA-P2MP approach in applications that are too demanding (e.g., in terms of transmission capacity and/or distance requirements) for the direction-detection TDMA-P2MP approach.

In the 5G network densification application described in Section 6.1.1, multiple DUs need to be connected with their corresponding centralized unit (CU) via a mid-haul link. As an example, each DU connects with 16 compact AAUs/RRUs with a total front-haul bit rate of 50 Gb/s per direction, which corresponds to a mid-haul bit rate of about 10 Gb/s (assuming a fivefold reduction in interface bit rate after the DU processing). Assuming there are 20 DUs in multiple remote locations, then the aggregated bit rate of the mid-haul would be 200 Gb/s per direction. In the uplink direction of the conventional point-to-point (P2P) network configuration, an electrical aggregation unit is needed to aggregate all the traffic from the front-haul links and send it to the CUs over the mid-haul link. In the downlink direction, the electrical aggregation unit de-aggregates the traffic from the mid-haul link for all the front-haul links. Fig. 6.16 illustrates the use of conventional P2P optical connections to connect N (e.g., 20) DUs with their corresponding CUs, requiring 2N subband optical transceivers (e.g., at 10 Gb/s), 2 full-band optical transceivers (e.g., at 200 Gb/s) and one electrical aggregation unit that consumes power.

With the use of FDMA-P2MP, each DU can be assigned with a frequency slot that contains a subchannel, and all the subchannels from the DUs are aggregated by a passive optical splitter/combiner for direct optical connection with a full-band optical transceiver at the CU site (e.g., a cloud data center), as illustrated in Fig. 6.17. Remarkably, the network configuration is much simplified by halving the number of subband and full-band optical transceivers and eliminating the electrical aggregation

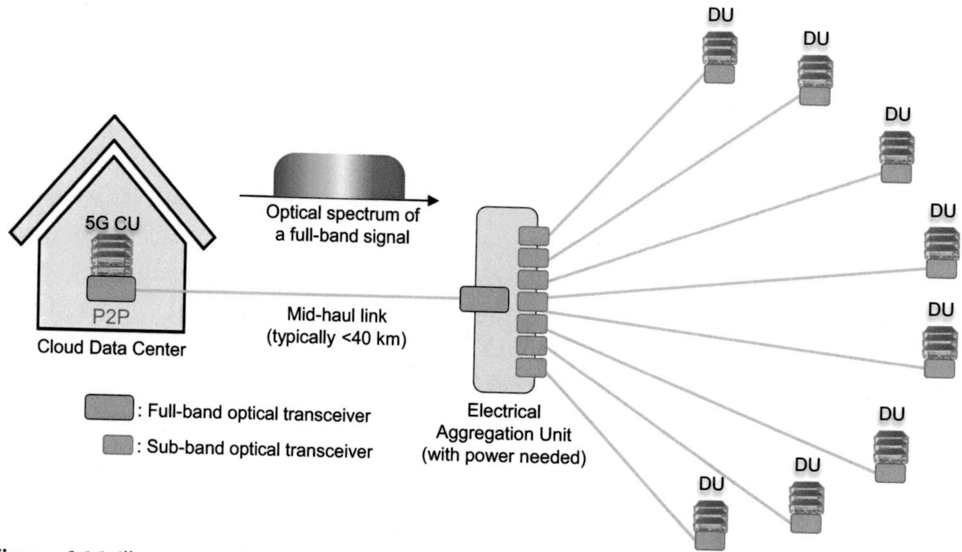

Figure 6.16 Illustration of the use of conventional point-to-point optical connections to connect multiple distributed units with their corresponding CUs.

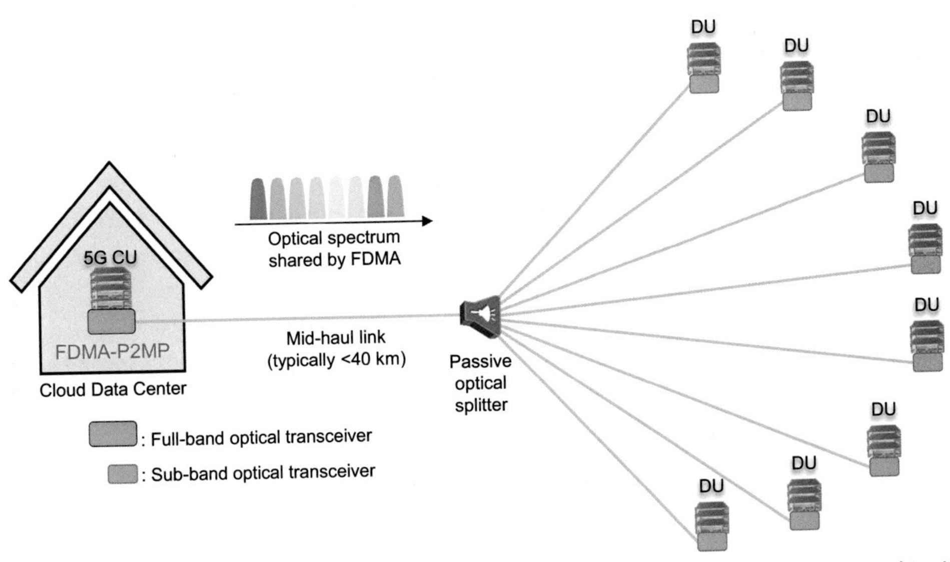

Figure 6.17 Illustration of the use of frequency-division multiple access-based point-to-multipoint optical connections to connect multiple distributed units with their corresponding CUs.

unit. After the simplification, there is only one full-band optical transceiver at the CU site, which can be afforded to be a high-end transceiver with spectrally efficient coherent in-phase and quadrature (IQ) modulation. For the subband transceivers, on the other hand, coherent IQ modulation is optional. In some cases, one may want to reduce the transmitter cost by using intensity modulation instead of coherent modulation. We will present two implementation examples of the subband transceivers in FDMA-based P2MP, one using intensity modulation and simplified coherent detection and the other using full-fledged coherent modulation and coherent detection in the Sections 6.2.2 and 6.2.3, respectively.

6.2.2 FDMA-P2MP with simplified subband transceivers

In cost-sensitive FDMA-P2MP deployment scenarios, it is desirable to simplify the subband transceivers to reduce the overall system cost. Digital coherent subband detection offers high receiver sensitivity, high dispersion tolerance, and fine wavelength selectivity. To reduce the complexity and cost of a digital coherent receiver, simplified low-cost coherent detection based on Alamouti-coding has been demonstrated and proposed for use in coherent optical access [37,38]. Recently, the combined use of Alamouti-coded multicarrier modulation at the core node and simplified subband coherent detection at the remote access nodes has been demonstrated to achieve high aggregated bandwidth, low latency, and low cost in a seamlessly integrated metro-access network [40]. The simplified subband coherent receiver front-end only requires one single-ended photodiode (PD) and a low-bandwidth ADC. Also used was a distributed feedback (DFB) laser that emits through both the front and back facets with two output ports, one connected for the OLO port of the coherent receiver and the other connected to a monolithically integrated electro-absorption modulator for upstream modulation [40]. This compact two-in-one laser device is referred to as the dual-side electro-absorption laser (DS-EML), which automatically ensures that the center frequency of the upstream signal is locked to that of the downstream signal (with a predetermined frequency offset). This scheme was experimentally demonstrated based on both offline DSP and real-time DSP. It was also used in a real-time 300-km communication link carrying live VR traffic, showing error-free performance with a low end-to-end latency of 1.5 Ms. More details on this demonstration are provided as follows.

The architecture of the proposed metro-access networking scheme is shown in Fig. 6.18. At the core node, an external cavity laser (ECL) operating at 1550 nm was used as the optical source for the integrated dual-polarization (DP) IQ modulator. The frequency spectrum of the downstream channel contained multiple modulated subcarriers, one for each of the access nodes. Multiple subcarriers were aggregated together and transmitted as an FDMA signal. In the proof-of-concept experiment based on offline-DSP, 16 subcarriers were generated with 0.625-baud QPSK per carrier and 3.125-GHz frequency spacing between two adjacent carriers. Nyquist pulse shaping with a roll-off factor of 0.1 was performed for each subcarrier. The bandwidth of each modulated carrier can be flexibly configured according to the actual needs of different users. Four electrical signals generated using four 80-GSa/s DACs were used to drive the DP-IQ modulator. The signal was then transmitted over a 300-km optical fiber link consisting of four 75-km SSMF spans, each followed by an EDFA.

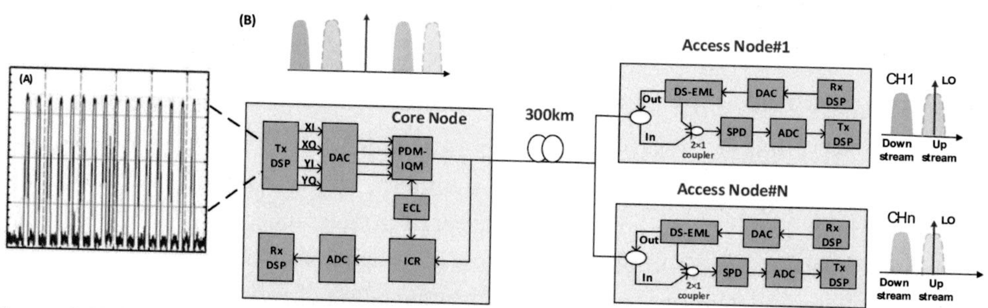

Figure 6.18 Experimental setup for generating a multicarrier Alamouti-coded signal at a core node and detecting these carriers via simplified subband coherent detection at multiple access nodes. *(After [40]).*

At each access node, the transmitted signal was detected by using a simplified heterodyne coherent receiver with a 2×1 coupler followed by only one single-ended PD with 10-GHz bandwidth. The received electrical analog signal was converted to a digital signal by an ADC. Each access node can selectively receive any given subband by tuning the OLO to a frequency that is suitably offset from the downstream signal. The accuracy of the frequency of the OLO can be control within ± 1.5 GHz by a thermo-electric cooler. In the receiver DSP, each modulated subcarrier was selectively received using low-pass filtering with 1-GHz bandwidth. The DSP functions included time recovery, frequency offset compensation, carrier-phase estimation, frame synchronization, and equalization. Finally, the errors of the bit were counted with the transmitted signal as reference sequence. In the upstream direction, the upstream signal was generated by the DS-EML with PAM4 modulation format. The upstream signal was then transmitted to the core node where it was received, alongside with the upstream signals from other access nodes, by a full-band coherent receiver.

The transmission performances were measured. At the typical hard-decision FEC BER threshold of 4×10^{-3}, the receiver sensitivity of the 0.625-GBaud QPSK subband signal was found to be better than -27 dBm. It was found that the DFB laser in the DS-EML had a large linewidth of a few MHz, and using it as the OLO for the subband coherent receiver caused a moderate sensitivity penalty of 1.2 dB as compared to a 100-kHz ECL. This moderate performance degradation can be accepted for the low-cost DS-EML based implementation in cost-sensitive scenarios.

Real-time demonstration of the simplified FDMA-P2MP scheme was also conducted with live VR traffic [40]. A server generated the real-time downstream Ethernet signal carrying live VR video and then sent it to the transmitter DSP in a field-programmable gate array (FPGA) to form a multicarrier signal, which was then sent through a 33-GSa/s DAC to drive a DP-IQ modulator. After 300-km SSMF transmission, the signal was detected by a simplified coherent receiver and recovered by the receiver DSP in another FPGA. The signal was then switched by an Ethernet switch and finally sent to a VR user equipment. The upstream signal was generated with the head movements by the VR equipment and then directly sent to the server to complete the bidirectional communication. The end-to-end latency of the 300-km transmission link was measured to be 1.5 Ms, almost entirely from the optical fiber propagation delay. The tested pre-FEC BER was about 10^{-5}, which was corrected to be error free by the FEC.

This proof-of-concept demonstration shows the feasibility of using multicarrier modulation and digital coherent subband detection to support the FDMA-P2MP architecture. With dramatic progresses made by our industry in reducing the power consumption and form factor of full-fledged coherent transceivers such as the

400G-ZR transceivers [41,42], it is becoming more attractive to use full-fledged coherent transceivers at both the core node and the access nodes to achieve high spectral efficiency and aggregated transmission capacity. We will present FDMA-P2MP with fully coherent subband transceivers in the following section.

6.2.3 FDMA-P2MP with fully coherent subband transceivers

Thanks to the dramatic progresses made in the field of digital coherent transceivers, a 400G-ZR coherent transceiver based on 60-Gbaud polarization-division multiplexing (PDM)-16QAM can fit into a Quad Small Form-factor Pluggable – Double Density (QSFP-DD) format factor with a total power consumption of less than 15 W [41,42]. Such 400G-ZR type coherent transceivers can be used as the full-band transceivers in the core node [43,44]. Assuming an optical bandwidth of 64 GHz, 16 4-GHz-spaced subchannels can be supported by one 64-Gbaud FDMA channel, as shown in Fig. 6.19. Assuming Nyquist spectral shaping with a roll-off factor of 0.1, each subchannel can be modulated at 3.6 Gbaud and carry a raw data rate of 28.8 Gb/s via PDM-16QAM. Excluding the FEC overhead, payload data rate of each subchannel can reach 25 Gb/s, leading to an aggregated payload data rate per FDMA channel of 400 Gb/s.

At each access node, a fully coherent subband transceiver can be used to receive any one of the subchannels in the downstream (or downlink) direction and transmit a neighboring subchannel in the upstream (or uplink) direction, as shown in Fig. 6.20. In the fully coherent subband transceiver, a common laser is used as the OLO and the transmitter laser, and bidirectional transmission over a single fiber is enabled by the use of a circulator. When receiving a downlink subchannel, the OLO is accurately locked to a predetermined frequency offset from the center frequency of the downlink subchannel. Therefore the same laser can be modulated by a coherent IQ modulator to generate an uplink subchannel whose center frequency is locked to the received downlink subchannel with an intended frequency offset. As an example, when the intended downlink subchannel and uplink subchannel are next to each other in an access node, the laser frequency of the subband transceiver can be set in the middle of these two subchannels. Doing so minimizes the RF bandwidth requirement of the

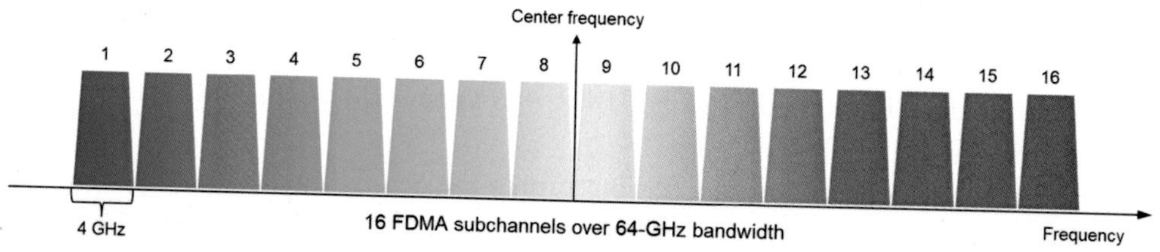

Figure 6.19 Illustration of the optical spectrum of a 400-Gb/s frequency-division multiple access channel consisting of sixteen 4-GHz-spaced subchannels each carrying 25-Gb/s payload data rate.

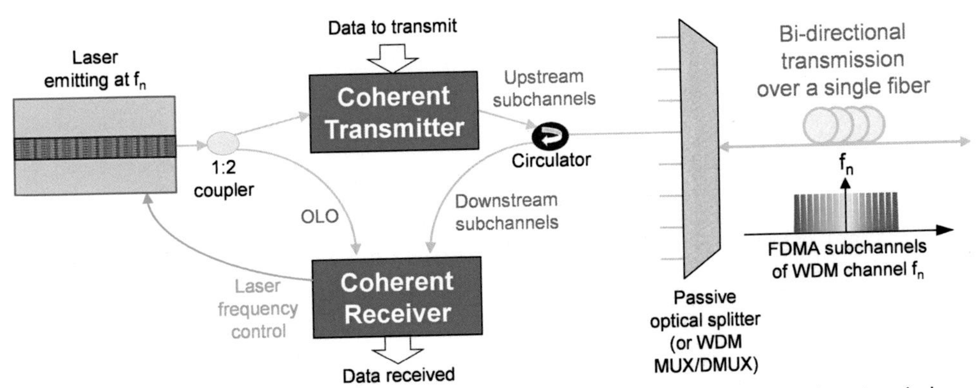

Figure 6.20 Illustration of a fully coherent subband transceiver supporting bidirectional downstream and upstream transmission with a common laser.

subband transceiver to the subchannel bandwidth, which is 4 GHz in this case shown in Fig. 6.20.

It is also feasible to allow the subband transceiver to transmit and receive more than one subchannel by increasing its RF bandwidth. For example, when the RF bandwidth of the subband transceiver is increased to 16 GHz, four adjacent downlink subchannels can be received and another four successive uplink subchannels can be transmitted at 100-Gb/s payload data rate per direction. This indicates the flexibility in the implementation of the subband transceiver to address different data rate requirements in the access nodes.

To increase the overall transmission capacity, multiple FDMA channels can be multiplexed via wavelength-division multiplexing (WDM), for example, on the ITU 100-GHz channel grid, as shown in Fig. 6.21. Different subchannel bandwidths can also be allocated to better match the traffic pattern. Moreover, the downlink traffic and uplink traffic can be carried by different WDM channels, further increasing the flexibility of the FDMA-P2MP technology. These WDM channels can be distributed to the edge nodes via a passive optical splitter or a WDM multiplexer/demultiplexer (MUX/DMUX). The use of passive optical splitter allows for colorless aggregation/de-aggregation at the expense of increased loss, while the use of WDM MUX/DMUX allows for low-loss aggregation/de-aggregation of FDMA channels on a predetermined channel grid.

Fig. 6.22 shows the schematic of a fully coherent subband transceiver supporting bidirectional downstream and upstream transmission with a pair of twin lasers having a predetermined frequency spacing. For example, the frequency spacing between the twin lasers can be set to 100 GHz by the design of the DFB laser gratings. The twin lasers can be fabricated on the same substrate and packaged together so that their frequency spacing is insensitive to ambient temperature, which helps make the upstream

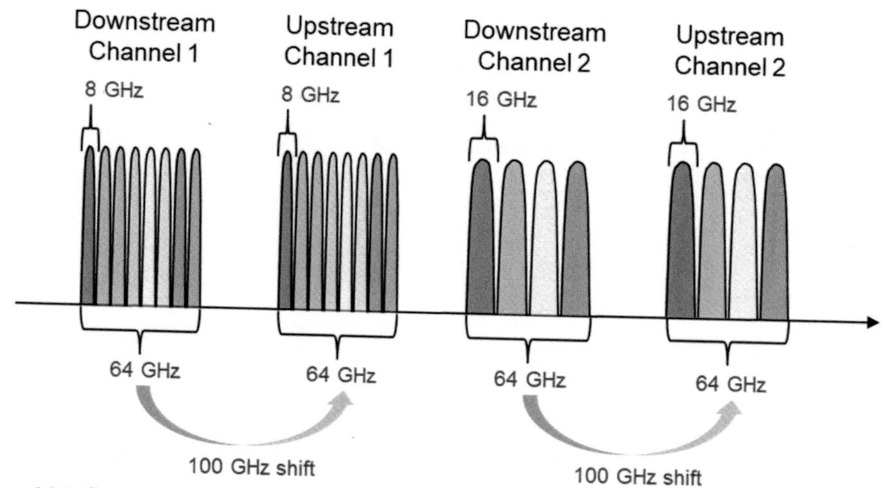

Figure 6.21 Illustration of a bidirectional transmission of frequency-division multiple access channels on a 100-GHz-spaced WDM grid with flexible allocations of subchannel bandwidth.

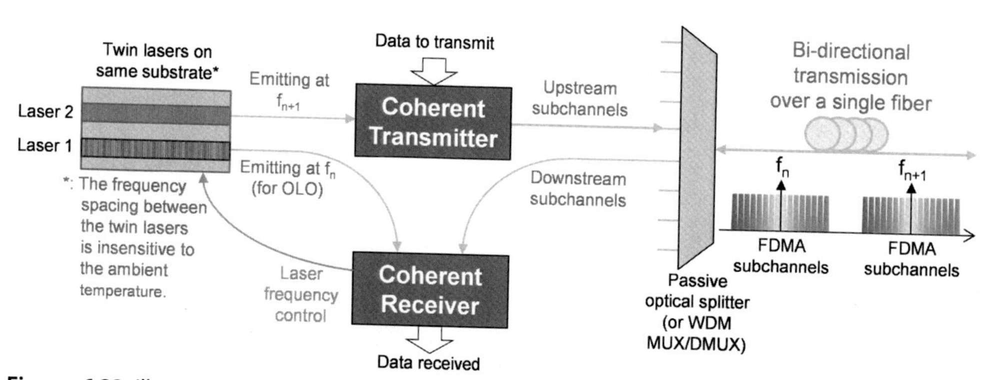

Figure 6.22 Illustration of a fully coherent subband transceiver supporting bidirectional downstream and upstream transmission with a pair of twin lasers having a predetermined frequency spacing.

subchannels from different access nodes to be accurately located in their intended frequency slots. The twin laser approach offers 3 dB more power to the transmitter and the OLO than the single-laser approach shown in Fig. 6.20. In addition, the RF bandwidth efficiency can be doubled. For example, to transmit (receive) four successive upstream (downstream) subchannels, the laser frequency can be set at the middle of the four subchannels, and the required RF bandwidth is 8 GHz, half of that required in the single-laser approach.

On the basis of the above discussion, we can summarize the key benefits of the fully coherent FDMA-P2MP technology as follows.

- High capacity: supporting high-capacity traffic aggregation from multiple access nodes to a core node.
- Long reach: supporting long-distance transmission without dispersion penalty.
- Flexible bandwidth allocation: supporting flexible and dynamic bandwidth allocations to match downstream and upstream traffic demands.
- High scalability: allowing the increase of aggregated capacity by only upgrading the equipment at the core node.
- Low total cost of ownership (TCO): reducing both capital expenditure (by halving the number of transceivers needed as compared to P2P) and operational expenses (by eliminating the need of active electrical aggregation and allowing for flexible and scalable bandwidth allocations).

6.3 Concluding remarks

On the basis of the discussions presented in the previous sections, we can draw some concluding remarks on the TDMA- and FDMA-based P2MP optical transmission technologies, as follows.

- The P2MP optical transmission is naturally suitable for distributing the downstream traffics from a central node (or core node) to multiple edge nodes (or access nodes), as well as aggregating upstream traffics from these edge nodes back to the central node.
- TDMA-based P2MP with intensity modulation/direct detection is suitable for short-reach aggregation scenarios such as the front-haul aggregation scenario for the 5G network densification. 50G-PON is well suited for aggregating multiple compact AAUs and RRUs.
- FDMA-based P2MP with coherent modulation and detection is suitable for medium-reach aggregation scenarios such as the mid-haul aggregation scenario for the 5G network densification. 400G-ZR type coherent transceivers are well suited to serve as the full-band transceivers in the central nodes and subband transceivers in the edge nodes.
- Fully coherent FDMA-P2MP offers benefits such as high capacity, long reach, flexible bandwidth allocation, high scalability, and low TCO and is expected to find various applications in metro, access, and data center networks.

Going forward in the 5G era, P2MP optical transmission technologies are expected to leverage advances on high-speed PON, such as 50G-PON and beyond [8–10,16,45], and high-speed coherent transceivers, such as 800G coherent transceivers [46–48]. Compared to well-established P2P coherent optical transmission technologies, the fully coherent FDMA-P2MP technology can be regarded as a disruptive technology [44]. To fully unleash its potential to transform the cost and complexity of certain aggregation segments of the optical network, industry-wide development and standardization efforts can be deemed necessary and rewarding.

References

[1] Gigabit capable passive optical network (G-PON). ITU-T recommendation G.984 series; 2008.
[2] 10 Gigabit capable passive optical network (XG-PON). ITU-T recommendation G.987 series; 2012.
[3] Wong E. Next-generation broadband access networks and technologies. J Light Technol 2012;30 (4):597–608.
[4] 10-gigabit-capable passive optical networks (XG-PON): general requirements. ITU-T recommendation G.987.1. https://www.itu.int/rec/T-REC-G.987.1/en; 2016.
[5] Liu X, Effenberger F. Emerging optical access network technologies for 5G wireless [Invited]. J Opt Commun Netw 2016;8:B70–9.
[6] Nesset D. PON roadmap. J Opt Commun Netw 2017;9(1):A71–6.
[7] Wey JS, Zhang J. Passive optical networks for 5G transport: technology and standards. J Light Technol 2019;37(12):2830–7.
[8] van Veen D, Houtsma V. Strategies for economical next-generation 50G and 100G passive optical networks [Invited]. IEEE/OSA J Opt Commun Netw 2020;12(1):A95–103.
[9] Zhang D, Liu D, Wu X, Nesset D. Progress of ITU-T higher speed passive optical network (50G-PON) standardization. IEEE/OSA J Opt Commun Netw 2020;12(10):D99–108.
[10] Houtsma V, Mahadevan A, Kaneda N, van Veen D. Transceiver technologies for passive optical networks: past, present, and future [Invited Tutorial]. IEEE/OSA J Opt Commun Netw 2021;13 (1):A44–55.
[11] Pizzinat A, Chanclou P, Saliou F, Diallo T. Things you should know about fronthaul. JLT 2015;33 (5):1077–83.
[12] Shibata N, Tashiro T, Kuwano S, Yuki N, Fukada Y, Terada J, et al. Performance evaluation of mobile front-haul employing ethernet- based TDM-PON with IQ data compression [Invited]. IEEE/OSA J Opt Commun Netw 2015;7(11):B16–22.
[13] Kani J. Solutions for future mobile fronthaul and access-network convergence. In: Proceeding OFC, Tutorial paper W1H.1, Anaheim; 2016.
[14] Chanclou P, Suzuki H, Wang J, Ma Y, Boldi MR, Tanaka K, et al. How does passive optical network tackle radio access network evolution? J Opt Commun Netw 2017;9:1030–40.
[15] Zhou S, Liu X, Effenberger F, Chao J. Low-latency high-efficiency mobile fronthaul with TDM-PON (mobile-PON). J Opt Commun Netw 2018;10:A20–6.
[16] Higher speed passive optical networks: requirements. ITU-T recommendation G.9804.1. https://www.itu.int/rec/T-RECG.9804.1-201911-I/en; 2019.
[17] Tashiro T, Kuwano S, Terada J, Kawamura T, Tanaka N, Shigematsu S, et al. A novel DBA scheme for TDM-PON based mobile fronthaul. In: Proceeding OFC, Tu3F.3, San Francisco; 2014.
[18] Shibata N, Tashiro T, Kuwano S, Yuki N, Terada J, Otaka A. Mobile front-haul employing Ethernet-based TDM-PON system for small cells. In: Proceeding OFC, Los Angeles, CA, M2J.1; 2015.
[19] Kobayashi T, Ou H, Hisano D, Shimada T, Terada J, Otaka A. Bandwidth allocation scheme based on simple statistical traffic analysis for TDM-PON based mobile fronthaul. In: Proceeding OFC, Anaheim, CA, paper W3C. 7; 2016.
[20] Zeng H, Liu X, Megeed S, Shen A, Effenberger F. Digital signal processing for high-speed fiber-wireless convergence [Invited]. J Opt Commun Netw 2019;11:A11–19.
[21] Houtsma V, van Veen D. A study of options for high-speed TDM-PON beyond 10G. J Light Technol 2017;35(4):1059–66.

[22] Tao M, Zhou L, Zeng H, Li S, Liu X. 50-Gb/s/λ TDM-PON based on 10G DML and 10G APD supporting PR10 link loss budget after 20-km downstream transmission in the O-band. In: Optical fiber communication conference, paper Tu3G.2; 2017.
[23] Zhang J, Wey JS, Yu J, Tu Z, Yang B, Yang W, et al. Symmetrical 50-Gb/s/λ PAM-4 TDM-PON in O-band with DSP and semiconductor optical amplifier supporting PR-30 link loss budget. In: 2018 Optical fiber communication conference, paper M1B.4; 2018.
[24] Torres-Ferrera P, Ferrero V, Valvo M, Gaudino R. Impact of the overall electrical filter shaping in next-generation 25 and 50 Gb/s PONs. J Opt Commun Netw 2018;10:493–505.
[25] Tao M, Zheng J, Dong X, Zhang K, Zhou L, Zeng H, et al. Improved dispersion tolerance for 50G-PON downstream transmission via receiver-side equalization. In: Optical fiber communication conference, paper M2B.3; 2019.
[26] Yi L, Liao T, Huang L, Xue L, Li P, Hu W. Machine learning for 100 Gb/s/λ passive optical network. J Light Technol 2019;37:1621–30.
[27] Torres-Ferrera P, Wang H, Ferrero V, Valvo M, Gaudino R. Optimization of band-limited DSP-aided 25 and 50 Gb/s PON using 10G-class DML and APD. J Light Technol 2020;38:608–18.
[28] Li B, Zhang K, Zhang D, He J, Dong X, Liu Q, et al. DSP enabled next generation 50G TDM-PON. J Opt Commun Netw 2020;12:D1–8.
[29] Effenberger FJ, Zeng H, Shen A, Liu X. Burst-mode error distribution and mitigation in DSP-assisted high-speed PONs. J Light Technol 2020;38:754–60.
[30] Reis JD, Shahpari A, Ferreira R, Ziaie S, Neves DM, Lima M, et al. Terabit + (192 × 10 Gb/s) nyquist shaped UDWDM coherent PON with upstream and downstream over a 12.8 nm band. J Light Technol 2014;32:729–35.
[31] Rohde H, Gottwald E, Teixeira A, Reis JD, Shahpari A, Pulverer K, et al. Coherent ultra dense WDM technology for next generation optical metro and access networks. J Light Technol 2014;32:2041–52.
[32] Cano IN, Lerín A, Polo V, Prat J. Direct phase modulation DFBs for cost-effective ONU transmitter in udWDM PONs. IEEE Photon Technol Lett 2014;26:973–5.
[33] Lavery D, Thomsen BC, Bayvel P, Savory SJ. Reduced complexity equalization for coherent long-reach passive optical networks [Invited]. J Opt Commun Netw 2015;7:A16–27.
[34] Presi M, Corsini R, Artiglia M, Ciaramella E. Ultra-dense WDM-PON 6.25 GHz spaced 8 × 1 Gb/s based on a simplified coherent-detection scheme. Opt Exp 2015;23:22706–13.
[35] Shahpari A. Coherent access: a review. J Light Technol 2017;35:1050–8.
[36] Presi M, Artiglia M, Bottoni F, Rannello M, Cano IN, Tabares J, et al. Field-trial of a high-budget, filterless, λ-to-theuser, UDWDM-PON enabled by an innovative class of low-cost coherent transceivers. J Light Technol 2017;35:5250–9.
[37] Erkılınç MS, Lavery D, Shi K, Thomsen BC, Savory SJ, Killey RI. Bidirectional wavelength-division multiplexing transmission over installed fibre using a simplified optical coherent access transceiver. Nat Commun 2017;8:1043.
[38] Erkılınç MS, Lavery D, Shi K, Thomsen BC, Killey RI, Savory SJ, et al. Comparison of low complexity coherent receivers for UDWDM-PONs (λ-to-the-user). J Light Technol 2018;36:3453–64.
[39] Teixeira A, Lavery D, Ciaramella E, Schmalen L, Iiyama N, Ferreira RM, et al. DSP enabled optical detection techniques for PON. J Light Technol 2020;38:684–95.
[40] Le Y, Zeng H, Zhou X, Deng N, Megeed S, Shen A, et al. Real-time demonstration of an integrated metro-access network carrying live VR traffic based on multi-carrier modulation and simplified sub-band coherent detection. In: Optical fiber communication conference (OFC) 2019, paper Th3F.3; 2019.
[41] Optical Internetworking Forum (OIF). Implementation agreement 400ZR [Online]. https://www.oiforum.com/wp-content/uploads/OIF-400ZR-01.0_reduced2.pdf; 2020 March.
[42] Filer M. The coherent 400 Gb/s revolution. Lightreading optical networking digital symposium; May 28, 2020.
[43] Rashidinejad A, Nguyen A, Olson M, Hand S, Welch D. Real-time demonstration of 2.4Tbps (200Gbps/) bidirectional coherent DWDM-PON enabled by coherent

nyquist subcarriers. In: Optical fiber communication conference (OFC) 2020, paper W2A.30; 2020.

[44] Welch DF. Disruption cycles for optical networks: how point to multi-point coherent optics can transform the cost and complexity of the optical network. In: European conference on optical communications 2020, Invited Talk Tu1C-1; 2020.

[45] Borkowski R, Straub M, Ou Y, Lefevre Y, Jeliü Ž, Lanneer W, et al. World's first field trial of 100 Gbit/s flexible PON (FLCS-PON). In: European conference on optical communications 2020, PDP2.2; 2020.

[46] See for example, https://www.lightreading.com/optical-ip/the-long-distance-view-of-ofc-800g-and-more!-/d/d-id/758149.

[47] Sun H, et al. 800G over 1000 km enabled by real-time DSP ASIC employing probabilistic shaping and digital sub-carrier multiplexing. In: 2020 22nd International conference on transparent optical networks (ICTON), Paper Tu.C5.3, Bari, Italy; 2020.

[48] See for example, https://www.oiforum.com/oif-launches-800g-coherent-and-co-packaging-framework-ia-projects-elects-new-board-members-positions-officers-and-working-group-chairs/.

CHAPTER 7

Cloud data center optics

7.1 Overview of the global cloud data center infrastructure

5G and cloud services are supported by the global cloud data center (DC) infrastructure for important functions such as content generation and delivery, data processing, storage, and routing. The global cloud DC infrastructure consists of a large number of DCs distributed all over the world and a global optical network infrastructure for interconnecting these DCs [1–5]. The key elements of the global DC infrastructure are as follows [6,7].

- DC: A DC is a physical facility that performs data processing, storage, and routing needed for cloud services. DCs have various scales, with hyperscale DCs each having a space of over 1 million square foot and capable of operating over 1 million servers. Smaller DCs are distributed closer to the end users to provide low-latency cloud services such as those required by 5G Cloud Radio Access Network (C-RAN).
- Availability zone (AZ): An AZ consists of multiple discrete DCs with redundant power, networking, and connectivity to provide highly available, fault-tolerant, and scalable cloud services. Fault tolerance is achieved by using reductant DCs that are connected with high-capacity inter-DC interconnects (DCIs). There are typically two to five AZs in a region with a public cloud infrastructure.
- Region: A region is a geographic area that contains multiple AZs. In a public cloud infrastructure, a region can cover multiple provinces or even multiple countries. In a private cloud infrastructure, a region may only cover one equipment room or a few equipment rooms within a short distance of tens of kilometers.
- Geography: Geography is a discrete geographic area, typically containing one or more regions, that preserves data residency and compliance boundaries. Geographies allow customers with specific data residency and compliance needs to keep their data and applications in their own areas.
- DCI: DCIs support communication among DCs for functions such as active–active operation, remote storage, virtual machine migration, disaster recovery and backup. There are hundreds of hyperscale DCs over the globe and thousands of small-scale DCs close to the end users; thus DCIs form the essential communication network that supports the global DC infrastructure.
- DC optics: Each DC contains numerous servers and routers/switches that need to be networked, and fiber-optic connections are widely used for the same. Various

Optical Communications in the 5G Era
DOI: https://doi.org/10.1016/B978-0-12-821627-9.00003-6

types of optical transceivers with different requirements on speed (e.g., from 100 to 400 Gb/second), reach (from 50 m to 10 km), power consumption, and form factor are used to support the intra-DC communication needs.

Fig. 7.1 illustrates the global DC infrastructure, a representative large-scale DC campus, and the interior of a large-scale DC. As of 2021, there are about 600 hyperscale DCs deployed globally, of which approximately 40% are located in the United States [7]. Amazon, Microsoft, and Google collectively account for over 50% of the world's largest DCs across the globe to accommodate the ever-increasing demand for cloud services [7]. In addition to the expansion of the footprint of hyperscale DCs, small-scale DCs are fast growing to support cloud services close to large population, industry, and enterprises. With the availability of DCs in close proximity, dedicated services can be provided as follows.

- Point of delivery (POD): A POD is located on a Layer 2 network and is used to differentiate virtual resource pools (including computing, storage, and network resources). A POD can exist only in one DC, while one DC may have multiple PODs. With an open and standard cloud computing platform, infrastructure-as-a-service can be readily provided in both public and private clouds where dedicated virtual servers and other resources are made available to the end users.
- Local zone: A local zone allows the end users to run applications that demand low latency locally. Local zones are well suited for hosting 5G core processing, such as 5G enhanced mobile broadband user plane processing that demands a one-way latency of <4 ms (as described in Chapter 4: Optical Technologies for 5G X-haul). Local zones are also well suited for use cases such as media and entertainment content creation, real-time gaming, and live video streaming [6].
- Wavelength zone: A wavelength zone uses a dedicated wavelength channel to support a given application with a deterministically low latency. The use of a dedicated wavelength channel enables an on-hop connection from the end user to the nearby DC and thus avoids the latency associated with data routing/processing at intermediate nodes. The wavelength zone is well suited for mobile edge computing (MEC) applications that demand a one-way latency of <0.5 ms to achieve 5G ultra-reliable and low-latency communications, as described in Chapter, Optical Technologies for 5G X-haul.
- Outpost: An outpost is a cloud processing facility that is closest to the end users for ultralow-latency applications and applications that demand localized data processing and storage [6]. An outpost can be in the form of a central office (CO) or an equipment room. It is well suited for hosting 5G distributed units (DUs) in the front-haul segment of C-RAN that requires a one-way latency of <0.1 ms, as described in Chapter 4, Optical Technologies for 5G X-haul.

Inside the global cloud DC infrastructure, optical communication technologies are essential in supporting both inter- and intra-DC communication needs. For inter-DC

Figure 7.1 Top: Illustration of the global data center (DC) infrastructure; Middle: A representative DC campus built by Microsoft in Middenmeer, Holland; Bottom: The interior of a hyperscale DC. *Available from: https://www.google.com/about/datacenters/*

connections, optical DCIs are deployed to provide the needed communication capacity, latency, reliability, flexibility, and scalability. For intra-DC connections, a massive amount of optical transceivers and fibers are used to provide the needed connections with low cost and power per bit. In this chapter, we will describe the overall DC communication infrastructure in Section 7.2, 400G-ZR for DCI in Section 7.3, intra-DC connections and 400-Gb/second pluggable optical transceivers in Section 7.4, and multi-Tb/second DC switches and their evolution in Section 7.5. Finally, some concluding remarks will be provided in Section 7.6. A discussion on long-haul, ultra-long-haul, and undersea transmission for national and international backbone optical networks will be presented in Chapter 8, High-Capacity Long-Haul Optical Fiber Transmission.

7.2 Data center communication infrastructure

Fig. 7.2 shows the schematic of the connection between a regional DC campus and a global optical network. As a representative example, the regional DC campus has three AZs for fault-tolerant support of cloud services in the corresponding region. Each AZ has two DCs to host sufficient resources for data computing, storage, and routing. Two regional network gateways (RNGs) are used to aggregate the data traffic from all the six DCs in the region to the global optical network with high fault tolerance. In

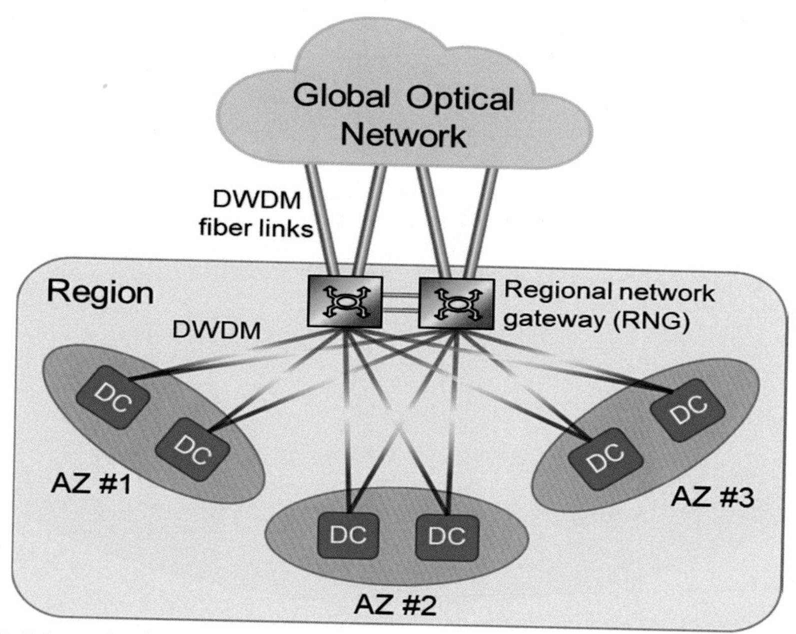

Figure 7.2 Schematic of the connection between a regional data center campus and a global optical network.

parallel, the RNGs also de-aggregate the data traffic from the global optical network to all the six DCs. Assuming that the regional DC campus hosts 100,000 servers each having a 50-Gb/second connection and ~23% of the total traffic is the north–south traffic [2], the total communication bandwidth between the RNGs and the global optical network is on the order of ~1 Pb/second (10^{15} bits per second). Thus high-capacity dense-wavelength-division multiplexing (DWDM) is needed for connections between the RNGs and the global network. Assuming the typical C-band DWDM transmission with 80 50-GHz-spaced 100-Gb/second wavelength channels, each optical fiber offers a transmission capacity of 8 Tb/second, and over 100 optical fiber links are needed (per direction) to support the connection between the RNGs and the global optical network. This calls for continued technological advances in increasing the transmission capacity per fiber, for example, to 24 Tb/second by using 120 50-GHz-spaced 200-Gb/second wavelength channels in the super-C band [8,9].

Similarly, high-capacity DWDM transmission is needed for the connections between the RNGs and the six DCs. The transmission distances for these connections are typically within 80 km, corresponding to the ZR reach defined by the IEEE. To address this important demand, the optical internetworking forum (OIF) published the first implementation agreement for a 400G-ZR coherent optical interface on April 29, 2020 [10]. This 400G-ZR interface supports DWDM transmission with distances up to 120 km. More detailed description on the 400G-ZR interface will be provided in the next section.

Fig. 7.3 shows the schematic of the communication infrastructure inside a DC. To facilitate the description of different optical transceiver types with different data rates, reaches, and implementations, we will use the popular IEEE Ethernet physical layer (PHY) terminology defined as follows.

- *x*GBase-*y*R*z* (or simply *x*G-*y*R*z*)

where "*x*" is the data rate in Gb,

"Base" stands for baseband,

"*y*" is one of the six letters, *S*, *D*, *F*, *L*, *E*, and *Z*, which represent:

S: Short reach (SR) up to 100 m,

D: DC reach (DR) up to 500 m

F: Far reach (FR) up to 2 km,

L: Long reach (LR) up to 10 km,

E: Extended reach (ER) up to 40 km,

Z: Reach up to 80 ~ 120 km (ZR), and

"*z*" is the parallelism order to indicate the number of fibers or wavelengths used.

- For example, we have
  - 400G-SR16 for 400-Gb/second transmission of up to 100 m with 16 parallel multimode fibers (MMFs)
  - 400G-DR4 for 400-Gb/second transmission of up to 500 m with four parallel single-mode fibers (SMFs)

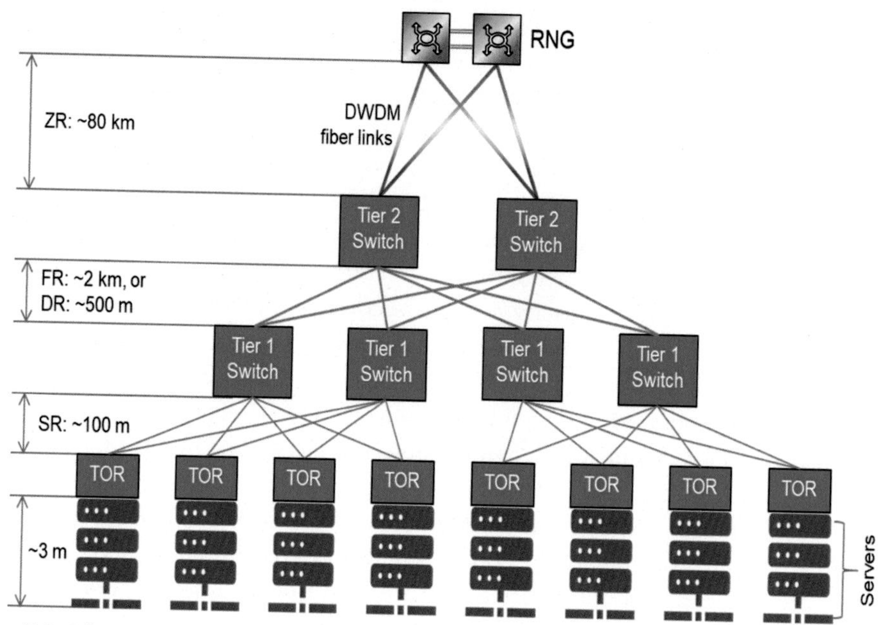

Figure 7.3 Schematic of the communication infrastructure inside a data center.

– 400G-FR4 for 400-Gb/second transmission of up to 2 km with four coarse wavelength-division multiplexing (CWDM) channels
– 400G-LR8 for 400-Gb/second transmission of up to 10 km with 8 local area network wavelength division multiplexing (LAN-WDM) channels
– 400G-ER8 for 400-Gb/second transmission of up to 40 km with 8 LAN-WDM channels
– 400G-ZR for 400-Gb/second transmission of up to 80–120 km with one dual-polarization 16-state quadrature amplitude modulation (DP-16QAM) channel, also commonly referred to as polarization-division-multiplexed 16-QAM (PDM-16QAM) channel,

As shown in Fig. 7.3, the communication infrastructure inside a DC can be divided into the following segments:

- Server to top-of-rack switch (TOR), which has a reach of up to 3 m and can be supported by a direct attach cable at 50 Gb/second;
- TOR to Tier-1 switch, which has a typical reach of up to 100 m (SR) and can be supported by 400G-SR16;
- Tier-1 switch to Tier-2 switch, which has a reach of up to 500 m (DR) or up to 2 km (FR) and can be supported by 400G-DR4 or 400G-FR4; and
- Tier-2 switch to RNG, which has a reach of up to 80–120 km (ZR) and can be supported by 400G-ZR.

Owing to the high demand for various high-speed optical transceivers to support the abovementioned intra-DC communication needs, the optical communication industry has devoted much effort to develop and standardize these high-speed optical transceivers. Fig. 7.4 shows the state-of-the-art 400-Gb/second-class optical transceivers for different reaches and their corresponding IEEE and OIF standards [10–15]. There are certain key considerations for the choices of the fiber type and the operating wavelength window and are enumerated as follows.

- For SR, the modal and chromatic dispersion effects of MMF are sufficiently small to allow the use of low-cost multimode vertical-cavity surface-emitting lasers (VCSELs) emitting at 850 nm and low-cost multimode photodiodes for 25-Gb/second non-return-to-zero (NRZ) [11,16]. Thus 400G-SR16 with 16 parallel MMFs can be realized.
- For DR, SMF is used to avoid the modal dispersion and the CWDM channel 3 (1304.5 – 1317.5 nm), which is in the zero-dispersion window of the standard G.652 fiber, is used to minimize the chromatic dispersion, thus supporting 100-Gb/second PAM4 per fiber [12]. So, 400G-DR4 with four parallel SMFs can be realized.
- For FR, WDM is used to save fiber resource and cost. Eight LAN-WDM wavelengths each modulated with 50-Gb/second PAM4 for an aggregated payload data rate of 400 Gb/second can be supported [12]. To reduce the cost of optical transceivers, four CWDM wavelengths each modulated with 100-Gb/second PAM4 for the same aggregated payload data rate of 400 Gb/second can be supported [13]. Thus 400G-FR8 and 400G-FR4 can be realized via 8-channel LAN-WDM with 50-Gb/second PAM4 and 4-channel CWDM with 100-Gb/second PAM4, respectively.
- For LR, the fiber dispersion effect becomes stronger and the dispersion-induced penalty needs to be carefully assessed. 400G-LR8 based on 8-channel LAN-WDM with 50-Gb/second PAM4 can still be supported [12], while the reach of 400G-LR4 based on 4-channel CWDM with 100-Gb/second PAM4 is slightly shorter than 10 km [13]. The LR transceivers are useful in large DC campuses that span between 2 and 10 km.

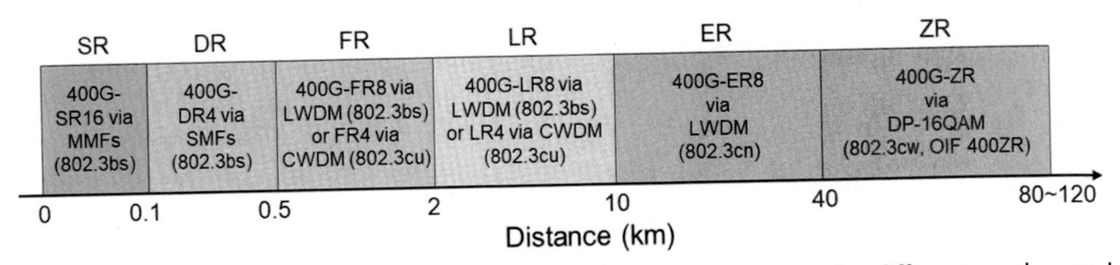

Figure 7.4 The state-of-the-art 400-Gb/second-class optical transceivers for different reaches and their corresponding IEEE and optical internetworking forum standards [10–15].

- For ER, the fiber dispersion effect becomes even more pronounced and 400G-ER8 based on 8-channel LAN-WDM with 50-Gb/second PAM4 can be marginally supported [14]. The ER reach class is not commonly used for DC applications, but it may be useful for some 5G mid-haul applications.
- For ZR, the fiber dispersion effect becomes too strong to be overcome by intensity-modulation and direct detection (IM/DD) transceivers, and therefore single-carrier 400-Gb/second DP-16QAM with digital coherent detection, which is capable of digitally compensating linear transmission impairments, such as chromatic dispersion and polarization-mode dispersion (PMD), is selected [10,15]. The 400G-ZR signal is modulated at about 60 Gbaud and can be transmitted on a 75-GHz DWDM grid. Using the extended C-band with 4.8 THz bandwidth, 64 400G-ZR channels can be carried on each fiber, providing a remarkably high capacity of 25.6 Tb/second per fiber [10].

A more elaborate discussion on 400G-ZR will be provided Section 7.4, and more in-depth study on the impact of dispersion penalty will be presented in Section 7.5. For easy use in real-world applications, these optical transceivers are made into pluggable form factors such as quad small form-factor pluggable - double density (QSFP-DD) [17] and octal small form-factor pluggable (OSFP) [18], which will be also described in more depth in Section 7.5.

7.3 400G-ZR for data center interconnects

Given the importance of 400G-ZR in connecting DCs to global optical networks, OIF published the first implementation agreement on 400ZR in 2020 [10], with the aim to create an interoperable, low-cost 400-Gb/second coherent interface for two main applications:

- Amplified, point-to-point DWDM links with reaches of 120 km or less
- Unamplified, single-wavelength links with a loss budget of 11 dB

The OIF 400ZR DWDM implementation is normatively based on 48 100-GHz-spaced wavelength channels with a channel plan shown in Table 7.1. Optionally, the OIF 400ZR DWDM implementation can be based on 64 75-GHz-spaced wavelength channels with a channel plan showed in Table 7.2. Alternatively, flexible DWDM

Table 7.1 Optical internetworking forum 400ZR dense-wavelength-division multiplexing channel plan for 48 100-GHz spaced channels.

Channel index	1	2	3	...	46	47	48
n (from G.694.1[a]) Frequency (THz)	30 196.1	29 196.0	28 195.9	...	− 15 191.6	− 16 191.5	− 17 191.4

[a]Based on the channel index *n* defined in ITU-T G.694 Section 7.6 "Fixed grid nominal central frequencies for dense WDM systems" [10].

Table 7.2 Optical internetworking forum 400ZR dense-wavelength-division multiplexing channel plan for 64 75-GHz spaced channels.

Channel index	1	2	3	...	62	63	64
n (from G.694.1)	120	117	114		− 63	− 66	− 69
Frequency (THz)	196.100	196.025	195.950	...	191.525	191.450	191.375

Table 7.3 Key channel specifications based on optical internetworking forum 400ZR dense-wavelength-division multiplexing system.

Parameter	Minimum value	Maximum value	Unit
Channel frequency	191.3	196.1	THz
Target reach	80	120	km
Fiber type	SMF with a default choice of G.652 fiber.		
Dispersion		2400	ps/nm
Instantaneous differential group delay (DGD)[a]		28	ps
Polarization-dependent loss		2	dB
Polarization rotation speed		50	krad/s
Ripple in the passband		2	dB
Interchannel crosstalk		− 8	dB
Interferometric crosstalk		− 35	dB

[a]The ratio of the maximum instantaneous DGD to the mean DGD is defined as 3.3, corresponding to the probability of exceeding the maximum DGD being 4.1×10^{-6} [10].

grids defined in ITU-T G 694 Section 7 "Flexible DWDM grid definition" can be used [10]. The key optical channel specifications based on the OIF 400ZR DWDM system are shown in Table 7.3.

The forward error correction (FEC) used in 400ZR is based on a concatenated FEC (C-FEC) that combines a hard-decision (HD) staircase FEC (255,239) outer code and an inner double-extended soft-decision (SD) FEC (128,119) Hamming code, leading to a total FEC overhead of ~14.8% [10]. With the C-FEC, the post-FEC bit error ratio (BER) can reach 10^{-15} at a pre-FEC BER of 1.25×10^{-2}, corresponding to a gross coding gain of 10.99 dB and a net coding gain (NCG) of 10.39 dB, which is ~2 dB away from the Shannon limit for NCG at 14.8% overhead. Table 7.4 compares two commonly used FEC codes, the KP4 code based on Reed–Solomon (544,

Table 7.4 Comparison of commonly used forward error correction codes.

FEC	KP4 RS(544, 514)	Staircase FEC (255, 239)	C-FEC used in OIF 400ZR
Overhead	5.84%	6.69%	14.8%
PreFEC BER (for 10^{-15} post-FEC BER)	2.3×10^{-4}	4.5×10^{-3}	1.25×10^{-2}
NCG	6.9	9.4	10.4
Power consumption	1 ×	~2.5 ×	~3 ×
Processing latency	1 ×	~100 ×	~100 ×

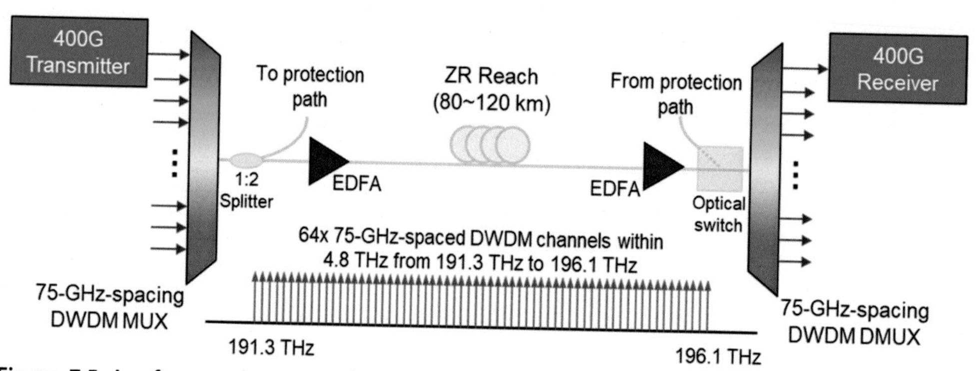

Figure 7.5 A reference dense-wavelength-division multiplexing system design for 400G-ZR based on the optical internetworking forum 400ZR implementation agreement [10].

514) and the staircase FEC (255,239), with the C-FEC code. The KP4 code is extensively used in the IEEE standards for Pulse Amplitude Modulation 4-level (PAM4)-based optical transceivers [12–14], while the staircase FEC is used in The International Telecommunication Union Telecommunication Standardization Sector (ITU-T) standards for 100-Gb/second coherent links [19]. The relative power consumptions and processing latencies of the three FEC codes are also available [20], displaying the competent overall performance of the C-FEC.

Fig. 7.5 shows a reference system design for 400G-ZR using 64 75-GHz-spaced DWDM channels in the extended C-band, achieving a total transmission capacity of 25.6 Tb/second per fiber. Each 400G-ZR channel is modulated with DP-16QAM at a symbol rate of 59.84375 Gbaud, corresponding to a raw bit rate of 478.75 Gb/second due to the overheads for C-FEC and optical transport network framing [10]. At the transmitter side, the 64 400-Gb/second DP-16QAM wavelength channels are multiplexed by a DWDM multiplexer (MUX). A 1:2 passive optical splitter is used for the 1 + 1 protection. A booster Erbium-doped fiber amplifier (EDFA) is used to boost the power to 3 dBm per channel. After transmission over a standard SMF with a reach of up to 120 km, the DWDM channels are preamplified by another EDFA before being demultiplexed by a DWDM demultiplexer (DMUX). A 2:1 optical switch is inserted between the pre-EDFA and the DMUX for the purpose of 1 + 1 protection. Each demultiplexed 400-Gb/second DP-16QAM channel is then received by a digital coherent receiver. For bidirectional transmission, a similar fiber link is used in the reverse direction. A 400-Gb/second DP-16QAM transceiver can be used for simultaneously transmitting and receiving at the same wavelength, thus requiring only one laser per transceiver.

Table 7.5 shows the margin allocation table assuming practical optical component losses. Silicon photonics (SiP) has been well adopted in compact coherent transceivers [21–28]. To accommodate SiP-based transmitters with relatively low output powers (due to the coupling loss between the SiP modulator and the external laser), the minimum transmitter power is assumed to be −10 dBm. The EDFA noise figure (NF) is assumed to be 6 dB. The optical signal-to-noise ratio (OSNR) due to optical amplification can be calculated by [29]

$$\mathrm{OSNR(dB)} = 58 + \mathrm{P_{in}(dBm)} - \mathrm{NF(dB)} \quad (7.1)$$

where OSNR(dB) is the OSNR (in dB) defined with the conventional 0.1-nm noise bandwidth, P_{in}(dBm) is the input signal power to the EDFA in dBm, and NF(dB) is the EDFA noise figure in dB. After the booster EDFA, the minimum OSNRs is calculated to be 32.5 dB. Similarly, the pre-EDFA causes an OSNR of 29 dB. The final OSNR at the receiver (RX) is then

$$\mathrm{OSNR_{RX}(dB)} = -10\log_{10}\left(10^{-\frac{32.5}{10}} + 10^{-\frac{29}{10}}\right) = 27.4\ \mathrm{dB} \quad (7.2)$$

Table 7.5 A reference margin allocation table for the 400G-ZR dense-wavelength-division multiplexing transmission over 120-km standard single-mode fiber.

Optical component	Loss	Channel power	OSNR
400ZR transmitter (478.75-Gb/s DP-16QAM)	Variable	−10 dBm (min.)	
DWDM MUX	5.5 dB	−15.5 dBm	
1:2 Splitter	4 dB	−19.5 dBm	
Booster EDFA	−23.5 dB	3 dBm	32.5 dB
120-km Fiber	24 dB	−21 dBm	
Connectors	2 dB	−23 dBm	
PreEDFA	−24 dB	1 dBm	27.4 dB
3 dB OSNR penalty due to PDL, ripple, crosstalk etc.			24.4 dB
Theoretical OSNR requirement at 1.25×10^{-2} pre-FEC BER			20.4 dB
OSNR margin (for nonideal implementations, aging, etc.)			4 dB

Assuming 3 dB OSNR margin for transmission impairments such as polarization dependent loss (PDL), ripple, and crosstalk, the OSNR at the receiver is adjusted down to 24.4 dB.

To assess the final OSNR margin, we need to know the required OSNR at 1.25×10^{-2} pre-FEC BER for the 400G-ZR transceiver operating at 478.75 Gb/second. Let us first investigate the theoretical OSNR requirement. The BER of a binary phase shift keying (BPSK) or quadrature phase shift keying (QPSK) signal in an additive white Gaussian noise (AWGN) channel can be expressed as [30–32]

$$\mathrm{BER}_{\mathrm{BPSK,QPSK}} = \frac{1}{2}\mathrm{erfc}\left(\sqrt{\mathrm{SNR}_b}\right), \tag{7.3}$$

while the BER of a M-point quadrature amplitude modulation (M-QAM) signal (with Gray mapping) can be expressed as

$$\mathrm{BER}_{\mathrm{M-QAM}} = \frac{2\left(1 - \frac{1}{\sqrt{M}}\right)}{\log_2(M)}\,\mathrm{erfc}\left(\sqrt{\frac{3\log_2(M)}{2(M-1)}\mathrm{SNR}_b}\right), \tag{7.4}$$

where erfc() is the complementary error function and SNR_b is the signal-to-noise per bit, which is related to OSNR as

$$\mathrm{OSNR(dB)} = 10\log_{10}\left(\frac{\mathrm{SNR}_b \cdot R}{2 \cdot 12.5\mathrm{GHz}}\right), \tag{7.5}$$

where R is the raw bit rate of the signal. Fig. 7.6 shows the theoretical BER versus SNR_b for BPSK, QPSK, 16-QAM, and 64-QAM. Evidently, BPSK and QPSK require the same SNR_b for a given BER, while 16-QAM and 64-QAM require higher SNR_b. Fig. 7.7 shows the theoretical BER versus OSNR for 100-Gb/second PDM-BPSK, 200-Gb/second PDM-QPSK, 400-Gb/second PDM-16QAM, and 600-Gb/second PDM-64QAM. Table 7.6 summarizes the key results. At a raw BER of 1.25×10^{-2}, 478.75-Gb/second PDM-16QAM requires an OSNR of 20.4 dB, thus leaving an OSNR margin of 4 dB to accommodate implementation imperfections, such as nonideal transceiver implementations and component aging.

In 2020, SiP-based 400G-ZR transceivers have been demonstrated with the required OSNR at the C-FEC BER threshold (1.25×10^{-2}) of <24 dB [20], indicating a moderate transceiver implementation penalty of ~3 dB. With further development of the 400G-ZR ecosystem, the transceiver performance can be further improved, and the goal of supporting up to 120 km transmission over standard SMF with 64 75-GHz-spaced 400G-ZR wavelength channels, delivering a remarkable per-fiber transmission capacity of 25.6 Tb/second, can be comfortably achieved.

It is desirable to make the 400G-ZR transceivers to be of pluggable form factors, such as QSFP-DD and OSFP, so that they can be directly connected to routers via the front panels. Both QSFP-DD and OSFP form factors support 36 pluggable ports on one rack

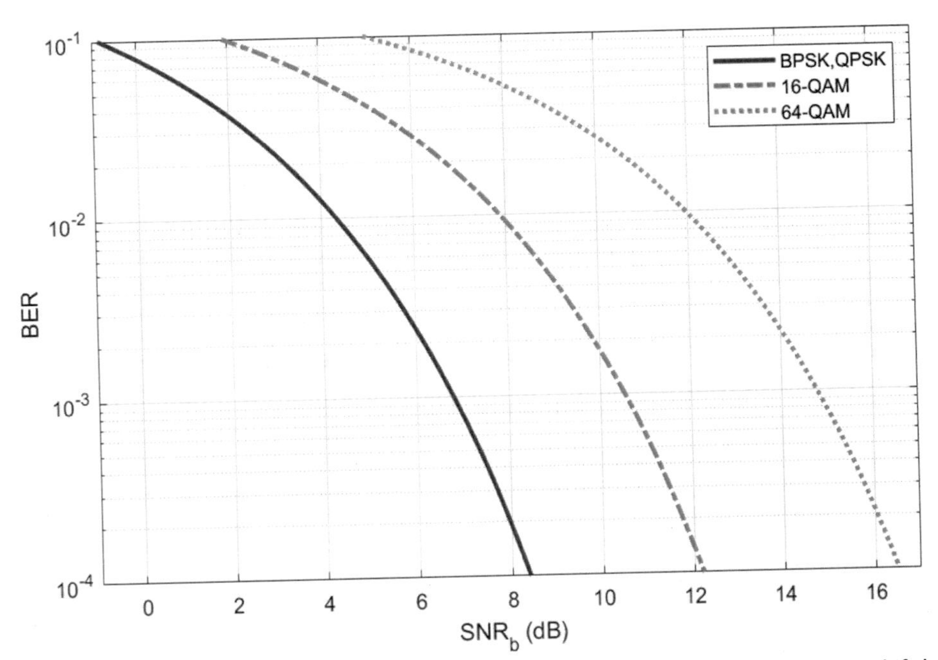

Figure 7.6 Theoretical bit error ratio versus signal-to-noise ratio per bit for binary phase shift keying, quadrature phase shift keying, 16- quadrature amplitude modulation (QAM), and 64-QAM.

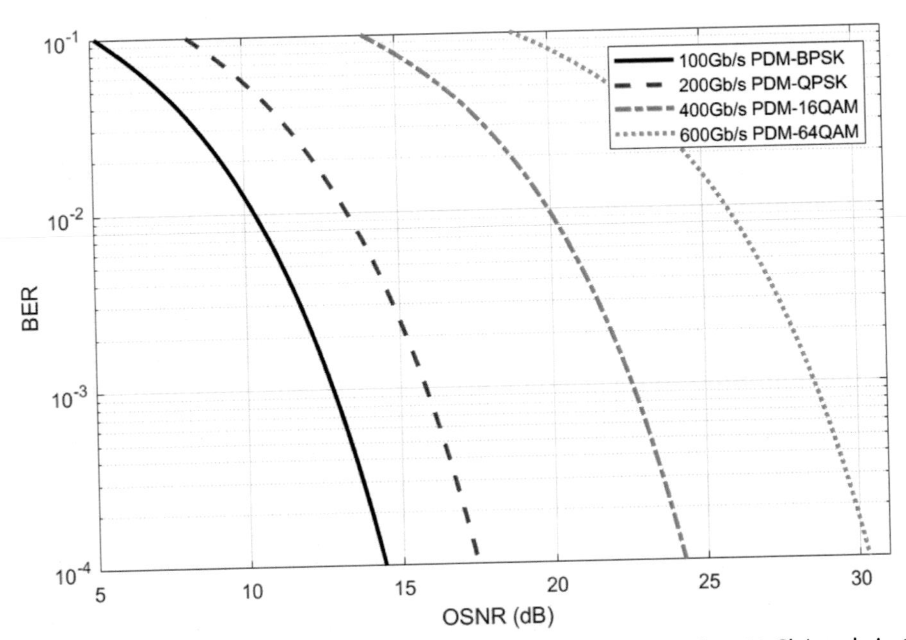

Figure 7.7 Theoretical bit error ratio versus optical signal-to-noise ratio for 100-Gb/s polarization-division-multiplexed (PDM) - binary phase shift keying, 200-Gb/s PDM - quadrature phase shift keying, 400-Gb/s PDM-16 quadrature amplitude modulation (QAM), and 600-Gb/s PDM-64QAM.

Table 7.6 Theoretical signal-to-noise ratio/optical signal-to-noise ratio requirements for different modulation formats.

	BPSK	QPSK	16-QAM	64-QAM
SNR at 10^{-3} BER (dB)	6.8	9.8	16.5	22.6
SNR at 10^{-2} BER (dB)	4.3	7.3	13.9	19.7
OSNR at 10^{-2} BER for 50-Gbaud PDM (dB)	10.3 (100 Gb/s)	13.3 (200 Gb/s)	19.9 (400 Gb/s)	25.7 (600 Gb/s)
OSNR at 1.25×10^{-2} BER for 59.84375-Gbaud PDM (dB)	10.8 (~120 Gb/s)	13.8 (~239 Gb/s)	20.4 (478.75 Gb/s)	26.2 (~718 Gb/s)

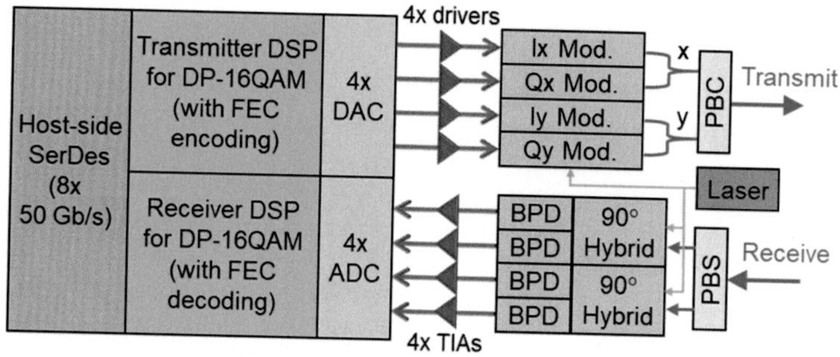

Figure 7.8 Schematic of the overall layout of a 400G-ZR–type coherent transceiver.

unit (1U) front panel, which has a height of 1.75 inches and a typical width of 19 inches, enabling 14.4 Tb/second overall capacity per 1U. Much effort has been made by the optical communication industry to reduce the size and power consumption of 400ZR-type coherent transceivers to fit the QSFP-DD and OSFP form factors [33]. In order to achieve this goal, both the overall transceiver layout and the digital signal processing (DSP) design have been refined. Fig. 7.8 shows the overall layout of a 400G-ZR–type transceiver. Eight 50-Gb/second serializer/deserializer (SerDes) lanes are used to communicate between the host and the transceiver. A DSP chip performs transmitter- and receiver-side DSP functions. It also consists of four digital-to-analog converters (DACs) and four analog-to-digital converters (ADCs). The four DAC outputs drive four Mach–Zehnder modulators (MZM) through four modulator drivers to generate the in-phase (I) and quadrature (Q) components of two orthogonal polarizations (x and y) of the optical carrier, which is provided by an external laser. The I and Q components of each polarization are combined with a 90-degree phase offset. Then a polarization beam combiner combines the x and y polarizations for transmission. For the receiver side, the incoming optical signal is divided into two orthogonal polarizations before mixing with the optical local oscillator (OLO) in two 90-degree hybrids. The OLO is provided by the same laser in the transceiver to save cost, size, and power consumption. The two 90-degree hybrids have four pairs of complementary outputs, which are detected by four balanced photodiodes

(BPDs). Four transimpedance amplifiers (TIAs) then convert the current outputs of the BPDs to voltages, which are sampled by the four ADCs.

Fig. 7.9 shows the flow diagram of the main DSP functions in a 400G-ZR–type digital coherent receiver. The four sampled signal components from the four ADCs are used to construct two complex waveforms representing two orthogonal polarizations of the received optical signal. Chromatic dispersion compensation is applied to each polarization. A 2×2 multiple-input-multiple-output equalizer then performs PMD compensation and polarization demultiplexing to recover the x and y polarization components of the optical signal. Each polarization component then goes through carrier frequency recovery and phase recovery, followed by SD- and HD-FEC decoding in accordance with the C-FEC design. Finally, the payload data bits are recovered. More details on the DSP functions used in digital coherent receivers can be found in a number of classical references [34–39].

Leveraging advances in 7-nm complementary metal oxide semiconductor or CMOS technology and photonic integration, 400-ZR–type coherent optical transceivers can fit into QSFP-DD form factor with a total power consumption of <15 W [40]. Fig. 7.10 shows the rough estimates of the power consumptions of the key components in a 400G-ZR–type coherent optical transceiver. The DSP, FEC, DAC, and ADC collectively consume approximately half of the power [20,39], while the laser, electrical-to-optical (E/O) and optical-to-electrical (E/O) converters, SerDes, and other accessories consume roughly equally the other half of the power. With continued advances in this field, higher data rates such as 800 Gb/second and longer reaches (beyond 120 km) are expected to be supported in pluggable form factors, which we will briefly discuss in Section 7.6.

7.4 Intra-DC connections and 400-Gb/second pluggable optical transceivers

For intra-DC connections, the transmission distance is typically shorter than 2 km, corresponding to the FR class and simple IM/DD is used instead of coherent

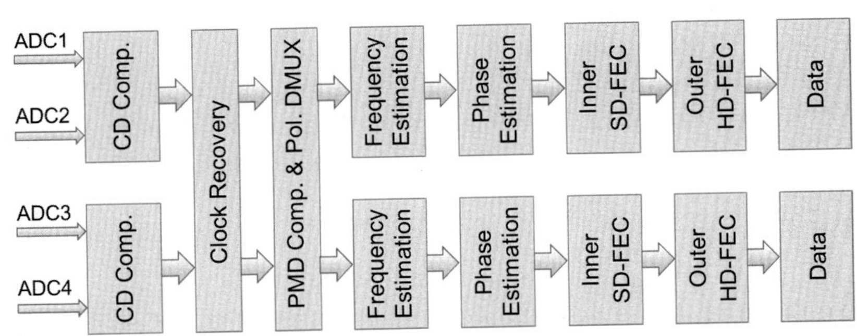

Figure 7.9 Flow diagram of the main digital signal processing functions in a 400G-ZR–type digital coherent receiver. CD comp.: chromatic dispersion compensation.

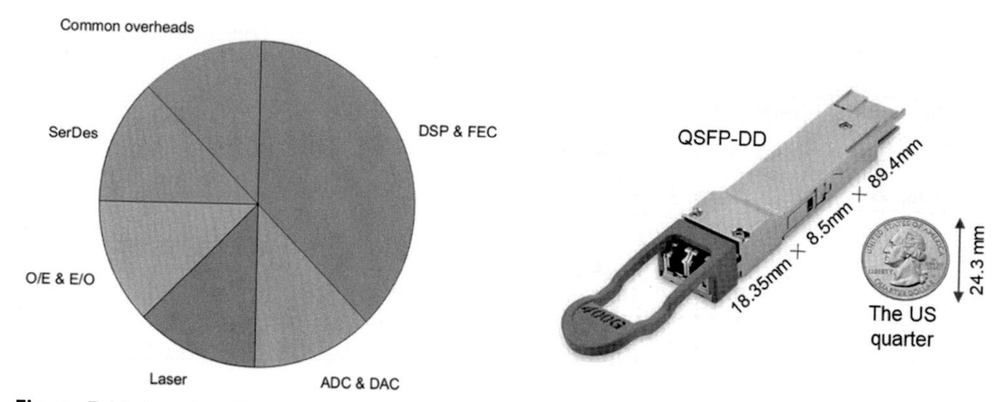

Figure 7.10 Rough estimates of the power consumptions of key components in a 400G-ZR–type coherent optical transceiver that fits into a quad small form-factor pluggable - double density form factor.

Table 7.7 Typical features of the first four generations of intra-DC transceivers.

Generation	First	Second	Third	Fourth
Net bit rate (Gb/s)	10	40	100	400
Modulation	NRZ	NRZ	NRZ	PAM4
Baud (GHz)	10	10	25	25 or 50
Multiplexing scheme	No	CWDM	CWDM	LAN-WDM or CWDM
Typical form factor	SFP	QSFP	QSFP	QSFP-DD, OSFP
Electrical lanes	1 × 10 Gb/s	4 × 10 Gb/s	4 × 25 Gb/s	8 × 50 Gb/s

modulation/detection [41]. Table 7.7 shows the first four generations of pluggable optical transceivers used for intra-DC connections. The current generation (the fourth generation) FR-class pluggable optical transceivers are based on PAM4 at 25 Gbaud using eight LAN-WDM wavelengths or 50 Gbaud using four CWDM wavelengths. Table 7.8 and Table 7.9 show the wavelength plans of the 8-channel LAN-WDM and 4-channel CWDM schemes.

Fig. 7.11 shows the overall layout of a 400-Gb/second IM/DD transceiver with four CWDM channels. Evidently, the IM/DD transceiver and the 400G-ZR coherent transceiver share some common components such as eight 50-Gb/second SerDes lanes, four DACs, four ADCs, and a DSP chip. The main difference is in the optics and the DSP algorithms. At the transmitter side, there are four externally modulated laser (EMLs) emitting at the first four CWDM wavelengths. Each EML consists of a laser followed by an electro-absorption modulator modulated with 100-Gb/second PAM4. The outputs of the four EMLs are wavelength-multiplexed to achieve an

Table 7.8 LAN-WDM wavelength plan used in intra-DC connections.

LAN-WDM channel index	Center frequency (THz)	Center wavelength (nm)	Passband wavelength range (nm)
L0	235.4	1273.54	1272.55–1274.54
L1	234.6	1277.89	1276.89–1278.89
L2	233.8	1282.26	1281.25–1283.28
L3	233.0	1286.66	1285.65–1287.69
L4	231.4	1295.56	1294.53–1296.59
L5	230.6	1300.05	1299.02–1301.09
L6	229.8	1304.58	1303.54–1305.63
L7	229.0	1309.14	1308.09–1310.19

Table 7.9 Coarse wavelength-division multiplexing wavelength plan used in intra-DC connections.

CWDM channel index	Center wavelength (nm)	Passband wavelength range (nm)
Ch1	1271	1264.5–1277.5
Ch2	1291	1284.5–1297.5
Ch3	1311	1304.5–1317.5
Ch4	1331	1324.5–1337.5

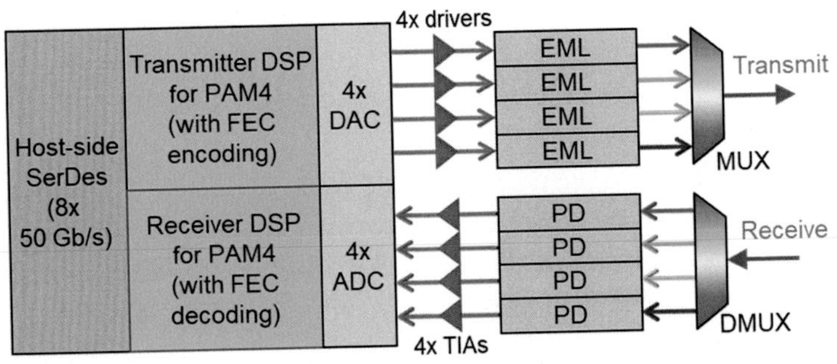

Figure 7.11 Schematic of the overall layout of a 400-Gb/s intensity-modulation and direct detection transceiver with four coarse wavelength-division multiplexing channels.

aggregated data rate of 400 Gb/second. At the receiver side, the four CWDM wavelength channels are demultiplexed and received by four photodiodes (PDs). Four TIAs then convert the current outputs of the PDs to voltages, which are sampled by the four ADCs. Fig. 7.12 shows the flow diagram of the main DSP functions in a 400G direct-detection receiver. The four sampled signal components from the four ADCs are processed in parallel through DSP stages such as clock recovery, multitap feed-forward equalization (FFE), maximum likelihood sequence estimation (MLSE), and HD-FEC decoding. Finally, the payload data bits are recovered. Further details on the

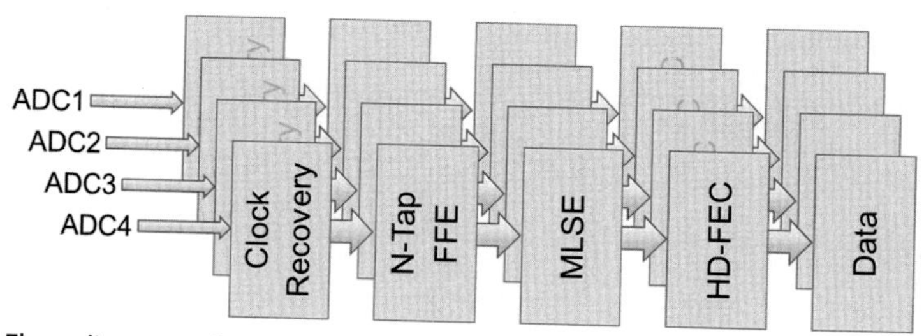

Figure 7.12 Flow diagram of main digital signal processing functions used in direct-detection receivers.

DSP functions used in direct-detection PAM4 receivers can be found in references such as [5,20,42–44].

The use of FFE and MLSE increases the receiver tolerance to intersymbol interference (ISI) caused by transceiver bandwidth limitation and fiber dispersion. However, unlike coherent-detection receivers, direct-detection receivers generally have limited dispersion tolerances because of the loss of the optical phase information upon direct direction of signal intensity, which prevents the full compensation of dispersion-induced ISI. It is thus important to carefully assess the dispersion-limited reach of multichannel 400-Gb/second IM/DD transmission in the O-band. For the commonly used standard SMF (G.652), the fiber dispersion coefficient (in ps/nm/km) can be expressed as [12–14]

$$D(\lambda) \approx \frac{S_0 \lambda}{4}\left(1 - \frac{\lambda_0^4}{\lambda^4}\right), \tag{7.6}$$

where λ is the operating wavelength (in nm), λ_0 is the zero-dispersion wavelength, and S_0 is the zero-dispersion slope. Owing to fiber manufacturing uncertainties, we have $1300\ \text{nm} \leq \lambda_0 \leq 1324\ \text{nm}$ and $S_0 \leq 0.093\ \text{ps}/(\text{nm}^2 \cdot \text{km})$, therefore the minimum and maximum dispersion coefficients (in ps/nm/km) can be expressed as [12–14]

$$D_{\text{min}}(\lambda) \approx 0.02325\lambda\left(1 - \frac{(1324\ \text{nm})^4}{\lambda^4}\right), \quad D_{\text{max}}(\lambda) \approx 0.02325\lambda\left(1 - \frac{(1300\ \text{nm})^4}{\lambda^4}\right). \tag{7.7}$$

As discussed in Chapter 4, Optical Technologies for 5G X-haul, the dispersion penalty depends on many factors such as signal modulation symbol rate, modulation format, transmitter chirp, receiver-side equalization, and the reference BER at which the dispersion penalty is measured. The chirp of the EML can also be slightly adjusted. Fig. 7.13 shows the estimated worst-case dispersion penalty for a moderately chirped 100-Gb/second PAM4 signal after 10-km transmission over the standard SMF in the O-band.

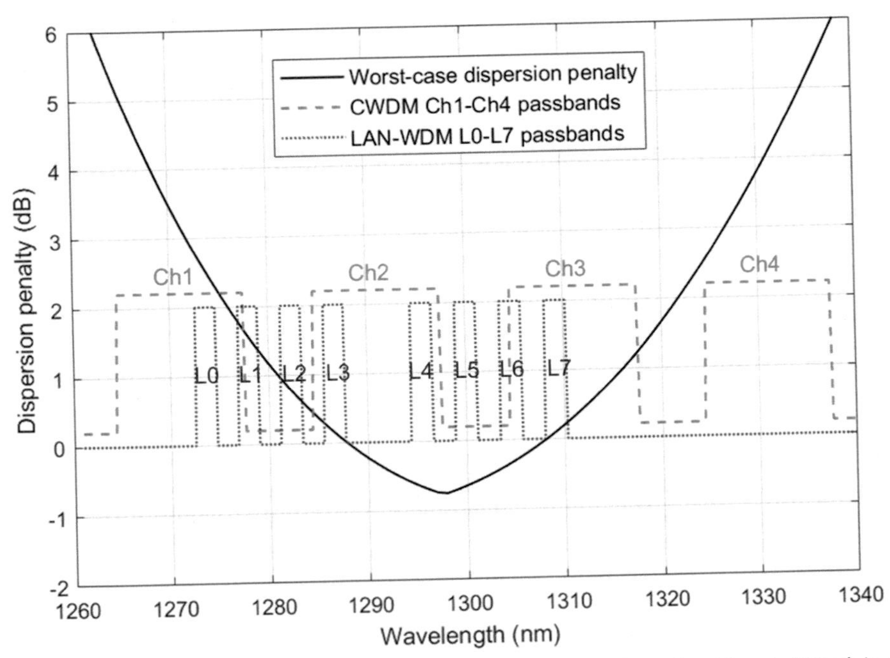

Figure 7.13 Estimated worst-case dispersion penalty for a moderately chirped 100-Gb/s pulse amplitude modulation 4-level signal after 10-km standard single-mode fiber transmission.

In Fig. 7.13, the passbands of the eight-channel LAN-WDM and the four-channel CWDM are also plotted. The dispersion penalty of the 100-Gb/second PAM4 signal can be much larger than 3 dB at the two outer edges of the four-channel CWDM passbands, making it difficult to support 400 GBase-LR4. For this reason, the IEEE 802.3 cu standard reduced the reach requirement to 6 km and introduced the 400 GBase-LR4−6 variant [13]. In addition, 400 GBase-FR4 and 400 GBase-DR4 have been standardized in the IEEE 802.3 bs standard [12].

When LAN-WDM is used instead of CWDM, the worst-case dispersion penalty of the 100-Gb/second PAM4 signal in the eight LAN-WDM passbands is less than 3 dB. Since the dispersion tolerance of a given signal quadruples when the modulation symbol rate is halved, a 50-Gb/second PAM4 signal can transmit over 40 km in standard SMF in LAN-WDM passbands. Thus 400 GBase-ER8 is feasible and has been standardized in the IEEE 802.3 cn standard [14]. In addition, 400 GBase-LR8 and 400 GBase-FR8 have been standardized in the IEEE 802.3 bs standard [12]. Table 7.10 summarizes key IEEE standards on 400 GBase Ethernet physical layer transceivers (PHYs) [11−15], including specifications such as fiber medium, transmitter type and transceiver form factor, optical modulation symbol rate and format, and reach.

There are four common pluggable form factors used in the 400GBase PHYs, SFP28, QSFP28, DSFP-DD, and OSFP [5], which are described in Table 7.11. These

Table 7.10 IEEE standards on 400GBase formats.

400GBase PHY	IEEE standards	Fiber medium	Transmitter type (form factor)	Gigabaud per lane	Reach
400G-SR16	802.3 bs	16 parallel MMFs	VCSEL at 850 nm (SFP28)	26.5625 (NRZ)	70(100) m in OM3(4)
400G-SR8	802.3 cm	8 parallel MMFs	VCSEL at 850 nm (QSFP)	26.5625 (PAM4)	100 m in OM5
400G-DR4	802.3 bs	4 parallel SMFs	DML at CWDM CH3: 1311 nm (QSFP28)	53.125 (PAM4)	500 m
400G-FR4 400G-LR4–6	802.3 cu 802.3 cu	SMF	4-Channel CWDM at 1271, 1291, 1311, and 1331 ± 6.5 nm (QSFP-DD or OSFP)	53.125 (PAM4) 53.125 (PAM4)	2 km >6 km
400G-FR8 400G-LR8 400G-ER8	802.3 bs 802.3 bs 802.3 cn	SMF	8-Channel LAN-WDM in 1272.55 − 1310.19 nm (QSFP-DD or OSFP)	26.5625 (PAM4) 26.5625 (PAM4) 26.5625 (PAM4)	2 km 10 km 40 km
400G-ZR	802.3 cw	SMF	DWDM (QSFP-DD or OSFP)	59.84375 (DP-16QAM)	80 km

Table 7.11 Common pluggable form factors used in 400GBase Ethernet physical layer transceivers.

SFP	Electrical lanes (nominal rate)	Dimensions (width × height × length)	Power consumption
SFP28	1 × 25 Gb/s	13.4 × 8.5 × 56.5 mm	<1 W
QSFP28	4 × 25 Gb/s	18.35 × 8.5 × 72.4 mm	<3.5 W
QSFP-DD	8 × 50 Gb/s	18.35 × 8.5 × 89.4 mm	<15 W
OSFP	8 × 50 Gb/s	22.58 × 13.0 × 100.4 mm	<20 W

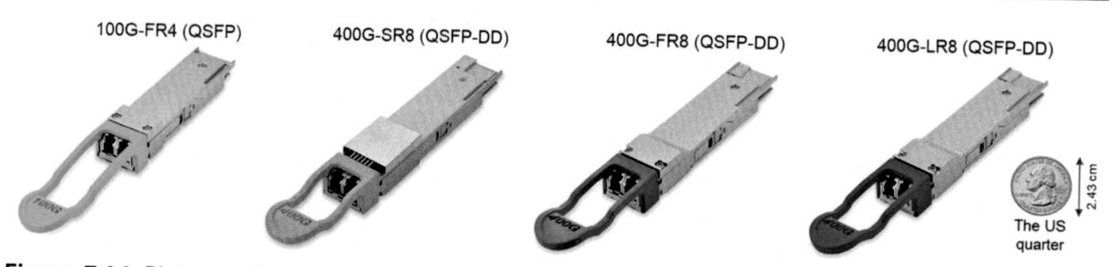

Figure 7.14 Pictures of representative color-coded pluggable format factors used in intra-data center connections.

pluggable form factors are usually color-coded on their handles (or tabs) to represent their reach specifications. For example, SR, DR, FR, LR, ER, and ZR are represented as beige, yellow, green, blue, red, and white, respectively. Fig. 7.14 shows some representative color-coded pluggable format factors in intra-DC connections.

The term "double density (DD)" in QSFP-DD refers to the doubling of the number of high-speed electrical interfaces (lanes) that the module supports as compared to the regular QSFP28 module. In addition, the nominal rate per electrical lane in QSFP-DD is doubled to 50 Gb/second, thereby offering a throughput that is four times as high as that of QSFP. A good feature of QSFP-DD is that its width and height are identical to those of QSFP and it is backward compatible with QSFP. The only mechanical difference is that QSFP-DD is ~20% longer than QSFP to accommodate the doubled number of electrical lane contacts. QSFP-DD supports the same system port count density as QSFP, for example, 36 pluggable ports per 1U front panel, enabling an overall capacity of 14.4 Tb/second per 1 U.

Compared to QSFP-DD, OSFP has a slightly larger size, but it can also support 36 pluggable ports per 1 U front panel. OSFP can accommodate more power consumption, which is helpful for achieving even higher throughputs such as 800 Gb/second per module [5]. Both QSFP-DD and OSFP form factors are under further development to support higher throughput and power consumption by industry-wide multisource agreement (MSA) groups such as the QSFP-DD MSA group and the OSFP MSA group [17,18].

7.5 Multi-Tb/second DC switches and their evolution

Driven by the ever-increasing demand in cloud services, both the east–west data traffic inside DCs and the north–south data traffic entering and leaving DCs are continuously increasing. According to the Cisco Global Cloud Index 2015–20 report, the compound annual growth rate (CAGR) has been about 27% over the last few years [2]. As disaggregated computing and storage become more popular, the east–west data traffic inside DCs further increases, and most of the DC traffic stays within DCs. For the estimated global DC traffic by destination in 2020, the intra-DC, DC-to-DC, and DC-to-user traffics account for 77%, 9%, and 14% of the total traffic, respectively [2].

To support the routing of the massive amount of intra-DC traffic, the routing capacities of DC switches are exponentially increasing as well. As shown in Fig. 7.3, a full-mesh topology is often used between the Tier-1 switches on the "leaf layer" and the Tier-2 switches on the "spine layer." In effect, the leaf layer contains access switches that connect to servers via the TOR switches, while the spine layer forms the spine of the network for interconnecting all the leaf switches. This type of "flattened" network architecture helps improve network efficiency and reduce latency. In addition, the use of multiple spine switches in the DC network increases the network resilience against failures in spine switches. Moreover, the leaf-spine topology is highly scalable. Network capacity can be readily increased by adding new spine switches and leaf switches. Fig. 7.15 shows two interior pictures of typical modern DCs with

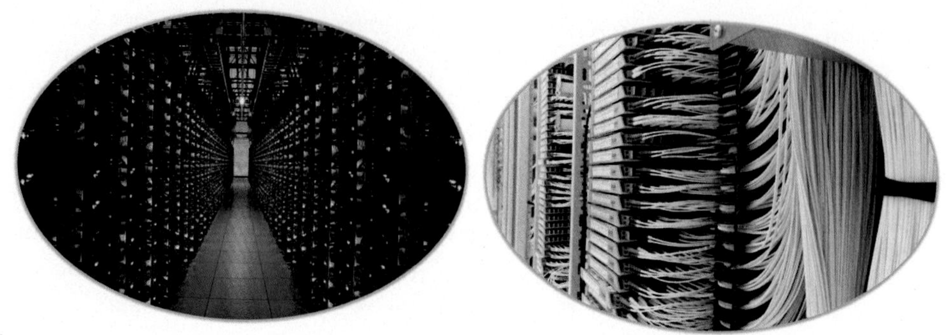

Figure 7.15 Data center interior pictures (available from the internet) showing racks full of servers (left) and fiber cables (right).

Table 7.12 Scaling of data center switch capacity.

Optical module speed	Number of optical ports	Switch capacity	Module form factor	Total optical transceiver power
100 Gb/s	32	3.2 Tb/s	QSFP28 (~3.5 W)	~112 W
	64	6.4 Tb/s		~224 W
	128	12.8 Tb/s		~448 W
400 Gb/s	32	12.8 Tb/s	QSFP-DD (~13 W)	~416 W
	64	25.6 Tb/s		~832 W
	128	51.2 Tb/s		~1664 W
800 Gb/s	32	25.6 Tb/s	OSFP or QSFP-DD (~20 W)	~640 W
	64	51.2 Tb/s		~1280 W
	128	102.4 Tb/s		~2560 W

densely packed servers, switches, and fiber cables, which are essential for data processing and networking.

As DC traffic scales up, the leaf-and-spine topology expands by connecting more switches and servers, and the switch capacity and port count are desired to scale up accordingly. Table 7.12 shows the scaling of DC switch capacity with the increase of optical transceiver module speed. With the third generation intra-DC optical transceivers operating at 100 Gb/second, switch capacities of 3.2, 6.4, and 12.8 Tb/second were respectively realized with the use of 32, 64, and 128 ports in the late 2010s. With the fourth generation intra-DC optical transceivers operating at 400 Gb/second, switch capacities of up to 51.2 Tb/second are expected in the early 2020s. However, with the power consumption of ~13 W per 400-Gb/second optical transceiver (in QSFP-DD format factor) [45], a 128-port 51.2-Tb/second switch would require a power consumption of ~1664 W for the needed optical transceivers alone. Together with the powers needed for data routing, processing, fans, and peripherals, the total power consumption of such a 51.2-Tb/second switch would exceed 2500 W, which is very challenging to realize [46].

The fifth generation intra-DC optical transceivers are expected to operate at 800 Gb/second. Extensive research efforts have been devoted to IM/DD-based 800-Gb/second transmission [5,47–52]. With 800-Gb/second intra-DC optical transceivers, switch capacities of up to 102.4 Tb/second could be realized in the mid-2020s [5]. However, with the power consumption of ~20 W per 800-Gb/second optical transceiver in OSFP or QSFP-DD format factor [53], a 128-port 102.4-Tb/second switch would require a power consumption of ~2560 W for the needed optical transceivers alone. Thus continued innovations are needed to reduce the power consumption of intra-DC optical transceivers.

A promising evolution path to scale up the switch capacity and port count with acceptable power consumption is via on-board optics and copackaged optics. The Consortium for On-Board Optics (COBO) was established in 2015 to explore this evolution path [54]. The COBO member companies work together to develop embedded optical module form factor specifications with multigeneration evolution path, broader technology capability, and enhanced thermal and signal integrity properties. Fig. 7.16 illustrates an evolution path of optical interface for increased capacity and port density. First, the use of on-board optics (OBO) allows the optical transceivers to be much closer to the switch application-specific integrated circuit (ASIC), thereby increasing the bandwidth of the electrical lanes that connect the transceivers and the ASIC and reducing their power consumptions. In addition, the port density can be increased because of the absence of pluggable packages. Second, the switch ASIC and the optical transceivers can be copackaged in a multichip module (MCM), via co-packaged optics (CPO), to further increase the electrical connection bandwidth and reduce the connection-associated power consumption. The MCM can further evolve from a 2.5D architecture to a 3D architecture. Finally, monolithic integration between the optical chip (for the optical transceivers) and the electrical chip (for the switch ASIC), via optical-electrical integrated circuit (OEIC), could provide the ultimate capacity and port density.

7.6 Concluding remarks and future trends

Data centers are essential for supporting the 5G networks and cloud services. At the beginning of the 5G era in the early 2020s, 400-Gb/second-class pluggable optical transceivers are the key enablers for DC interconnects and intra-DC connections. For connecting DCs with global optical networks via RNGs, 400 G-ZR has become a primary solution. For intra-DC connections with various reach requirements, 400 GBase PHY transceivers such as 400 GBase-SR8/DR4/FR4 are the main

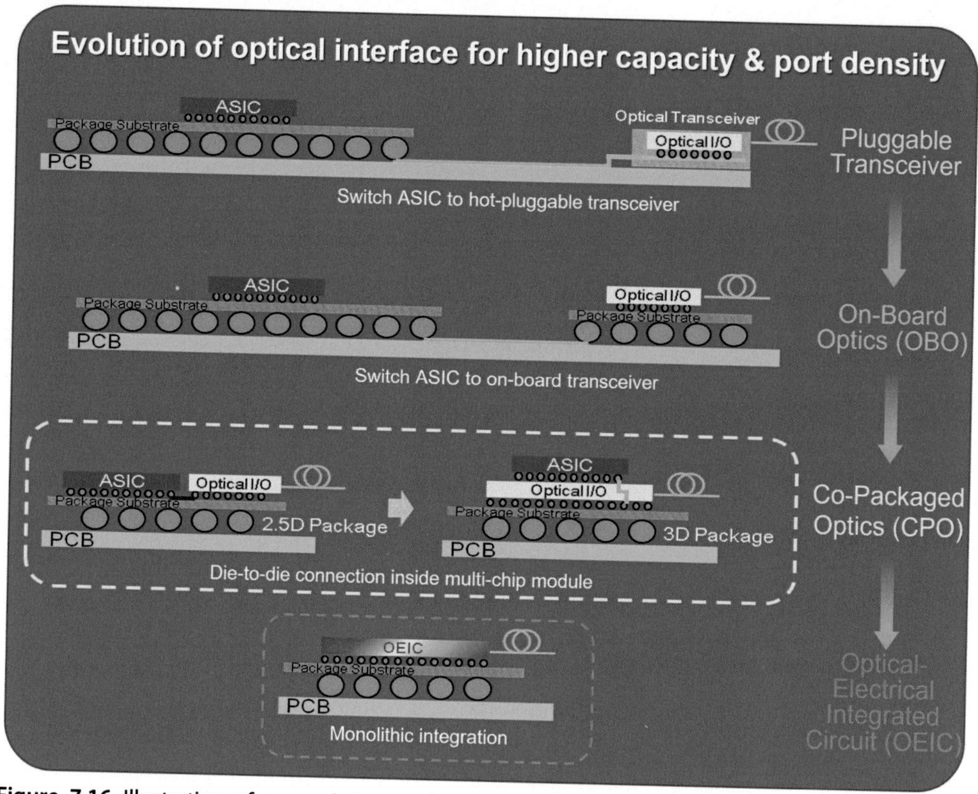

Figure 7.16 Illustration of an evolution path of optical interface for increased capacity and port density.

solutions. With 400-Gb/second-class pluggable optical transceivers, switch capacity can be scaled up to 51.2 Tb/second with 128 ports.

With a continual increase in DC traffic, the optical interface bit rate is expected to be further increased from 400 to 800 Gb/second or even 1.6 Tb/second [5] during the 2020s. At the same time, the DC switch capacity needs to be increased accordingly. However, we need to be mindful about DC-related energy consumptions. According to a recent report [55], US-based DCs alone used up more than 90 billion kilowatt-hours of electricity in 2017, which accounted for ~2.3% of the total US electricity consumption of 3.9 trillion kilowatt-hours [56]. Thanks to the technological advances in this field, the DC energy consumption increased at a rate of ~4% between 2010 and 2020 [55] even though the DC traffic increased at a much higher rate of ~27% [2]. Going forward, continued innovations are required to sustain the increase in the optical interface bit rate and the switch capacity with acceptably low power consumptions.

For the evolution of intra-DC networks, it is desirable to have technological advances in multiple areas, some of which are listed as follows.

- Increasing the electrical interface speed to 100 Gb/second and beyond, as being studied in the IEEE 802.3ck group [57]
- Increasing the power dissipation of QSFP-DD to 20 W and beyond to support 800 Gb/second optical interface speed, as being studied by the QSFP-DD MSA group [53]
- Reducing the cost and power consumption of coherent optics for 1.6 Tb/second and beyond transmission [5]
- Progressing copackaged optics toward monolithic integration of switch ASIC and optical transceivers [54]
- Optimizing the leaf-spine network architectures, for example, by network flattening with a high radix of 512 [58]

For the evolution of inter-DC networks, 800G-ZR is expected to be the next step after 400G-ZR for edge DCI [59]. In addition, ZR+ class pluggable optical transceivers are being developed to reach beyond 120 km for regional DCI and metro applications [60,61]. The key techniques being explored are as follows [60]:

- More powerful FEC codes such as the O-FEC code that provide ~0.8 dB higher NCG than the C-FEC used in 400G-ZR, and
- Flexible modulation formats, such as 16QAM/8QAM/QPSK, to address a large range of reach requirements, for example, up to ~450 km for both DCI and metro network applications.

Going beyond edge and regional DCIs, longer reach and higher per-fiber transmission capacity are needed to effectively complete the global optical network connections. We will discuss long-haul, ultra-long-haul, and undersea transmission aspects in Chapter 8, High-Capacity Long-Haul Optical Fiber Transmission. Advanced flexible-grid WDM transmission, superchannel transmission, and wavelength routing via optical cross-connects will be presented in Chapter 9, Superchannel Transmission And Flexible-Grid Wavelength Routing.

References

[1] Kachris C, Kanonakis K, Tomkos I. Optical interconnection networks in data centers: recent trends and future challenges. IEEE Commun Mag 2013;51(9):39–45.
[2] Cisco Global Cloud Index 2015–2020, Cisco knowledge network (CKN) session, November 2016 [Online]. Available from: https://www.cisco.com/c/dam/m/en_us/service-provider/ciscoknowledgenetwork/files/622_11_15-16-Cisco_GCI_CKN_2015-2020_AMER_EMEAR_NOV2016.pdf.
[3] Chang F. Datacenter connectivity technologies: principles and practicer. River Publishers; 2018.
[4] Nagarajan R, Filer M, Fu Y, Kato M, Rope T, Stewart J. Silicon photonics-based 100 Gbit/s, PAM4, DWDM data center interconnects. J Opt Commun Netw 2018;10:B25–36.

[5] Zhou X, Urata R, Liu H. Beyond 1 Tb/s intra-data center interconnect technology: IM-DD OR coherent? J Light Technol 2020;38(2):475–84 15.
[6] See for example, https://aws.amazon.com/about-aws/global-infrastructure/; https://www.google.com/about/datacenters/locations/; https://azure.microsoft.com/en-us/global-infrastructure/.
[7] See for example, https://www.crn.com/news/data-center/aws-google-microsoft-are-taking-over-the-data-center
[8] See for example, press release. Huawei's ON2.0 leads the commercial use of all-optical networks in partnership with operators worldwide (http://www.tagitnews.com/en/article/36282).
[9] Li J, Zhang A, Zhang C, Huo X, Yang Q, Wang J, et al. Field trial of probabilistic-shaping-programmable real-time 200-Gb/s coherent transceivers in an intelligent core optical network. In: Asia communications and photonics conference (ACP), PDP Su2C.1; 2018.
[10] Optical Internetworking Forum (OIF). Implementation agreement 400ZR [Online]. Available from: https://www.oiforum.com/wp-content/uploads/OIF-400ZR-01.0_reduced2.pdf, 2020 March.
[11] IEEE P802.3 cm, 400 Gb/s over multimode fiber task force. Available from: https://www.ieee802.org/3/cm/.
[12] IEEE P802.3 bs, 200 Gb/s and 400 Gb/s ethernet task force. Available from: https://www.ieee802.org/3/bs/.
[13] IEEE P802.3 cu, 100 Gb/s and 400 Gb/s over SMF at 100 Gb/s per wavelength task force. Available from: https://www.ieee802.org/3/cu/.
[14] IEEE P802.3 cn, 50 Gb/s, 200 Gb/s, and 400 Gb/s over greater than 10 km of SMF task force. Available from: https://www.ieee802.org/3/cn/.
[15] IEEE P802.3 cw, 400 Gb/s over DWDM systems task force. Available from: https://www.ieee802.org/3/cw/.
[16] Tatum JA, et al. VCSEL-based interconnects for current and future data centers. J Light Technol 2015;33(4):727–32.
[17] QSFP-DD Multi-source Agreement Specification [Online]. Available from: http://www.qsfp-dd.com/specification/.
[18] OSFP Multi-source Agreement Specification [Online]. Available from: https://osfpmsa.org/specification.html.
[19] Smith BP, Farhood A, Hunt A, Kschischang FR, Lodge J. Staircase codes: FEC for 100 Gb/s OTN. J Light Technol 2012;30(1):110–17.
[20] Nagarajan R. and Lyubomirsky I. Low-complexity DSP for inter-data center optical fiber communications. In: 2020 European conference on optical communications (ECOC), Tutorial paper SC04; 2020.
[21] Soref RA. The past, present and future of silicon photonics. IEEE J Sel Top Quant Electron 2006;12(6):1678–87.
[22] Miller DAB. Device requirements for optical interconnects to silicon chips. Proc IEEE 2009;97 (7):1166–85.
[23] Reed GT, Mashanovich G, Gardes FY, Thomson DJ. Silicon optical modulators. Nat Photonics 2010;4(8):518–26.
[24] Asghari M, Krishnamoorthy AV. Energy-efficient communication. Nat Photonics 2011;5 (5):268–70.
[25] Doerr CR, Buhl LL, Baeyens Y, Aroca R, Chandrasekhar S, Liu X, et al. Packaged monolithic silicon 112-Gb/s coherent receiver. IEEE Photonics Technol Lett 2011;23(12):762–4.
[26] Vivien L, Pavesi L. Handbook of silicon photonics. Taylor & Francis; 2013.
[27] Dong P, Liu X, Chandrasekhar S, Buhl LL, Aroca R, Chen Y. Monolithic silicon photonic integrated circuits for compact 100^{+}Gb/s coherent optical receivers and transmitters. IEEE J Sel Top Quant Electron 2014;20(4):150–7.
[28] Nagarajan R, Doerr C, and Kish F. Semiconductor photonic integrated circuit transmitters and receivers. In: Kaminow I, Li T, Willner A, editors. Optical fiber telecommunications VIA: components and subsystems, 2013, pp. 25–88.
[29] Kaminow I, Li T, Willner AE, editors. Optical fiber telecommunications V. Elsevier; 2007.

[30] Proakis JG. Digital communications. 4th Edition McGraw-Hill; 2000.
[31] Ho K-P. Phase-modulated optical communication systems. New York: Springer; 2005.
[32] Liu X, Chandrasekhar S, Leven A. Self-coherent optical transport systems. In: Kaminow IP, Li T, Willner AE, editors. (Chapter 4) in Optical fiber telecommunications V B. Academic Press; 2008.
[33] See for example, https://www.oiforum.com/press-room/oif-in-the-news/.
[34] Taylor MG. Coherent detection method using DSP for demodulation of signal and subsequent equalization of propagation impairments. IEEE Photon Technol Lett 2004;16:674–6.
[35] Savory SJ. Digital filters for coherent optical receivers. Opt Express 2008;16:804–17.
[36] Ip E, Lau APT, Barros DJF, Kahn JM. Coherent detection in optical fiber systems. Opt Express 2008;16(2):753–91.
[37] Winzer PJ. High-spectral-efficiency optical modulation formats. J Light Technol 2012;30:3824–35.
[38] Laperle C, O'Sullivan M. Advances in high-speed DACs, ADCs, and DSP for optical coherent transceivers. J Light Technol 2014;32:629–43.
[39] Fludger C. Performance orientated DSP design for flexible coherent transmission. In: Optical fiber communication conference (OFC) 2020, Tutorial paper Th3E.1; 2020.
[40] Filer M. The coherent 400 Gb/s revolution. LightReading optical networking digital symposium; 2020 May 28.
[41] Cole C. Beyond 100G client optics. IEEE Commun Mag 2012;50(2):s58–66.
[42] Liu X, Effenberger F. Emerging optical access network technologies for 5G wireless [invited]. IEEE/OSA J Opt Commun Netw 2016;8(12):B70–9.
[43] Liu GN, Zhang L, Zuo T, Zhang Q. IM/DD transmission techniques for emerging 5G fronthaul, DCI, and metro applications. J Light Technol 2018;36(2):560–7.
[44] Zhong H, Zhou X, Huo J, Yu C, Lu C, Lau APT. Digital signal processing for short-reach optical communications: a review of current technologies and future trends. J Light Technol 2018;36:377–400.
[45] See for example, http://100glambda.com/specifications; https://ii-vi.com/product/400gbase-fr8-qsfp-dd-optical-transceiver/.
[46] Rump Session on When will co-packaged optics replace pluggable modules in the datacenter? In: Organized by Chris Cole and Dan Kuchta, Optical fiber communication conference (OFC) 2020; 2020 March 10.
[47] Lange S, Wolf S, Lutz J, Altenhain L, Schmid R, Kaiser R, et al. 100 GBd intensity modulation and direct detection with an InP-based monolithic DFB laser Mach–Zehnder modulator. J Light Technol 2018;36:97–102.
[48] Shen Y, Meng X, Cheng Q, Rumley S, Abrams N, Gazman A, et al. Silicon photonics for extreme scale systems. J Light Technol 2019;37:245–59.
[49] Zhu Y, Zhang F, Zhang L, Ruan X, Li Y, Chen Z. Towards single lane 200G optical interconnects with silicon photonic modulator. J Light Technol 2020;38:67–74.
[50] Jacques M, Xing Z, Samani A, El-Fiky E, Li X, Xiang M, et al. 240 Gbit/s silicon photonic Mach–Zehnder modulator enabled by two 2.3-Vpp drivers. J Light Technol 2020;38:2877–85.
[51] Pang X, Ozolins O, Lin R, Zhang L, Udalcovs A, Xue L, et al. 200 Gbps/lane IM/DD technologies for short reach optical interconnects. J Light Technol 2020;38:492–503.
[52] Zhang H, Li M, Zhang Y, Zhang D, Liao Q, He J, et al. 800 Gbit/s transmission over 1 km single-mode fiber using a four-channel silicon photonic transmitter. Photonics Res 2020;8(11):1776–82.
[53] QSFP-DD MSA Whitepaper on Optimizing QSFP-DD Systems to Achieve at Least 25 Watt Thermal Port Performance. January 2021. Available from: http://www.qsfp-dd.com/wp-content/uploads/2021/01/2021-QSFP-DD-MSA-Thermal-Whitepaper-Final.pdf.
[54] The Consortium for On-Board Optics (COBO), https://www.onboardoptics.org/.
[55] See for example, https://www.vxchnge.com/blog/growing-energy-demands-of-data-centers.
[56] See for example, https://www.eia.gov/energyexplained/electricity/use-of-electricity.php.
[57] IEEE P802.3ck, 100 Gb/s, 200 Gb/s, and 400 Gb/s electrical interfaces task force. Available from: https://www.ieee802.org/3/ck/.
[58] Filer M. The race to 800G: a reality check. COBO Presentation on January 13, 2021. Available from: https://www.onboardoptics.org/race-to-800g-a-reality-check-micros.

[59] See for example, https://www.oiforum.com/oif-launches-800g-coherent-and-co-packaging-framework-ia-projects-elects-new-board-members-positions-officers-and-working-group-chairs/.
[60] Trowbridge S. Hot topics in optical transport networks. In: ITU-T study group 15 presentation at Optical fiber communication conference (OFC) 2019. Available from: https://www.itu.int/en/ITU-T/studygroups/2017-2020/15/Documents/OFC2019-3-otn.pdf.
[61] See for example, https://www.lightwaveonline.com/optical-tech/transmission/article/14188934/understanding-400zropenzr400zr-optics.

CHAPTER 8

High-capacity long-haul optical fiber transmission

8.1 Introduction

High-capacity long-haul optical fiber transmission is important in forming the global optical network that supports communication services such as 5G and cloud services. In this chapter, we theoretically explore the capacity limit of optical fiber transmission in Section 8.2. Common coherent modulation and detection techniques including dispersion compensation, polarization-mode dispersion compensation, carrier frequency offset, compensation and phase compensation, are reviewed in Section 8.3. Advanced digital signal processing (DSP) schemes for high-speed coherent transmission such as multicarrier modulation, multiband dispersion compensation, nonlinearity mitigation and compensation, capacity-approaching forward-error correction (FEC), and constellation shaping (CS) are described in Section 8.4. A real-time demonstration of probabilistic constellation shaping that shows a clear performance improvement is presented in Section 8.5. Advanced fiber and Erbium-doped fiber amplifier (EDFA) technologies are discussed in Section 8.6. Ultra-long-haul undersea transmission is discussed in Section 8.7. Finally, a summary with a few concluding remarks will be given in Section 8.8.

8.2 Exploring the capacity limit of optical fiber transmission

In information theory, there is a well-known Shannon–Hartley theorem that states that the maximum rate at which information can be transmitted over a communication channel (C) of a specified bandwidth (B) in the presence of noise is [1]

$$C = \text{B}\log_2\left(1 + \frac{S}{N}\right) \tag{8.1}$$

where S is the average power of the received signal over the bandwidth and N is the average power of the noise over the bandwidth. S/N is the signal-to-noise ratio (SNR) of the communication signal to the noise at the receiver.

Assuming that the signal distortions are Gaussian-distributed, which is a reasonably accurate assumption for dispersion-uncompensated high-speed transmission links [2,3],

Optical Communications in the 5G Era
DOI: https://doi.org/10.1016/B978-0-12-821627-9.00014-0

we can define an effective SNR as the ratio between the signal power and the sum of the linear noise resulting from ASE and the nonlinear noise, given by [4,5]

$$\text{SNR} = \frac{P}{\sigma_L^2 + \sigma_{NL}^2} = \frac{1}{aP^{-1} + bP^2} \tag{8.2}$$

where P, δ_L^2, and δ_{NL}^2 are respectively the signal power, linear noise power, and nonlinear noise power, and a and b are parameters that depend on link conditions. The SNR is maximized to

$$\text{SNR}_{\text{max}} = a^{-2/3} b^{-1/3} / (2^{1/3} + 2^{-2/3}) \tag{8.3}$$

at the optimum power $P_{\text{opt}} = \left(\frac{a}{2b}\right)^{1/3}$. Evidently, the SNR penalty because of fiber nonlinearity (as compared to the case without fiber nonlinearity) at the optimal signal launch power is $-10\log(2/3) \approx 1.76$ dB. This is in a good agreement with that obtained through numerical simulations for dispersive coherent transmission [2,3]. We have the following key observations:

1. Reducing the linear noise power coefficient a down by 3 dB causes P_{opt} to decrease by 1 dB and SNR_{max} to increase by 2 dB;
2. Reducing the nonlinear noise power coefficient b down by 3 dB causes P_{opt} to increase by 1 dB and SNR_{max} to increase by 1 dB;
3. At the optimal power P_{opt}, the linear and nonlinear noise powers are 2/3 and 1/3 of the total noise power, respectively.

Assuming that the transmission link consists of uniform fiber spans with the same physical characteristics and seamless multiplexing of signals with the same modulation format, the maximum effective SNR can be further expressed as [4]

$$\text{SNR}_{\text{max}} = \frac{(8\pi\alpha|\beta_2|)^{1/3}}{3\left[3n_0^2\gamma^2 N_s h_e \ln(B/B_0)\right]^{1/3}} \tag{8.4}$$

where α, β_2, and γ are the fiber coefficients for loss, dispersion, and nonlinearity, respectively, N_s is the number of fiber spans of the optical link, n_0 is the power spectral density of the optical amplified spontaneous (ASE) noise, which is proportional to $N_s \cdot 10^{\alpha L_s}$ (where L_s is the length of each fiber span), B is the bandwidth of the wavelength-division multiplexing (WDM) signals, B_0 is related to the chromatic dispersion (CD)-induced walk-off bandwidth, and h_e is the nonlinear multispan noise enhancement factor. The maximum SNR is obtained at the optimum signal power spectral density

$$I_{\text{max}} = \left(\frac{8n_0\pi\alpha|\beta_2|}{3\gamma^2 N_s h_e \ln(B/B_0)}\right)^{1/3} \tag{8.5}$$

For dispersion-uncompensated optical transmission (DUMT) or dispersion compensation module (DCM)-free transmission, which is a popular approach enabled by the capability of electronic dispersion compensation in digital coherent receivers (DCR)s, $h_e \approx 1$, we have

$$\mathrm{SNR}_{\max,\mathrm{DUMT}} \propto \frac{\left[\alpha|\beta_2|/\ln(B/B_0)\right]^{1/3}}{N_s 10^{2\alpha L_s/3}\gamma^{2/3}} \tag{8.6}$$

and

$$I_{\max,\mathrm{DUMT}} \propto 10^{\alpha L_s/3}\left[\alpha|\beta_2|/\ln(B/B_0)\right]^{1/3}\gamma^{-2/3} \tag{8.7}$$

The maximum signal qualify factor (Q^2) of a signal with n-state quadrature-amplitude modulation (n-QAM) format has the following approximate dependence on link parameters

$$Q^2_{\max,\mathrm{DUMT}} \propto \frac{\mathrm{SNR}_{\max.\mathrm{DUMT}}}{n} \propto \frac{\left[\alpha|\beta_2|/\ln(B/B_0)\right]^{1/3}}{nN_s 10^{2\alpha L_s/3}\gamma^{2/3}} \tag{8.8}$$

The maximum achievable spectral efficiency (SE) obtained by using polarization-division multiplexing (PDM) n-QAM for the link is achieved by setting the maximum signal quality factor equal to the quality factor corresponding to the underlying FEC threshold, Q^2_{FEC},

$$\mathrm{SE}_{\max,\mathrm{DUMT}} = 2\log_2(n) = 2\log_2\left(\frac{s\left[\alpha|\beta_2|/\ln(B/B_0)\right]^{1/3}}{N_s 10^{2\alpha L_s/3}\gamma^{2/3} Q^2_{\mathrm{FEC}}}\right) \tag{8.9}$$

where the first number 2 is for PDM, and s is a scaling factor. By setting the maximum signal qualify factor $Q^2_{\max}$ to the FEC threshold Q^2_{FEC}, we can express the achievable transmission distance ($L_{\mathrm{Achievable}}$) for an n-QAM signal in given fiber link as

$$L_{\mathrm{Achievable}} = N_s L_s = \frac{s(\alpha|\beta_2|)^{1/3} L_s}{n 10^{2\alpha L_s/3}\gamma^{2/3} Q^2_{\mathrm{FEC}}} \tag{8.10}$$

The above analytical results suggest the following key transmission characteristics in DUMT:

1. To increase spectral efficiency by 2 bit/second/Hz, one could increase the dispersion coefficient ($|\beta_2|$) by a factor of 8, reduce the nonlinear coefficient (γ) by a factor of $2^{1.5}$ (or 2.83), reduce the loss per span ($10^{\alpha L_s}$) by a factor of $\sim 2^{1.5}$, or reduce the number of fiber spans (N_S) by a factor of 2. These analytical findings are in reasonable agreement with those obtained through numerical simulation.

2. The optimum signal power spectral density is dependent on fiber parameters but is independent of the transmission distance and the modulation format [2].
3. The optimum signal quality factor Q^2 scales inversely proportional to the transmission distance. The physical reason behind this is that in DUMT, the nonlinear noises (or distortions) generated by the fiber spans are de-correlated, so the accumulation of the nonlinear noise during transmission scales the same way as the linear noise resulting from the ASE.
4. Reducing the FEC threshold (Q^2_{FEC}) by 3 dB doubles the transmission distance, demonstrating the importance of using powerful FEC codes with higher coding gains, for example, soft-decision FEC (SD-FEC) codes, to enable longer transmission distance.
5. For a given fiber link, the achievable transmission distance is inversely proportional to the constellation size (n). For example, if a PDM-QPSK signal can reach 4000 km in a dispersion-unmanaged fiber link, then the transmission distances of PDM-16QAM and PDM-64QAM would be reduced to about 1000 and 250 km, respectively.

The above findings from the analytical study are in a reasonably good agreement with the results obtained by experiments and numerical simulations for DUMT of superchannels based on optical orthogonal frequency division multiplexing (O-OFDM) and Nyquist-WDM [2]. It is useful to find the SE or modulation format that provides the minimum cost per bit (CPB) of a transmission system of a given reach. When the system cost is dominated by the optical transponders or the optical-electrical-optical (O-E-O) conversion, the minimum CPB is achieved when the SE-distance product (SEDP) of the transmission system is maximized. The effective SEDP of a maximum system length L_{Link} using PDM-n-QAM modulation can be expressed as [4]

$$\text{SEDP} = 2\log_2(n)\cdot\min(L_{\text{Link}}, L_{\text{Achievable}}) \tag{8.11}$$

Fig. 8.1 shows the calculated SEDP as a function of SE in systems using seamless WDM based on PDM-n-QAM modulation and 100-km ultra-large-effective-area fiber (ULAF) spans [4,5]. The calculation is based on Eqs. (8.10) and (8.11), and the scaling factor s in Eq. (8.10) was obtained using the results reported in Ref. [6]. The FEC bit error ratio (BER) threshold and the overall spectral overhead (for FEC and multiplexing) are assumed to be 2×10^{-2} and 29%, respectively [6]. The key implications from the results are as follows.

- There is an optimum SE, SE_{opt}, at which the system SEDP is maximized (for lowest CPB).
- The SE_{opt} depends on the system size L_{link}.
- For long-haul systems where $L_{\text{link}} > 1000$ km, the SE_{opt} is smaller than 10.

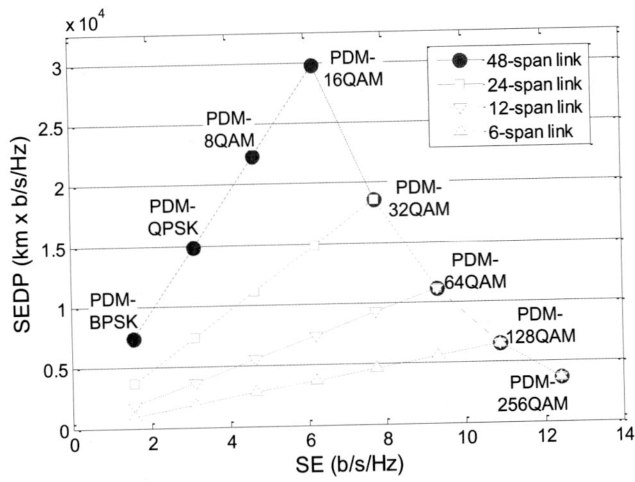

Figure 8.1 Calculated SEDP as a function of SE in systems using seamless WDM and 100-km ULAF spans. The FEC BER threshold and the overall spectral overhead (for FEC and multiplexing) are assumed to be 2×10^{-2} and 29%, respectively. (After [4]).

- When the implementation penalty of higher-level QAM and practical system margin are taken into consideration, the optimum system SE is expected to be pushed downward, for example, to be between 4 and 8 b/second/Hz in long-haul systems.

As can be seen later in this chapter, commercial high-capacity long-haul transmission systems demonstrated so far indeed follow the above predications. The above analysis is based on several assumptions, such as homogenous fiber spans and negligible signal-to-noise nonlinear interaction. More general and comprehensive studies on the ultimate capacities of fiber transmission systems can be found in Refs. [7–12].

8.3 Optical modulation and detection technologies

Generally, there are three basic physical attributes of a lightwave that can be modulated to carry information, its amplitude, phase, and polarization. As shown in Fig. 8.2, there are three popular modulation and detection schemes, intensity-modulation and direct-detection (IM/DD), differential phase keying (DSPK), and coherent optical modulation and detection. For IM/DD, the most common optical modulation format is on-off-keying (OOK), in which the light intensity is turned on and off to represent the "1" and "0" states of a digital signal. OOK is widely used in early optical fiber communication systems because of its simple transmitter and receiver configuration; however, its receiver sensitivity and SE are fairly limited as compared to most other modulation formats.

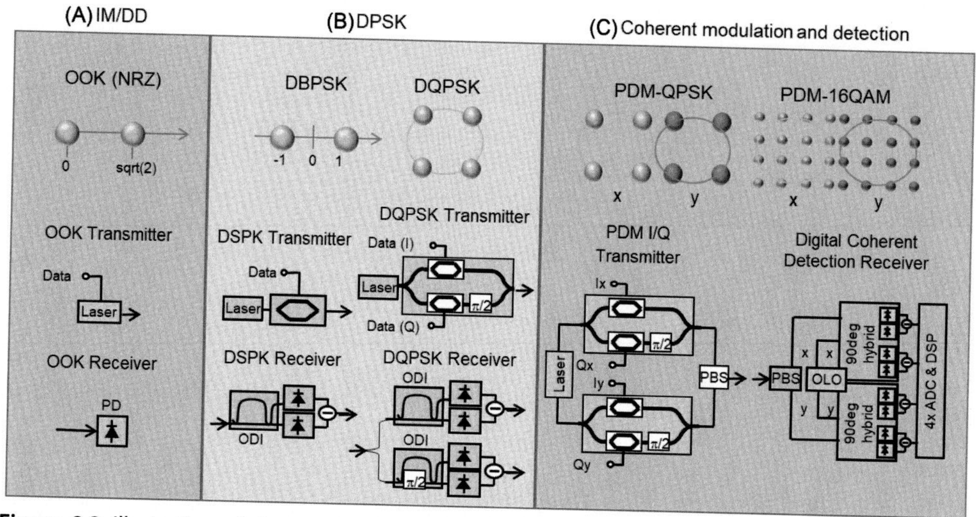

Figure 8.2 Illustration of three commonly used modulation and detection schemes, (A) intensity-modulation and direct-detection (IM/DD), (B) differential phase-shift keying (DPSK), and (C) coherent optical modulation and digital coherent detection. (After [5]).

For DSPK, the commonly used formats are differential binary-phase-shift keying (DBPSK) and differential quadrature-phase-shift keying (DQPSK), which utilizes the phase difference between adjacent symbols to carry information bits. For DBPSK detection, an optical delay interferometer (ODI) is used to compare the signal with a delayed copy of itself and convert the phase modulation into intensity-modulation that can be directly detected by square-law detectors. This detection scheme can be extended to DQPSK that carries 2 bits per symbol by using two ODIs with a 90-degree phase offset. DBPSK and DQPSK had been well used in 40-Gb/second-per-wavelength transmission systems in the mid-2000s [13—15].

For coherent optical modulation, both the in-phase (I) and quadrature (Q) components of each of the two orthogonal polarizations of an optical carrier can be modulated. With high-speed digital-to-analog converters (DACs), flexible n-QAM modulation can be realized with the same hardware. With PDM of two orthogonal polarization components of the same wavelength, doubled SE can be achieved. For coherent optical detection, the coherent optical signal to be received mixes with a reference optical carrier that is provided by an optical local oscillator (OLO). Polarization-diversity and phase diversity, both, are applied to allow for the retrieval of the four orthogonal components of the received optical signal, that is, I and Q components of two orthogonal polarization states. High-speed analog-to-digital converters (ADCs) are used to sample the four retrieved orthogonal components for a followup DSP that recovers the original information bits carried by the coherent optical signal. This type of receiver, that is, DCR, and the DSP used in recovering the optical signal are sometimes referred to as oDSP, where "o" indicates the optical nature of the signal being processed. As the full electromagnetic field of the optical signal is reconstructed by the DCR, all the linear effects experienced by the optical signal during fiber transmission, such as CD, polarization rotation, and polarization-mode dispersion (PMD), can be fully compensated by using linear digital filters [16–26]. This is a very powerful technology that has enabled today's high-speed long-haul transmission at a per-wavelength data rate of 100 Gb/second and above.

With access to the full optical field information at both transmitter and receiver, modern coherent optical communications have benefited from many powerful DSP techniques originally developed for wireless communication. However, data rates in optical communication are usually several orders of magnitude higher than in wireless communication, necessitating efficient DSP techniques tailored specifically for optical communications, indeed reflecting a statement by Haykin [27], "*Signal processing is at its best when it successfully combines the unique ability of mathematics to generalize with both the insight and prior information gained from the underlying physics of the problem at hand.*" For optical communications in single-mode fiber, channel equalization can be efficiently realized by separating slowly varying transmission effects such as CD from rapidly varying effects such as polarization rotations and PMD.

Fig. 8.3 shows the block diagrams of common DSP modules implemented in a polarization-diversity DCR. First, compensation of CD is conducted by using a linear filter that is relatively static, as the CD of a given fiber link is virtually fixed. Second, clock recovery is performed to extract the signal modulation symbol rate. Third, PMD compensation and polarization-demultiplexing are realized by a dynamic 2 × 2 linear filter, as polarization state of the received signal may vary very quickly due to vibrations and temperature changes occurred along the fiber link. This dynamic filter

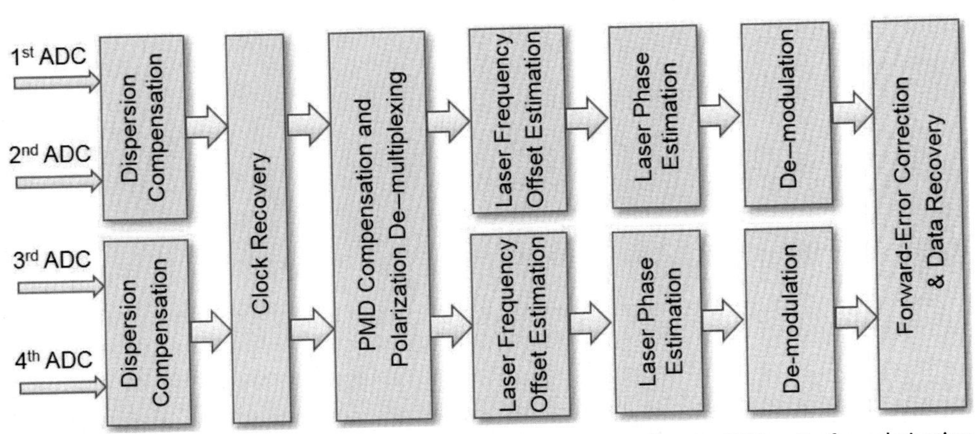

Figure 8.3 Block diagram of common DSP modules implemented in the DSP unit of a polarization-diversity digital coherent optical receiver. (After [5]).

is usually based on the constant modulus algorithm (CMA) for typical QPSK and n-QAM modulation formats. Fourth, laser frequency offset estimation (FOE) is performed for each polarization component of the optical signal in order to digitally compensate for the frequency offset between the center frequency of the received signal and the OLO. Fifth, laser phase estimation (PE) is performed to align the phase between the optical carrier of the signal and the OLO so that the signal constellation can be recovered. The ability to digitally perform the FOE and PE allows the OLO to be free running, without the need for a sophisticated optical phase-locked loop (OPLL), thereby enabling the practical implementation of coherent detection. Sixth, demodulation is conducted to recover the information bits carried by the optical signal. Finally, FEC is applied to increase the fidelity of the recovered information bits. As the digitally recovered signal constellation has a resolution of multiple bits in each quadrature, SD-FEC decoding can be applied to achieve better error correction performance than previously used hard-decision (HD) FEC decoding, showing another major benefit of digital coherent detection.

In the following, we will describe the basic DSP algorithms used in digital coherent detection in more depth.

8.3.1 Dispersion compensation

Fiber CD can be compensated via time-domain equalization (TDE) or frequency-domain equalization (FDE). For compensation of a large dispersion effect that needs a large number of equalizer taps, FDE has been demonstrated to offer lower computational complexity than TDE [28]. The main FDE technique for dispersion compensation is the overlap FDE (OFDE) technique based on fast Fourier transform (FFT) and inverse fast Fourier transform (IFFT) [29]. In the OFDE technique, an N-point FFT is first performed on a block of N received signal samples. The FDE then multiplies the received signal block in the frequency domain by a transfer function that reverses the dispersion effect as follows

$$g(f) = \exp\left(-j\frac{\pi cLDf^2}{f_c^2} \right) \tag{8.12}$$

where j is the imaginary unit, f is the frequency with respect to the center frequency of the optical signal, c is the speed of light, L is the transmission distance, D is the dispersion coefficient of the optical fiber, and f_c is the center frequency of the optical signal. Then, the frequency domain block is transformed back to the time domain by an N-point IFFT. The first M samples and the last M samples in the block are removed, and the remaining ($N-2M$) samples are extracted. Zero interblock interference is achieved by setting $2M$ to be larger than the dispersion-induced pulse broadening. The above process is repeated for the next block of N received signal samples that are shifted forward by $(N-2M)$. There is an *overlap* of $2M$ samples between two adjacent processing blocks, which is the reason behind the name "*overlap*-FDE" [29]. Thanks to the use of FFT and IFFT, OFDE is computational efficient and can be readily implemented in high-speed digital coherent receivers.

8.3.2 PMD compensation and polarization-demultiplexing via 2 × 2 MIMO

In a polarization-division multiplexed transmission system, two signals x_1 and x_2 are launched into two orthogonal polarization states that are linearly oriented. To compensate for PMD and perform polarization-demultiplexing, a 2×2 MIMO equalizer can be used [24] to recover the two signals as z_1 and z_2:

$$\begin{pmatrix} z_1(k) \\ z_2(k) \end{pmatrix} = \sum_{l=-B/2}^{B/2} \begin{pmatrix} h_{11}(l) & h_{12}(l) \\ h_{21}(l) & h_{22}(l) \end{pmatrix} \begin{pmatrix} y_1(k+l) \\ y_2(k+l) \end{pmatrix} \tag{8.13}$$

where y_1 and y_2 are two orthogonal polarization components received by the coherent receiver, k is the sampling time index, and h_{ij} is the coefficients of the 2×2 MIMO finite impulse response (FIR) filter H. To effectively compensate for PMD, the FIR filter length $B+1$ of the equalization filters $h_{ij}(l)$ should be longer than the maximum differential group delay (DGD) divided by the sampling time T_S. More desirable would be a method that can adaptively estimate the coefficients $h_{ij}(l)$ without the need of training sequences. Such an estimation algorithm is called a blind estimator. One commonly used algorithm for this purpose is the so-called Goddard algorithm or CMA. Originally proposed in [30] and independently in [31], this algorithm, in its original form, provides a criterion to reduce intersymbol interference for channels with impulse responses longer than the sampling time. It is designed to operate for modulation formats having a constant modulus, that is, a constant amplitude, as phase-shift keyed modulation formats. It has been shown though that the same algorithm is even applicable for some nonconstant modulus modulation formats, for example, QAM. In [32], this algorithm was proposed for source separation in a MIMO system. A polarization multiplexed transmission system is essentially a MIMO system as well, or more precisely a two-in, two-out system. A comprehensive review of CMA can be found, for example, in [33,34].

CMA algorithm tries to penalize deviations from a constant modulus of the equalized signals and adapt the equalization filter coefficients to minimize a cost function

$$\begin{aligned} \varepsilon_1 &= E\left\{\left(1-|y_1|^2\right)^2\right\} \\ \varepsilon_2 &= E\left\{\left(1-|y_2|^2\right)^2\right\} \end{aligned} \tag{8.14}$$

$E\{\cdot\}$ denotes the expectation operator. Using an iterative gradient-type optimization algorithm, coefficients $h_{ij}(l)$ can be found that minimize Eq. (8.14) for instance as follows:

$$\begin{aligned} h_{11}(k+1,l) &= h_{11}(k,l) + \mu\, \delta\varepsilon_1\, z_1(k) y_1^*(k+l) \\ h_{12}(k+1,l) &= h_{12}(k,l) + \mu\, \delta\varepsilon_1\, z_1(k) y_2^*(k+l) \\ h_{21}(k+1,l) &= h_{21}(k,l) + \mu\, \delta\varepsilon_2\, z_2(k) y_1^*(k+l) \\ h_{22}(k+1,l) &= h_{22}(k,l) + \mu\, \delta\varepsilon_2\, z_2(k) y_2^*(k+l) \end{aligned} \tag{8.15}$$

with constant μ being the step size of the algorithm. The parameters $\delta\varepsilon_1$ and $\delta\varepsilon_2$ are derivatives of the cost functions expressed in Eq. (8.14), which can be estimated by replacing the expectation values by their respective instantaneous values:

$$\begin{aligned} \delta\varepsilon_1 &= 2\left(1 - |y_1|^2\right) \\ \delta\varepsilon_2 &= 2\left(1 - |y_2|^2\right) \end{aligned} \tag{8.16}$$

8.3.3 Frequency estimation

In digital coherent (or intradyne) optical receivers, there is no frequency locking between the OLO and the received optical signal. Therefore frequency offset compensation has to be incorporated into the receiver DSP. Frequency offset compensation can be performed by estimating the phase shift $\Delta\varphi = 2\pi\Delta\nu T_s$ between two consecutive samples, a_k and a_{k+1}, that is caused by a frequency offset $\Delta\nu$ between the OLO and the transmit laser frequency, with T_s sampling time [24]. The operation can be described as follows. First, the received symbol is multiplied with the complex conjugate of the previous symbol. This results in a complex number whose phase is equal to the difference in phase of the two symbols. Then, any information that is encoded in the signal phase has to be removed. This can easily be performed for M-ary PSK signals by taking the m-th power of the complex symbol, for example, for QPSK, the fourth power has to be taken. Taking the m-th power of a complex number is equivalent to taking the m-th power of its magnitude and multiplying its phase by m. With proper phase wrapping, for a PSK-modulated signal, this operation will yield in a constellation where all constellation points lie on top of each other. The result is then summed up over a large number of samples. This summation essentially performs a *long-range* averaging operation in order to find the mean frequency difference between the OLO and the signal in the presence of laser noise. The phase then has to be divided by four to correct for the earlier power-of-four operation. The result is an estimate for the phase difference $\Delta\varphi$ between consecutive samples. To correct for the frequency offset, an accumulated phase offset $\varphi_k = k\Delta\varphi$ has to be subtracted from the phase of each symbol a_k, with k being a running symbol index, to obtain a corrected symbol [24].

8.3.4 Phase estimation

The primary task of the phase estimation is to establish a phase reference to decode the phase-encoded data of a PSK or DPSK modulated signal. The phase difference between the local oscillator laser and the transmit laser changes slowly with respect to the signal baud rate due to the phase noises of the two lasers, as well as optical noise from the ASE noise, etc. We thus need to establish a phase reference with high immunity to noise. This can be done by averaging the estimated phases over a number of consecutive symbols. One method is based on a feed-forward scheme in which the phase modulation of an M-ary PSK signal is removed by raising the received signal field to its m-th power [35]. Note that differential coding is needed to solve the phase ambiguity issue. A commonly used phase estimator is based on the Viterbi and Viterbi algorithm [36]. The operation of the phase estimator is as follows. In the first step, any coded information of the PSK-modulated signal has to be removed by the same m-th power operation as discussed for the frequency estimator. After phase modulation removal, the symbols can be passed through a digital low-pass filter, as discussed before, to establish a phase reference. In case of the Viterbi and Viterbi algorithm, filtering is performed by summing the complex symbols over a short sequence of symbols. The sequence length is known as the block length. The appropriate block length is determined by the amount of phase noise and nonlinear phase noise in the received signal and the laser linewidth.

Using the differential coding causes the double error issue and leads to performance degradation. Furthermore, differential coding is not conveniently compatible with SD-FEC. To avoid the use of the differential coding, pilots can be used to obtain the phase reference without ambiguity. Pilot-assisted PE can be based on schemes such as the decision-aided maximum-likelihood PE [37]. One issue with pilot-assisted PE is the presence of cycle slips. Techniques to detect and remove these cycle slips have been developed to make pilot-assisted PE reliable in practical systems [38–42].

8.4 Advanced techniques for high-speed transmission

8.4.1 Multicarrier modulation or digital subcarrier modulation

With the advances in high-speed DAC, ADC, and DSP, as well as high-speed optical modulators and detectors, coherent transceivers are operating at higher and higher modulation symbol rate. Increased speed in coherent transceivers offers benefits such as better meeting the capacity demand, reduced cost per bit by transmitting/receiving more bits with the set of same hardware, and reduced number of wavelengths to manage. 100-Gbaud-class coherent transceivers have recently been demonstrated [43–46]. On the other hand, there are three technical issues with high-baud transmission, namely:

- Large equalization-enhanced phase noise (EEPN)
- Large dispersion compensation complexity
- High nonlinear distortion

To address these issues, multicarrier modulation or digital subcarrier multiplexing (DSCM) in which a high-speed signal consists of multiple lower-speed signals that are multiplexed in the frequency domain, can be applied. In the following, we will describe these three technical issues and show how multicarrier modulation can help resolve these issues.

In 2008, Shieh and Ho found that unlike optical dispersion compensator, the electronic equalizer for dispersion compensation enhances the impairments from the laser phase noise. This EEPN imposes a tighter constraint on the phase noise of the OLO used in digital coherent receiver for transmission systems with high symbol rate and large electronically compensated CD [47]. The EEPN-induced penalty can be expressed as [47]

$$\Delta P_{\mathrm{EEPN}}(dB) \cong 4.343\pi c\left(2f_0^2\right)^{-1} DBf_{3\mathrm{dB}} \cdot (1 + \mathrm{SNR}_0) \qquad (8.17)$$

where c is the speed of light, f_0 is the laser center frequency, D is the CD to be electronically compensated at the receiver, B is the signal symbol rate, $f_{3\mathrm{dB}}$ is the 3-dB linewidth of the OLO, and SNR_0 is the effective SNR with the consideration of the EEPN. Evidently, the EEPN-induced penalty scales linearly with the signal symbol rate, the dispersion to be compensated, and the linewidth of the OLO. Thus for high-haul transmission over long dispersion-uncompensated fiber link, the EEPN-induced penalty may become too large for a typical OLO with 100-kHz linewidth.

Fig. 8.4 shows the calculated EEPN penalty for PDM-16QAM transmission over a large-effective-area fiber with a dispersion coefficient of 21 ps/nm/km with the OLO being a commonly used 100-kHz external cavity laser (ECL) and the SNR_0 being 13 dB (corresponding to a BER of $\sim 2 \times 10^{-2}$). With 96-Gbaud modulation, the EEPN-induced penalty can be larger than 0.2 dB after 1000 km transmission, thus limiting the performance of PDM-16QAM for ultra-long-haul transmission beyond 1000 km. With 64-Gbaud and 12-Gbaud modulations, the EEPN-limited reaches are increased by 1.5 and 8 times, respectively. When multicarrier modulation (or DSCM) is applied, each band (or subcarrier) is modulated as a low symbol rate and can thus reduce the EEPN-induced penalty. For example, a 96-Gbaud Nyquist-filtered single-carrier signal with a given roll-off factor may be replaced by a multicarrier signal that multiplexes eight 12-Gbaud Nyquist-filtered single-carrier signals with the same roll-off factor in the frequency domain. At the receiver, electronic dispersion compensation is performed on these lower-speed signals individually, thereby reducing the EEPN.

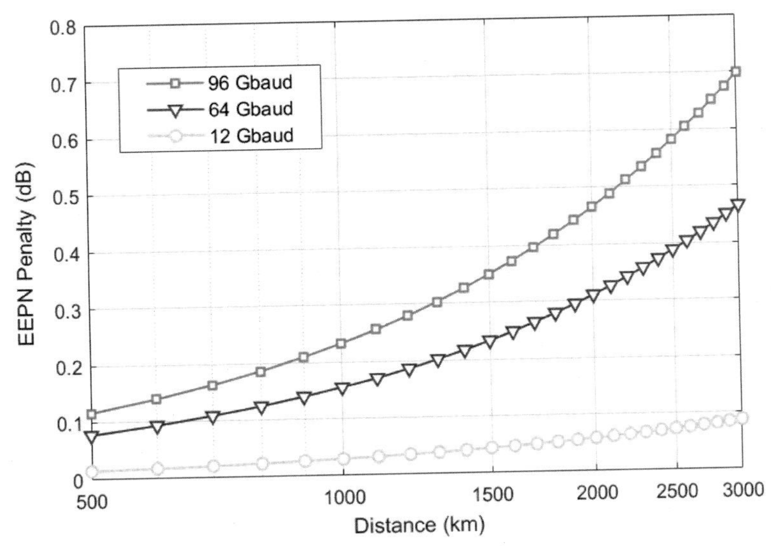

Figure 8.4 Calculated EEPN penalty for PDM-16QAM transmission over a large-effective-area fiber with a dispersion coefficient of 21 ps/nm/km with the OLO being a 100-kHz laser and the SNR_0 being 13 dB (corresponding to a BER of $\sim 2 \times 10^{-2}$).

With the reduced EEPN, the PDM-16QAM format can be used for ultra-long-haul undersea transmissions of over 6000 km.

8.4.2 Multiband electronic dispersion compensation

Digital coherent detection has enabled the compensation of linear optical transmission impairments such as CD. Efficient electronic dispersion compensation (EDC) has been demonstrated in the frequency domain through the use of the overlap-and-add scheme [29]. With the increase of modulation rates, the DSP complexity of EDC increases as the size of the CD equalizer scales *quadratically* with the signal bandwidth. Hence to reduce DSP complexity, subband CD equalizers, in which a signal is divided into multiple subbands and each subband is individually dispersion compensated, were proposed [48]. For Nyquist WDM, a filter-bank–based subband CD equalizer was demonstrated [49]. For coherent O-OFDM, overlap-and-add CD compensation can be applied prior to OFDM demultiplexing to reduce the needed guard interval (GI) and thus spectral and power overheads, but at the expense of increased DSP complexity [50]. A filter-bank–based subband CD equalizer was also demonstrated for multiband OFDM to reduce DSP complexity [51]. DFT-spread (DFTS) OFDM has recently attracted much attention owing to its lower peak-to-average-power-ratio as compared to conventional OFDM [52,53], and multiband DFTS-OFDM (MB-DFTS) equalization was introduced to realize subband CD equalization by allocating a sufficiently large GI to reduce the CD equalizer complexity [54]. Recently, a novel

MB-DFTS equalizer that combines overlap-and-add dispersion compensation with OFDM demultiplexing (so only one DFT is needed at the receiver) was demonstrated [55]. In a proof-of-concept experiment, it was shown that this MB equalizer offers ~36% reduction in the DSP complexity associated with multiplexing, equalization, and demultiplexing, as compared to the conventional single-band approach [55].

Fig. 8.5 shows the schematic of a single-band DFTS architecture with the EDC separated from the DFTS equalizer and demultiplexer (DFTS + EDC) and the accumulated fiber CD is compensated across the entire signal bandwidth. In comparison, the proposed MB-DFTS architecture is shown in Fig. 8.6. It combines an MB-EDC with the DFTS equalizer and demultiplexer (MB-DFTS-EDC). In either case, the EDC is based on the overlap-and-add scheme with an overlap length of L_{overlap}, which needs to be greater than the CD-induced channel memory length. Using the well-known radix-2 Cooley–Tukey algorithm, an m-point DFT (or IDFT) requires $m\,\log_2(m)/2$ complex multipliers [56]. For the DFTS + EDC scheme, the overall DSP complexity associated with multiplexing, equalization, and demultiplexing can be expressed in terms of the number of complex multipliers per input symbol (per polarization) as

$$\text{DSP}_{\text{SB-DFTS+EDC}} \approx \log_2(N) + L_1\log_2(L_1)/N + L_2\log_2(L_2)/N \tag{8.18}$$

where OSR is the oversampling ratio for the DFT spreading and $L_2 = N\cdot\text{OSR} + L_{\text{Overlap}}$ is the size of the DFT/IFFT used for the EDC. For the overlap-and-add EDC to be DSP-efficient, it is desirable to have $L_2 \geq 4L_{\text{Overlap}}$ [29]. For the MB-DFTS-EDC scheme, the overall DSP complexity using K subbands can be expressed in terms of the number of complex multipliers per input symbol (per polarization) as

$$\text{DSP}_{\text{MB-DFTS-EDC}} \approx \log_2\left(\frac{N}{K^2}\right)\Big/2 + L_1\Big/\left(\frac{N}{K}\right)\log_2(L_1)/2$$
$$+ L_2\Big/\left(\frac{N}{K}\right)\log_2(L_2)/2 + \left(\frac{N}{K^2} + L_o\right)\Big/\left(\frac{N}{K^2}\right)\log_2\left(\frac{N}{K^2}\right)\Big/2 \tag{8.19}$$

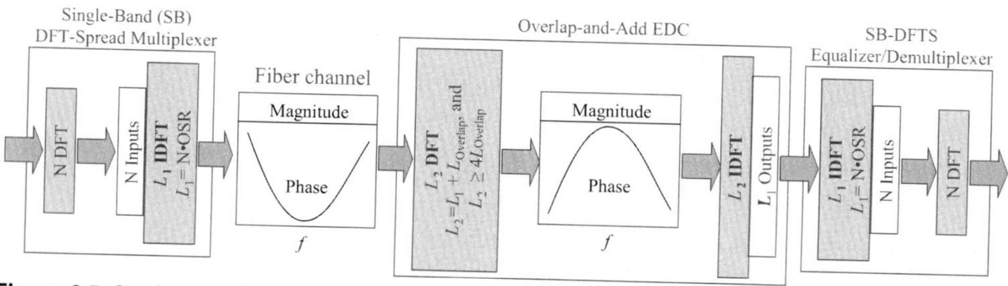

Figure 8.5 Single-band DFTS architecture with EDC separated from the DFTS equalizer and demultiplexer (DFTS + EDC). (After [55]).

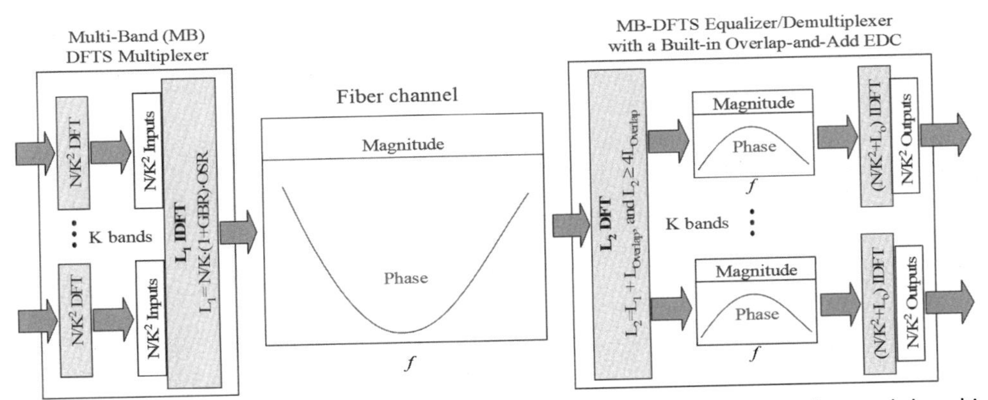

Figure 8.6 Multiband DFTS architecture with EDC combined with the DFTS equalizer and demultiplexer (MB-DFTS-EDC). (After [55]).

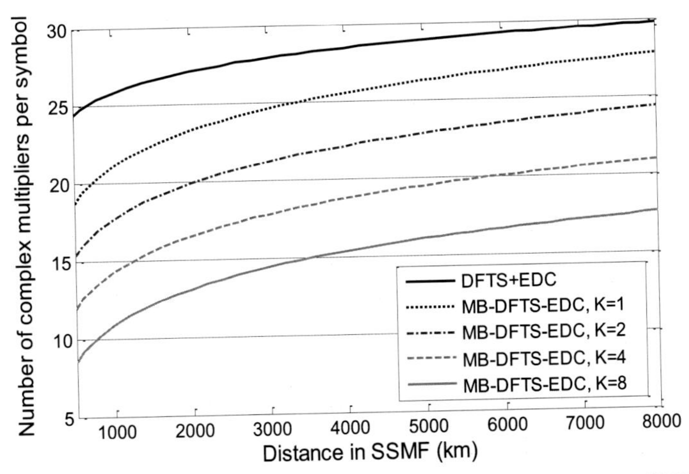

Figure 8.7 Calculated DSP complexity as a function of transmission distance in SSMF. (After [55]).

where $L_1 = N/K(1 + \text{GBR})\text{OSR}$, $L_2 = L_1 + L_{\text{Overlap}}$, $L_o = L_{\text{Overlap}}/[K(1 + \text{GBR})\text{OSR}]$, and GBR is the ratio between the guard-band and subband bandwidth. Note that the architectures shown in Fig. 8.6 can be straightforwardly extended to the case of PDM transmission by performing the 2×2 MIMO equalization in the receiver DSP. Note also that synchronization and channel estimation are not discussed here; these can be conducted in a parallel path at a much lower speed with a small additional DSP complexity. Fig. 8.7 shows the calculated DSP complexity as a function of transmission distance in a standard single-mode fiber (SSMF) with a CD coefficient of 17 ps/nm/km for the two schemes. In the calculation, we assumed $\text{OSR} = 5/4$, $\text{GBR} = 0.1$ (for quasiNyquist-WDM-like transmission), $L_2 = 4L_{\text{Overlap}}$, and an input

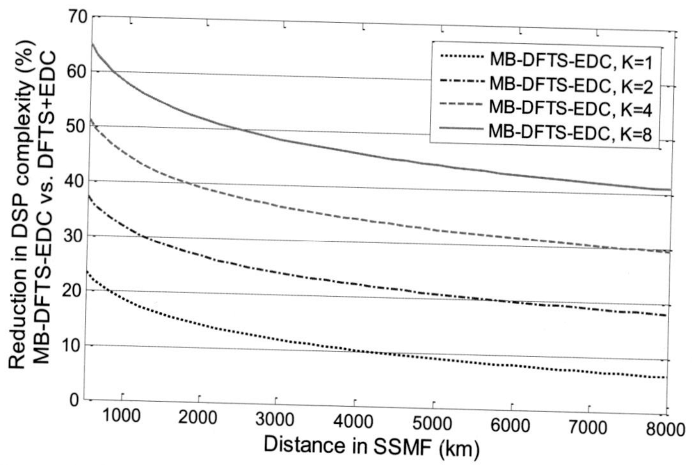

Figure 8.8 DSP complexity reduction obtained by MB-DFTS-EDC, as compared to DFTS + EDC, as a function of transmission distance. Signal symbol rate: 32.5 Gbaud. (After [55]).

total symbol rate of 32.5 Gbaud (suitable for 100-Gb/second transmission using PDM-QPSK and 20%-overhead FEC). For the DFTS + EDC case, $N=96$ was assumed. Evidently, the proposed MB-DFTS-EDC has less DSP complexity than the DFTS + EDC approach. The larger the number of subbands (K), the lower the DSP complexity of MB-DFTS-EDC. Thanks to the shared use of the L_2 DFT, even the 1-band MB-DFTS-EDC ($K=1$) offers a lower complexity compared to DFTS + EDC. Fig. 8.8 shows the reduction of DSP complexity offered by the MB-DFTS-EDC scheme as compared to the DFTS + EDC scheme. For long-haul dispersion-uncompensated SSMF transmission with a distance ranging from 2000 to 8000 km, the DSP complexity reduction using a four-band MB-DFTS-EDC ($K=4$) is between ~40% and ~30%.

Fig. 8.9 depicts the schematic of the experimental setup. Various DFTS signal waveforms defined in the previous section were generated offline and loaded into a field programmable gate array–based logic circuit feeding two 30-GS/second DACs driving an I/Q modulator. After PDM emulation, the signal was launched into a recirculating fiber loop consisting of four 80-km SSMF spans, each amplified by an EDFA. After 6400-km transmission, the signal was received by a digital coherent receiver with offline DSP. Fig. 8.10 shows the measured penalty in the Q^2 factor, derived from the BER, as a function of L_{Overlap} for DFTS + EDC and a four-band MB-DFTS-EDC. As expected, the smallest L_{Overlap} needed to ensure small equalization penalties for four-band MB-DFTS-EDC is only ~one-fourth of that for DFTS + EDC. Using the design rules shown previously, the four-band MB-DFTS-EDC can be implemented with a DSP complexity reduction of ~36% as compared to conventional DFTS + EDC.

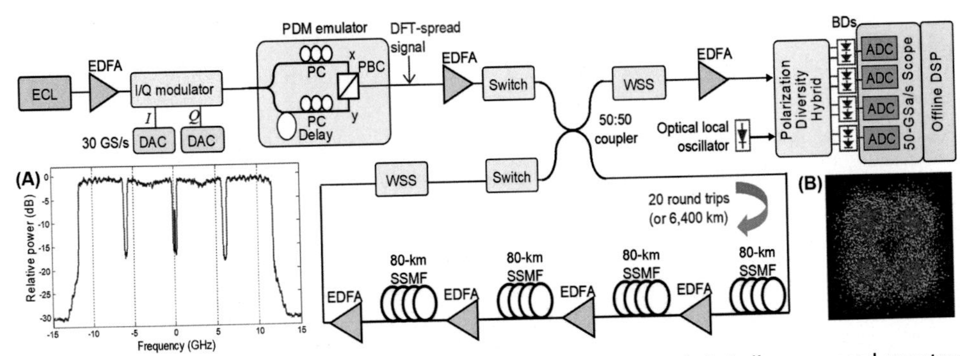

Figure 8.9 Schematic of the experimental setup. Insets: (A) a typical digitally recovered spectrum of a 22-Gbuad four-band DFTS signal; (B) equalized signal constellation after 6400-km transmission of SSMF. Signal launch power: −2 dBm. *ECL*, external cavity laser at 1550 nm. (After [55]).

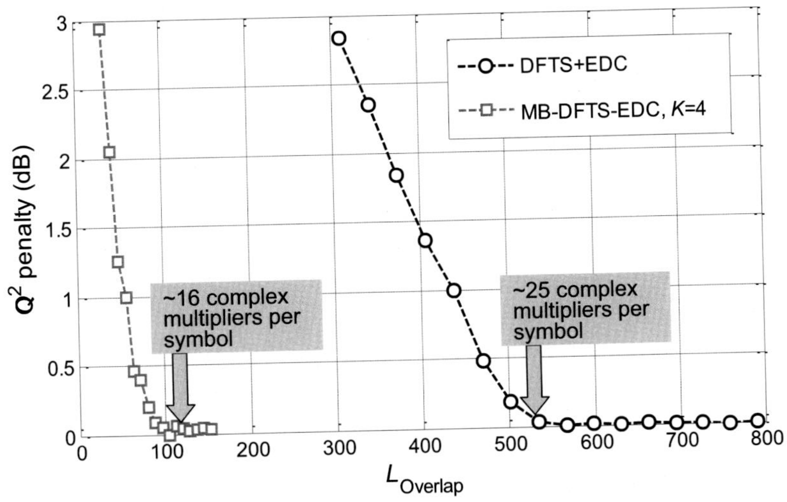

Figure 8.10 Measured Q^2 factor penalty as a function of the overlap length used in the EDC after 6400-km SSMF transmission. (After [55]).

Through this proof-of-concept experiment, we have shown that multiband EDC offers ~36% reduction in DSP complexity as compared to single-band EDC in the context of DFTS modulation. For high-baud-rate single-carrier modulation, the benefit of multiband EDC has also been well documented. In a more recent demonstration showing 800-Gb/second-per-wavelength transmission over 1000 km [57], the EDC complexity for a 95.6-Gbaud single-carrier signal is compared with that for a DSCM signal consisting of eight 11.95-Gbaud single-carrier signals. It was found that the DSCM approach would bring a complexity saving of 0.68 times for eight subcarriers versus single carrier, which leads to a noticeable power saving [57].

8.4.3 Nonlinearity-tolerant transmission and nonlinearity compensation

The theoretical capacity upper bound of a linear communication link is set by the Shannon limit. As optical fiber transmission capacity continues to increase, so does the needed optical signal power to ensure sufficient SNR, which results in signal distortions due to fiber Kerr nonlinearity. Fiber Kerr nonlinearity thus imposes an upper bound on the maximum effective SNR of a link and limits the achievable transmission performance. There has been extensive effort toward attempting to break the Kerr nonlinear limit through nonlinear compensation [58–63], which has limitations such as the inability to mitigate interchannel nonlinear impairments when the E-field evolution of other wavelength-division multiplexed channels is unknown to the compensator and exhibits impractically high complexity when numerous computation steps are needed to undo the nonlinear interactions, as in the case of dispersive transmission. Mid-link optical phase conjugation (ML-PC) is a well-known technique to compensate for CD [64], nonlinearity [65], and their combined effect [66–69], by performing phase conjugation on the E-field of each optical signal near the middle of a transmission link. However, ML-PC requires the transmission link to be modified, in order to insert a phase conjugator inside the link and requires near mirror-imaged power evolutions about the phase conjugator, which significantly reduces the flexibility of dynamically routing optical networks.

Recently, a method was proposed to realize the cancellation of nonlinear signal-to-signal interactions by transmitting a pair of mutually phase-conjugated twin waves (PCTW) together through a nonlinear medium and coherently superimposing them at the receiver site [70]. Modulating a pair of PCTW on two orthogonal polarizations of a same optical carrier via QPSK, we demonstrate nonlinear distortion reductions of greater than 8.5 dB when both dispersion and linear noise, for example, ASE, are negligibly small. For dispersive nonlinear transmission, the nonlinearity cancellation additionally requires a dispersion-symmetry condition that can be satisfied by appropriately predispersing the signals, for example, through electronic dispersion precompensation (preEDC).

To theoretically explain the PCTW-enabled nonlinearity cancellation, we can start with the coupled nonlinear Schrödinger equations [71] that govern the nonlinear optical propagation of a polarization-division multiplexed (PDM) vector wave $\vec{E} = \left(E_x, E_y\right)^T$ in an optical fiber, which, under the common assumption that the nonlinear interaction length is much greater than the length scale of random polarization rotations, can be reduced to the Manakov equation as [72]

$$\left[\frac{\partial}{\partial z} + \frac{\alpha(z) - g(z)}{2} + i\frac{\beta_2(z)}{2}\frac{\partial^2}{\partial t^2}\right] E_{x,y}(z,t) = i\frac{8}{9}\gamma\left(|E_x(z,t)|^2 + |E_y(z,t)|^2\right) E_{x,y}(z,t) \tag{8.20}$$

where i is the imaginary unit, z, α, g, β_2, and γ are respectively the propagation distance, the loss coefficient, the gain coefficient, the group-velocity dispersion (or second-order dispersion) coefficient, and the fiber nonlinear Kerr coefficient along a transmission link. Extending the previously reported perturbation approach [73–76], we can express the nonlinear distortions after transmission in the frequency domain (to first order) as

$$\delta E_{x,y}(L,\omega) = i\frac{8}{9}\gamma P_0 L_{eff}\int_{-\infty}^{+\infty} d\omega_1 \int_{-\infty}^{+\infty} d\omega_2 \eta(\omega_1\omega_2)\Big[E_{x,y}(\omega+\omega_1)E_{x,y}(\omega+\omega_2)E^*_{x,y}(\omega+\omega_1+\omega_2) + E_{y,x}(\omega+\omega_1)E_{x,y}(\omega+\omega_2)E^*_{y,x}(\omega+\omega_1+\omega_2)\Big] \quad (8.21)$$

where $E_{x,y}(\omega) = \int_{-\infty}^{\infty} E_{x,y}(0,t)e^{-i\omega t}dt/\sqrt{2\pi}$, $G(z) = \int_0^z [g(z') - \alpha(z')]dz'$, $L_{eff} = \int_0^L e^{G(z)}dz$ and $C(z) = \int_0^z \beta_2(z')dz'$ are the logarithmic signal power evolution, the effective length, and the cumulative dispersion along the link, respectively, and $\eta(\omega_1\omega_2)$ is the dimensionless nonlinear transfer function defined as

$$\eta(\omega_1\omega_2) = \int_0^L \exp[G(z) - i\omega_1\omega_2 C(z)]dz/L_{eff} \quad (8.22)$$

In the above derivations, we assume that the E-fields of the PCTW after transmission are normalized in power and postcompensated to be free of residual dispersion. When a symmetric dispersion map is applied such that $C(z) = -C(L-z)$, $G(z) = G(L-z)$, and $\eta(\omega_1\omega_2)$ become real-valued. Note that a symmetric dispersion map in a dynamic optical networking environment can be readily obtained by performing preEDC at the transmitter through techniques such as software-definable digital EDC. For PCTW-based transmission, we have the following relations

$$E_y(0,t) = E_x(0,t)^*, E_y(\omega) = E_x^*(-\omega) \quad (8.23)$$

Closely inspecting Eq. (8.21) and using $\eta(\omega_1\omega_2) = \eta(\omega_1\omega_2)^*$, we then have (for PCTW-based transmission with a symmetric dispersion map)

$$\delta E_y(L,\omega) = -[\delta E_x(L,-\omega)]^*, \delta E_y(L,t) = -[\delta E_x(L,t)]^* \quad (8.24)$$

The equations reveal that the nonlinear distortions experienced by the PCTW are *anticorrelated*. It is this anticorrelation that leads to the full cancellation of nonlinear distortions and restoration of the original signal field, $E(0,t)$, upon coherent superposition of the received PCTW

$$E_x(L,t) + E_y(L,t)^* = 2E(0,t) \quad (8.25)$$

A remarkable feature is that nonlinear signal distortions that can be canceled including those resulting from the interaction between Kerr nonlinearity and dispersion. In addition, as abundantly exploited in many areas of physics and engineering, the coherent superposition halves the variance of *linear noise* resulting from ASE, further improving the overall signal quality.

To quantify the benefit of PCTW under *stringent* nonlinear transmission conditions, an experiment was conducted for high SE superchannel transmission with eight pairs of closely spaced QPSK-modulated PCTW over TrueWave reduced slope (TWRS) fiber spans. The fiber dispersion and nonlinear Kerr coefficients are 4.66 ps/nm/km and 1.79 W/km, respectively. Fig. 8.11A shows the schematic of the experimental setup. The modulation speed of the PCTW is further increased to 32 Gbaud using 64-GSamples/second DACs. The channel spacing is 37.5 GHz, which is compatible with the emerging flexible-grid WDM architecture (ITU-T G.694.1). To achieve such close channel spacing, root-raised-cosine filter with a roll-off factor of 0.1 is used to confine the optical spectra of the signals. The raw data rate of the superchannel is 512 Gb/second. After allocating 23.46% overhead for FEC, including an inner SD and outer HD code, and 2% overhead for training symbols and pilots, the net data rate becomes 406.6 Gb/second. Fig. 8.11B shows the Q^2 factor of the center channel, derived from the measured BER versus the signal launch power after 8000-km (100 × 80 km) TWRS fiber transmission. At $P_{in} = -2$ dBm, the signal Q^2 factor obtained after digital coherent superposition (DCS) of the PCTW (DCS-PCTW) is 7.9 dB, which is ~5.2 dB higher than that of conventional PDM-QPSK. The improvement in terms of the optimal Q^2 factor is ~4 dB. Note that for the PCTW transmission, preEDC with $D_{pre} = -18{,}400$ ps/nm is applied to make the dispersion map nearly symmetric. It was also found that a change of the D_{pre} value by ±2% does not cause a noticeable change in the PCTW performance.

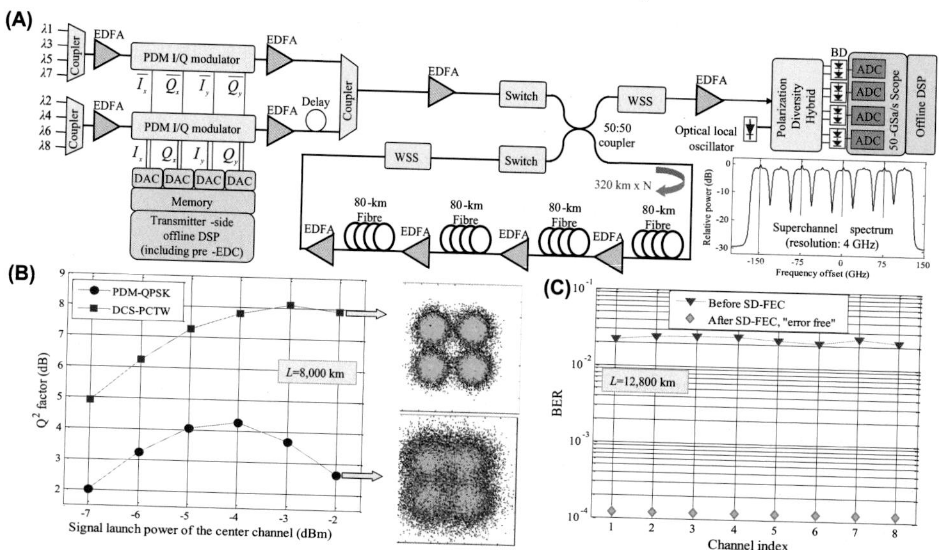

Figure 8.11 (A) Experimental setup, (B) Measured signal Q^2 factor as a function of the center channel signal launch power after 8000-km TWRS fiber transmission, and (C) Measured BER after 12,800-km TWRS transmission at $P_{in} = -3$ dBm with optimized preEDC in a superchannel transmission with improved nonlinear transmission performance brought by PCTW. (After [70]).

With the improved nonlinear transmission performance, the 406.6-Gb/second superchannel was then transmitted over a record distance of 12,800 km (160×80 km). Fig. 8.11C shows the transmission performance of all the eight channels comprising the superchannel performance after the 12,800-km TWRS transmission with $D_{pre} = -29,440$ ps/nm. The transmission penalty in terms of optical SNR is estimated to be ~ 3 dB. All the eight channels inside the superchannel have a raw BER below 2.5×10^{-2}, and error-free performance is obtained (with 10^6 bits processed per measurement) even before applying the outer HD-FEC whose threshold is $\sim 4 \times 10^{-3}$. The PCTW generated in the experiment can also be interpreted as a PDM-BPSK signal using the unitary transformation, and the performance improvement can be seen as the result of a beneficial nonlinear "noise" squeezing effect [70]. This demonstration further confirms the enhanced nonlinear transmission performance brought by PCTW under stringent transmission conditions with substantial interchannel nonlinear effects, PMD, and polarization-dependent loss.

The benefit of symmetric EDC in improving the nonlinearity tolerance was also found in PDM-QPSK systems, albeit with less performance gain [77,78]. Another method to increase the nonlinearity tolerance is to perform symbol-rate optimization (SRO) [78–84]. Various theoretical and experimental papers have reported on the effect of SRO. In terms of nonlinearity tolerance, the optimal symbol rate is inversely proportional to the square root of the total fiber link dispersion, and is between 2 and 6 Gbaud for typical dispersion-uncompensated long-haul transmission systems [78]. It was shown that the combined use of preEDC and SRO could bring additional improvement in nonlinearity tolerance [78]. In the recent 800-Gb/second-per-wavelength transmission demonstration, both preEDC and DSCM were implemented for added tolerance to fiber nonlinear effects [57].

DSP-based nonlinearity compensation (NLC) using digital back-propagation (DBP) has drawn much attention [62,85]. However, in dispersion-uncompensated high-speed transmission, the needed DSP complexity for DBP is usually impractically high. To reduce the DSP complexity, transmitter-side perturbation-based NLC, in which the signal nonlinear distortions could be computed in a single multiplier-free DSP step, has been introduced [86–91]. It has been underlined in Ref. [70] that "the perturbation analysis may be applied for digital compensation of fiber nonlinear effects in an efficient manner." Using the same perturbation theory as used in the derivation for PCTWs, one can precompute the nonlinear distortions and subtract them from the original signal at the transmitter through DSP [92]. The distortions can be expressed in the time domain as

$$\delta E_{x,y}(t) = \sum_{m=-N}^{N} \sum_{n=-N}^{N} C(m,n)[E_{x,y}(t+m)E_{x,y}(t+n)E^*_{x,y}(t+m+n) + E_{y,x}(t+m)E_{x,y}(t+n)E^*_{y,x}(t+m+n)] \tag{8.26}$$

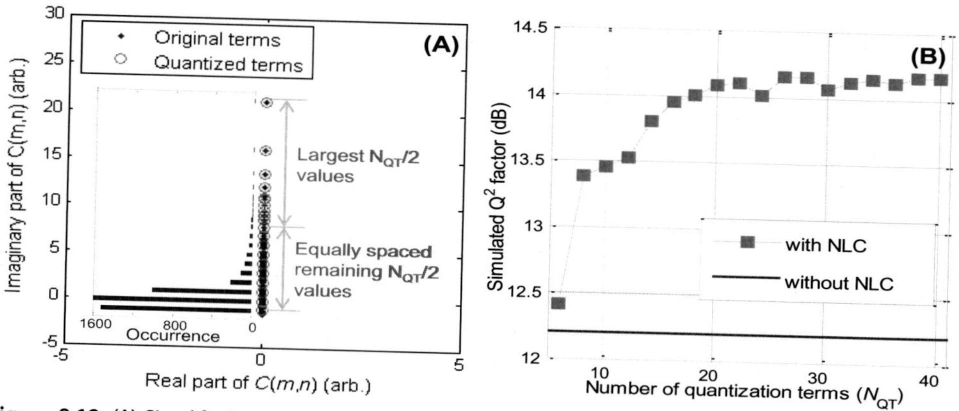

Figure 8.12 (A) Simplified quantization of $C(m,n)$ when symmetric-EDC is used, and (B) NLC performance versus N_{QT}, for 32-Gbaud PDM-QPSK transmission over 3200-km SSMF at $P_{in} = 2$ dBm. (After Ref. [88]).

where $C(m,n)$ is the nonlinear perturbation coefficient representing the four-photon-mixing (FPM) product of the m-, n-, and $(m+n)$-th symbols, generating nonlinear distortion on the 0-th symbol. The typically large number of terms has recently been shown to be significantly reduced by using the symmetric EDC [87]. When a symmetric dispersion map and a symmetric power map are assumed, $C(m,n)$ is purely imaginary [87]. This helps to further reduce the DSP complexity and power consumption of perturbation-based NLC because (1) the number of quantized $C(m,n)$ coefficients can be nearly halved and (2) the multiplication between imaginary-valued $C(m,n)$ and its corresponding FPM products is twice as efficient as that for complex-valued $C(m,n)$. Fig. 8.12A exemplifies $C(m,n)$ quantization for 32-Gbaud signal transmission over 40 80-km SSMF spans [88]. Fig. 8.12B shows the simulated performance as a function of the number of quantized terms (N_{QT}). Remarkably, only ~20 terms are needed to attain most of the NLC gain. The NLC scheme can be applied to other modulation formats. For 16-QAM, a QAM decomposition method enabling simple processing of perturbative nonlinearity mitigation was recently proposed [91]. After 1920-km SSMF, the simple approach exhibits only ~0.1-dB penalty compared with the complex conventional method. Note that the performance gain of single-channel NLC, however, is typically limited to well below 1 dB in fully loaded DWDM systems [93].

8.4.4 Capacity-approaching FEC

Forward-error correction is an important technology to enable a communication link to approach the Shannon limit. The use of FEC in optical fiber communication links has gone through three generations [94]

- The first generation of FEC codes appeared in the 1987–93 period, and the representative FEC code is Reed–Solomon (RS) code (255,239) with a FEC overhead of 6.7% and a net coding gain (NCG) of 5.8 dB at an output BER of 10^{-15}.

- The second generation of FEC codes in the 2000–04 period, and the representative FEC codes are the concatenated codes, showing an NCG of 9.4 dB at an FEC overhead of ~25%.
- The third generation of FEC codes started to be adopted in real systems around 2006, and the representative FEC codes are SD decoding enabled low-density parity check (LDPC) codes, turbo codes, etc., showing an NCG of over 10 dB at an FEC overhead of between ~15% and ~25%. The key feature of the third generation of FEC codes is the use of SD decoding [94].

A key performance indicator of FEC is its NCG, which is defined as

$$\mathrm{NCG(dB)} = 10\log_{10}(\mathrm{SNR}(\mathrm{BER}_{\mathrm{out}})) - 10\log_{10}(\mathrm{SNR}(\mathrm{BER}_{\mathrm{in}})) + 10\log_{10}(R) \quad (8.27)$$

where $\mathrm{BER}_{\mathrm{in}}$ is the maximally allowed input BER to the FEC to achieve a reference output BER of $\mathrm{BER}_{\mathrm{out}}$, $\mathrm{SNR}(x)$ is the SNR needed for a given modulation format to reach a BER of x without coding, and R is the FEC code rate. For BPSK modulation format, we have

$$\sqrt{\mathrm{SNR}_{\mathrm{BPSK}}(\mathrm{BER})} = \mathrm{erfc}^{-1}(2\mathrm{BER}) \quad (8.28)$$

where $\mathrm{erfc}^{-1}()$ is the inverse complementary error function. For QPSK modulation format, we have

$$\sqrt{\mathrm{SNR}_{\mathrm{QPSK}}(\mathrm{BER})} = \sqrt{2}\mathrm{erfc}^{-1}(2\mathrm{BER}) \quad (8.29)$$

For 16-QAM modulation format, we have

$$\sqrt{\mathrm{SNR}_{16\mathrm{QAM}}(\mathrm{BER})} = \sqrt{10}\mathrm{erfc}^{-1}\left(\frac{8}{3}\mathrm{BER}\right) \quad (8.30)$$

For high-speed transmission based on 16-QAM, multiple high-performance FEC codes have been studied. Table 8.1 shows some of the FEC codes and their performances. The first code is the concentrated FEC (CFEC) code adopted by the Optical

Table 8.1 High-performance FEC codes used for 16-QAM based high-speed transmission.

FEC code	$\mathrm{BER}_{\mathrm{in}}$ ($\mathrm{SNR}_{\mathrm{in}}$) for $\mathrm{BER}_{\mathrm{out}} = 10^{-15}$*	Code rate	NCG
CFEC [95,96]	1.22×10^{-2} (13.6 dB)	0.871	10.76 dB
CFEC + [97]	1.81×10^{-2} (12.9 dB)	0.871	11.45 dB
OFEC [98]	1.98×10^{-2} (12.7 dB)	0.867	11.60 dB
A 20%-OH LDPC [57]	$\sim2.76\times10^{-2}$ (12.0 dB)	0.833	12.12 dB
A 25%-OH LDPC [99]	$\sim3.45\times10^{-2}$ (11.5 dB)	~0.8	~12.5 dB

FEC, forward-error correction; *BER*, bit error ratio; *SNR*, signal-to-noise ratio; *NCG*, net coding gain; *LDPC*, low-density parity check.
*At the reference $\mathrm{BER}_{\mathrm{out}}$ of 10^{-15}, the corresponding SNR for 16-QAM is ~24.95 dB.

Internetworking Forum (OIF) for 400ZR [95,96]. Its required BER_{in} for a BER_{out} of 10^{-15} is 1.22×10^{-2}, and its code rate is 0.871, leading to an NCG of 11.76 dB. The second code is an enhanced version of CFEC, named as CFEC +, which offers an increased NCG of 11.45 dB [97]. The third code is referred to as open FEC (OFEC), which was adopted by the Open ROADM Multisource Agreement (MSA) and was proposed to the ITU project on 200 G/400 G FlexO-LR for 450 km black link applications [98]. The OFEC offers a further increased NCG of 11.6 dB.

In a recent 800-Gb/second-per-wavelength demonstration, a 20% overhead (OH) LDPC code was used, and a high NCG of 12.12 dB was achieved [57]. Finally, a 25%-OH FEC was demonstrated in real-time 200-Gb/second coherent transceivers, achieving a remarkably high NCG of 12.5 dB [99]. Similar performance has also been shown in the 800-Gb/second demonstration when the LDPC FEC OH was increased to 25% [57].

It is worthwhile to compare the above FEC performances with the theoretical limit. The ultimate NCG can be derived from the Shannon's capacity theorem. The channel capacity C of a binary symmetric channel with HD decoding is given as

$$C_{HD} = 1 + BER_{in} \cdot \log_2(BER_{in}) + (1 - BER_{in}) \cdot \log_2(1 - BER_{in}) \tag{8.31}$$

where BER_{in} is the input BER threshold and the channel capacity C_{HD} can be set to the FEC code rate R [94]. For a given FEC code rate R, we can calculate BER_{in}. Together with Eqs. (8.27–8.30), we can then calculate the NCG for a given modulation format at a given reference BER_{out}.

For SD decoding, the capacity of a binary symmetric channel, which has two possible inputs $X = A$ and $X = -A$, can be expressed as [35]

$$C_{SD} = \frac{1}{2}\int_{-\infty}^{+\infty} p(y|A)\log_2\left(\frac{p(y|A)}{p(y)}\right)dy + \frac{1}{2}\int_{-\infty}^{+\infty} p(y|-A)\log_2\left(\frac{p(y|-A)}{p(y)}\right)dy \tag{8.32}$$

where $p(y|A)$ is the conditional probability of getting y at the receiver when the input is A, and $p(y)$ is the probability of receiving y. Similar to the case with HD, for a given FEC code rate $R = C_{SD}$, we can calculate BER_{in}, from which we can then calculate the NCG for a given modulation format at a given reference BER_{out}. In the idealized case of SD decoding with infinite quantization bits, the NCG obtained by SD decoding is $\pi/2$ times as large as (or ~2 dB higher than) that obtained by HD decoding when R approaches zero.

Fig. 8.13 shows the Shannon limits of HD and SD NCGs for BPSK/QPSK, together with some recently demonstrated high-performance FEC codes. For HD decoding, KP4 and staircase FEC [100] are widely used in the optical communication industry. Remarkably, the staircase FEC offers a NCG of 9.41 dB, which is only 0.56 dB away from the HD Shannon limit [100]. For SD decoding, NCGs of 11.6

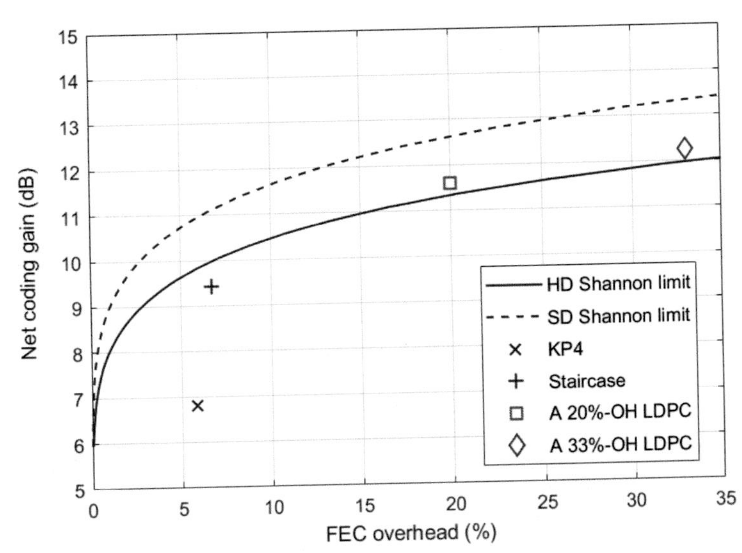

Figure 8.13 The Shannon limits of HD and SD NCGs for BPSK/QPSK and some recently demonstrated FEC codes [57,96,100]. The reference output BER (BER_{out}) is set at 10^{-15}.

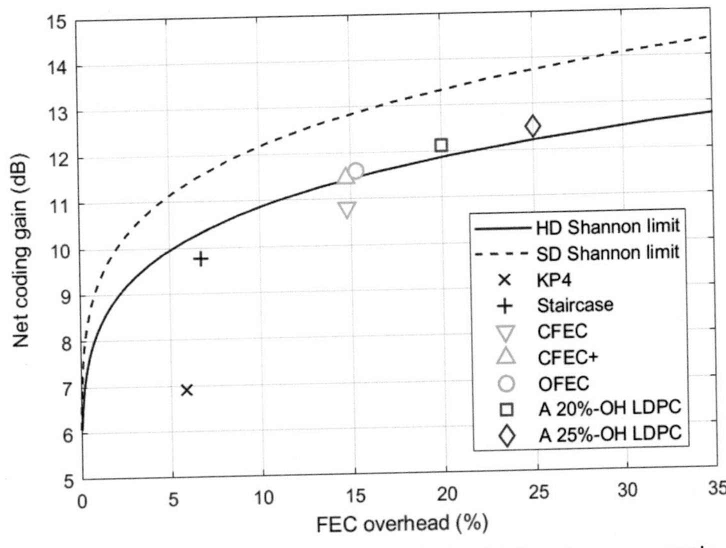

Figure 8.14 The Shannon limits of HD and SD NCGs for 16-QAM and some recently demonstrated FEC codes [57,96–99,100]. The reference output BER (BER_{out}) is set at 10^{-15}.

and 12.25 dB have been achieved with 20% and 33% OHs, respectively [57]. These SD NCGs are only about 1 dB away from the SD Shannon limit.

Fig. 8.14 shows the Shannon limits of HD and SD NCGs for 16-QAM, together with some recently demonstrated high-performance FEC codes. The Shannon limits of NCGs for 16-QAM are slightly higher than those for BPSK/QPSK. This can be understood by the slightly flatter BER curve of 16-QAM at high BER values as compared to BPSK/QPSK, as shown in Figure 7.6. For HD decoding, the staircase FEC is again only ~0.6 dB away from the HD Shannon limit. For SD decoding, the CFEC+, OFEC, 20%-OH LDPC, and 25%-OH LDPC described in Table 8.1 are all within 1.4 dB away from the SD Shannon limit. In terms of the absolute NCG, it increases as the OH increases. This offers the flexibility of adjusting the system performance and throughput based on link conditions.

The above analysis and review have shown the remarkable works done by the optical communication industry in approaching the Shannon limit via advanced FEC coding designs and implementations.

8.4.5 Constellation shaping and probabilistic shaping

To approach the Shannon capacity even more closely, signal constellation needs to be optimized. For the conventionally used m-QAM constellations, there is a theoretical performance gap from the Shannon capacity, which can be reduced via CS [101–120]. In the limit of a high SE, the performance gain that can be achieved via CS, or the shaping gain, approaches $\pi e/6$ or 1.53 dB. CS can be realized through geometric shaping (GS) or probabilistic shaping (PS). In this section, we first introduce the concept of CS via GS and then highlight some of the state-of-the-art results achieved via PS.

The most common approach to increase the SE of a communication system is to apply higher order modulation formats such as square-shaped Q-ary QAM. In order to exploit the available channel capacity up to its theoretical limit, we need to deviate from uniform square constellations. In the derivation of Shannon's fundamental equation of the channel capacity for the additive white Gaussian noise channel, Shannon stated that the codes need to be infinitely long and the distribution of the channel input symbols must be Gaussian. Therefore, for m-QAM, using a uniformly distributed signal constellation with equal probability of occurrence for each constellation point induces a gap of 1.53 dB to the fundamental Shannon limit [102].

The above statement is also applicable for OFDM modulation since the central limit theorem suggests that the use of the discrete Fourier transform (DFT) or inverse DFT (IDFT) in OFDM causes the recovered subcarriers to be corrupted by fairly Gaussian-distributed noise when the number of subcarriers is sufficiently large. Therefore in order to minimize this gap, Gaussian-distributed symbols by CS at the channel input can be used.

We first consider a shaping method based on optimized nonuniform signal constellations for providing a quasiGaussian source. Iterative polar quantization (IPQ) [104] was used to design shaped signal constellation. The constellation points of the obtained set are distributed over circles of radii determined by a Rayleigh distribution [104]

Using a Rayleigh distribution $p(r)$ originates from the fact that the envelope r of a two-dimensional Gaussian distribution is Rayleigh distributed. The optimized constellations based on IPQ are called iterative polar modulation (IPM) [104,105,107]. Fig. 8.15 shows the 256-IPM constellation obtained through the above procedure. The distribution of the 256-IPM constellation points is nonuniform and follows a discrete 2D Gaussian distribution with zero mean. An experimentally measured constellation is also shown. Fig. 8.16 shows the SEs of QAM and IPM, with 16, 64, and 256

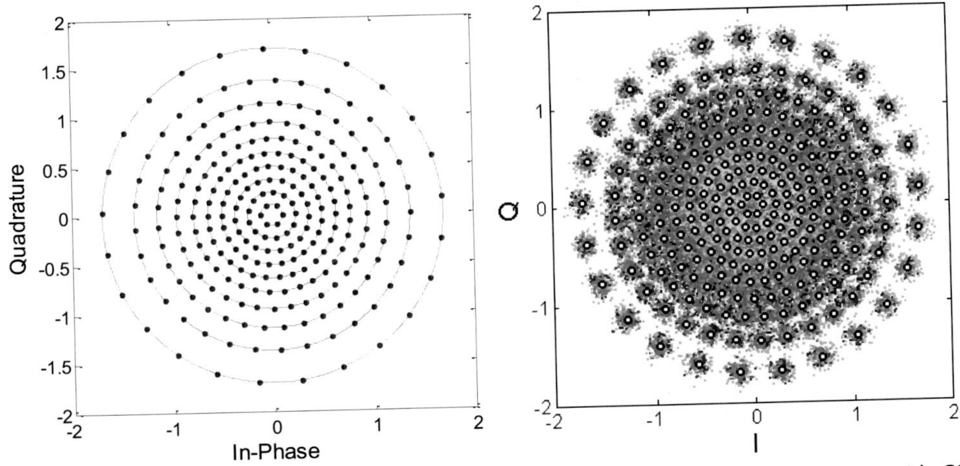

Figure 8.15 Calculated (left) and experimentally measured (right) 256-IPM constellations with GS. (After [107]).

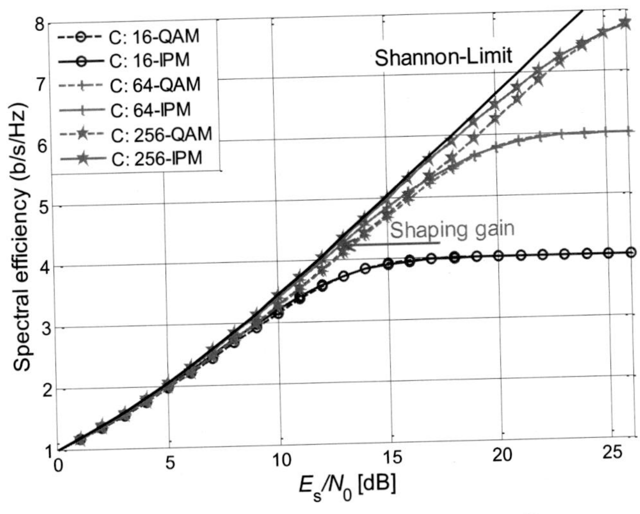

Figure 8.16 SEs of different IPM and square-QAM formats. (After [107]).

constellation points. Evidently, IPM signal constellations approach the Shannon limit better than the square-QAM constellations in the low and medium SNR regions. The gain of IPM over QAM can be expressed by examining the difference in SNR to obtain a certain SE. Fig. 8.17 shows the shaping gain of IPM over square-QAM as a function of the SNR. Obviously, for 256-IPM, the maximum theoretical shaping gain over 256-QAM is 0.88 dB, reached at an SNR of 18 dB.

A proof-of-concept experiment was conducted to verify the shaping gain due to GS [105]. Fig. 8.18 compares the measured back-to-back BER performance of 231.5 Gb/second PDM-OFDM-256-IPM and PDM-OFDM-256-QAM before and after inner SD-FEC processing. At the outer FEC threshold, 256-IPM outperforms 256-QAM by ~1.3 dB in the required optical SNR (OSNR). Remarkably, the observed OSNR improvement is more than twice the theoretical SNR improvement (0.51 dB) discussed in the previous section. We attribute this to the large contribution of quantization noise to the overall SNR, as a given reduction in the required overall SNR leads to a larger reduction in the required OSNR. This proof-of-concept demonstration shows the promise of using CS to achieve higher optical transmission performance than without.

The IPM-based GS scheme requires computationally intensive DSP, for example, iterative FEC decoding and demodulation, making it difficult for practical implementations at high speed [114]. Also, different sets of constellation points are needed to achieve different SEs, making the rate adaptation difficult. Moreover, other aspects such as binary gray labeling, bit accurate implementation, and effective equalization are also challenging to realize [57]. Aiming to achieve the shaping gain with

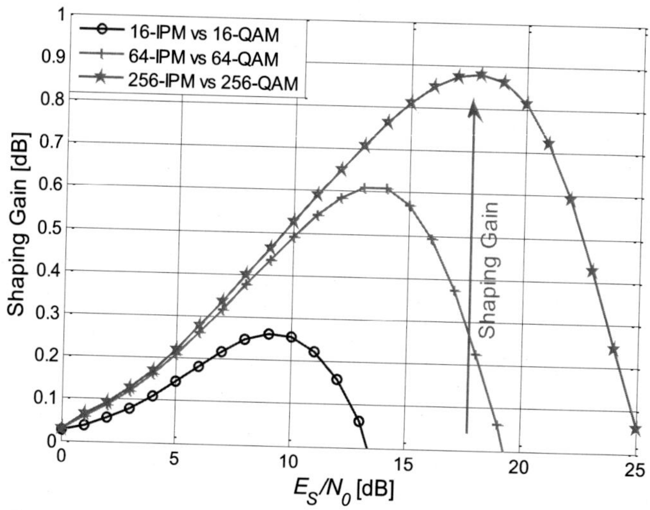

Figure 8.17 Shaping gains of different IPM formats over square-QAM. (After [107]).

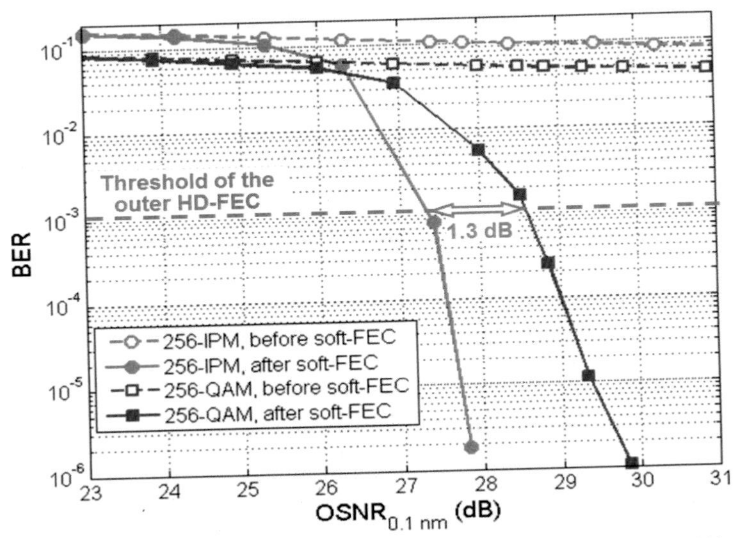

Figure 8.18 Measured BER performance comparison between 256-IPM and 256-QAM. (After [105]).

simple implementations, a new shaping method called probabilistic amplitude shaping (PAS) was introduced in 2014 [110]. The PAS scheme concatenates a distribution matcher (DM) for PS with a systematic binary encoder for FEC. At the receiver, bit-metric decoding is used without any iterative de-mapping. This PAS scheme directly applies to two-dimensional QAM constellations by mapping two real-valued M-ary PAM symbols to one complex M^2-QAM symbol. As compared to the IPM scheme, the newly developed PAS scheme offers the following advantages:

- Simple DSP implementation without iterative processes
- Compatibility with common QAM constellations, making modulation, gray mapping, equalization and demodulation straightforward
- Compatibility with common FEC encoding and decoding processes, enabling the use of already developed high-performance FEC codes
- Flexible rate adaptation by changing the distribution of the PAS before FEC encoding

Since the introduction of the PAS scheme, impressive experimental demonstrations have been reported, verifying its superior performance and efficient implementation [111,112,114,116,117]. Fig. 8.19 shows the architecture of the PAS scheme. For an M-ary PAM symbol set X, the probability of a constellation point $x \in X$ can be generated according to the Maxwell-Boltzmann (MB) distribution

$$P_X(x) = \frac{e^{-ax^2}}{\sum_{y \in X} e^{-ay^2}} \tag{8.33}$$

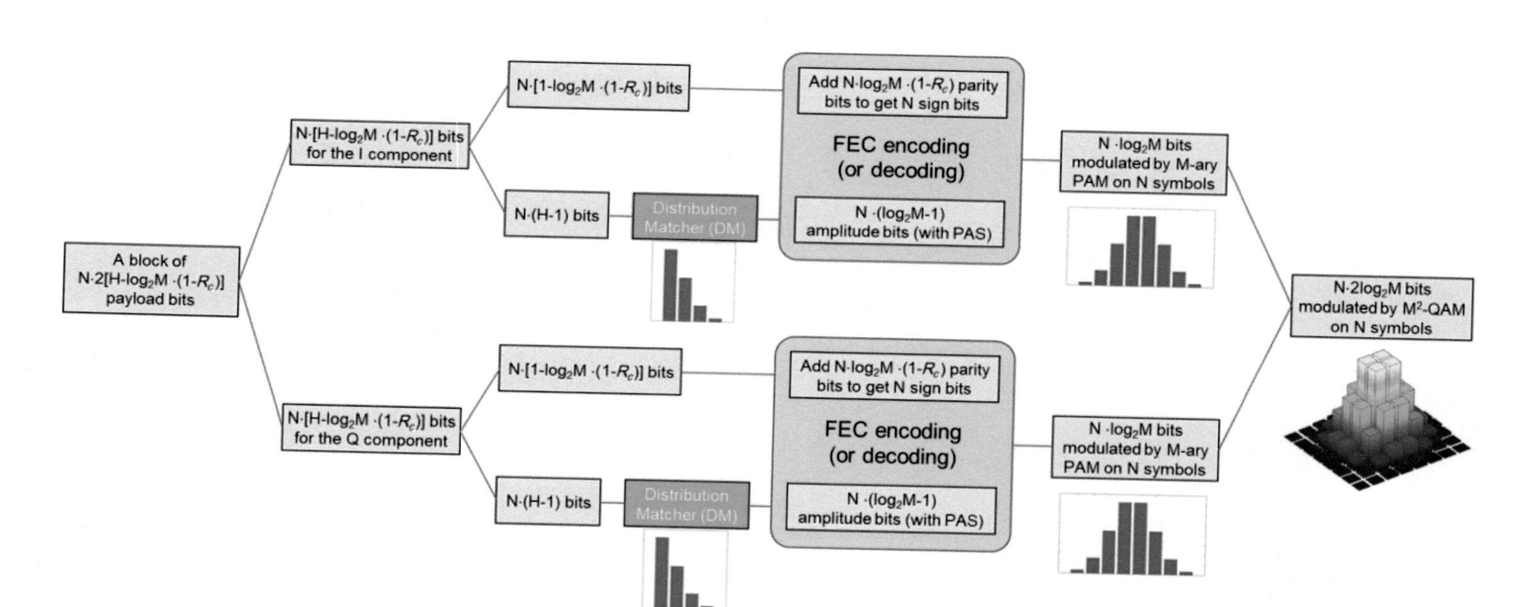

Figure 8.19 Illustration of the PAS-based PS architecture.

where $a \geq 0$ is the rate parameter, which controls the entropy rate (H) of the shaped M-ary PAM signal according to

$$H(X) = -\sum_{x \in X} P_X(x)\log_2 P_X(x) \tag{8.34}$$

When a FEC with a code rate of R_c is used, the information rate (IR) of the shaped M-ary PAM signal becomes

$$\mathrm{IR}(X) = H(X) - \log_2 M \cdot (1 - R_c) \tag{8.35}$$

At the transmitter, the PS codeword length is set to N. A block of $N \cdot 2[H - \log_2 M \cdot (1 - R_c)]$ payload bits are divided equally for the in-phase (I) component and the quadrature (Q) component. The payload bits of each component are further divided into two branches, with the first branch of $N \cdot [1 - \log_2 M \cdot (1 - R_c)]$ bits to be carried by the sign bits of the M-ary PAM and the second branch of $N \cdot (H - 1)$ bits sent to the DM, which generates $N \cdot (\log_2 M - 1)$ amplitude bits of the M-ary PAM according to the PAS distribution, as illustrated in Fig. 8.19. Then, the FEC encoder adds $N \cdot \log_2 M \cdot (1 - R_c)$ parity bits as the sign bits such that the total number of sign bits are exactly N, or one sign bit per PAM symbol. A key advantage of this scheme is that the FEC parity bits are added as sign bits, so that the PAS distribution is unchanged during the FEC encoding process. After the FEC encoding, there are $N \cdot \log_2 M$ bits, which are modulated onto N M-ary PAM symbols, as also illustrated in Fig. 8.19. Finally, the I and Q components are combined to form a M^2-QAM signal with the intended PS, as illustrated in the constellation distribution diagram in Fig. 8.19. At the receiver, the above processes are reversed. The FEC decoder will remove the parity bits and the DM will perform the reserve function of distribution matching.

Three exemplary constellation distribution diagrams of PS-64-QAM with entropies of 5.7, 5, and 4.3 bits per complexity symbol are shown in Fig. 8.20. Note that

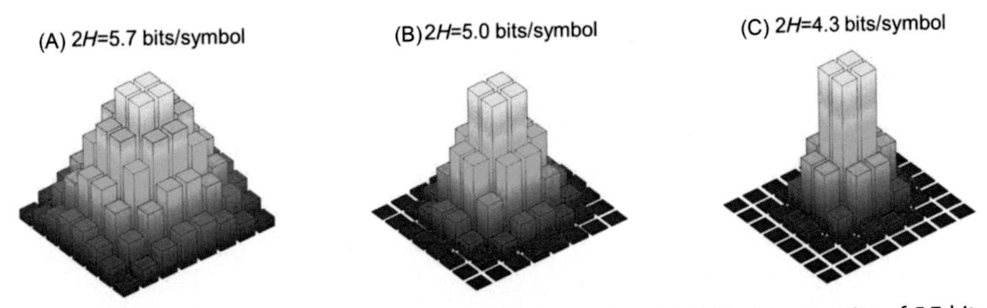

Figure 8.20 Exemplary constellation distribution diagrams of PS-64-QAM with entropies of 5.7 bits (A), 5 bits (B), and 4.3 bits (C) per complexity symbol.

the probability distribution can deviate from the MB distribution (or the Gaussian distribution). For example, superGaussian distribution was used to improve the nonlinearity tolerance [57].

As an example, we can have the following settings of PS:

- $M = 8$, leading to 8-ary PAM per quadrature and 64-QAM final constellation
- $H = 2.5$ bits per 8-ary PAM symbol (or $2H = 5$ bits per 64-QAM symbol)
- $R_c = 5/6$, leading to an IR of 2 bits per 8-ary PAM symbol
- The IR of the shaped 64-QAM: 4 bits per complex symbol
- The IR of the shaped PDM-64-QAM: 8 bits per PDM symbol

Regarding the PS codeword length N, it can be chosen to be ~ 1000 to achieve most of the theoretical shaping gain, while keeping the implementation complexity low [57]. Also, a fine SE step size of the order of a percent of a bit per symbol per polarization can be achieved. This fine SE granularity enables the transmission system to approach its maximal capacity for any given link condition, which is another important advantage of the PS scheme. Real-time demonstrations of PAS-based PS in commercial systems have been reported, and we will describe one such demonstration [117] in the following section.

8.5 Real-time demonstrations of PS

As the first real-time demonstration of PS in commercial optical transmission systems, a field trial on the use of PS-programmable real-time 200-Gb/second coherent transceivers in a deploying core optical network was reported in 2018, achieving a twofold increase in reach when the PS was activated [117].

Fig. 8.21A shows the schematic of an intelligent intent-driven optical network with a network cloud engine (NCE) consisting of intent, intelligence, automation, and analytics engines, and an intelligent optical layer consisting of reconfigurable optical add-drop multiplexers, advanced optical monitoring and PS-programmable optical DSP (oDSP). Fig. 8.21B illustrates the use of format-and-shaping-programmable coherent transceivers to achieve the tradeoff between performance and power consumption for a given application or intent.

The transmission performances of 200 G PDM-PS-16QAM, PDM-8QAM, and PDM-16QAM were first compared in a lab environment, as shown in Fig. 8.22. Evidently, PS-16QAM offers the best performance, and doubles the reach of 16QAM at a given OSNR penalty of 1.5 dB. Then the transmission performances in the field trial were compared [117]. Figs. 8.23 and 8.24 show representative recovered constellations, respectively, for 16-QAM after 571-km and PS-16QAM after 1142-km transmission.

Fig. 8.25 shows the corresponding BER curves. In the back-to-back case, the required OSNR values at the FEC threshold of 3×10^{-2} are 18 and 16.5 dB for

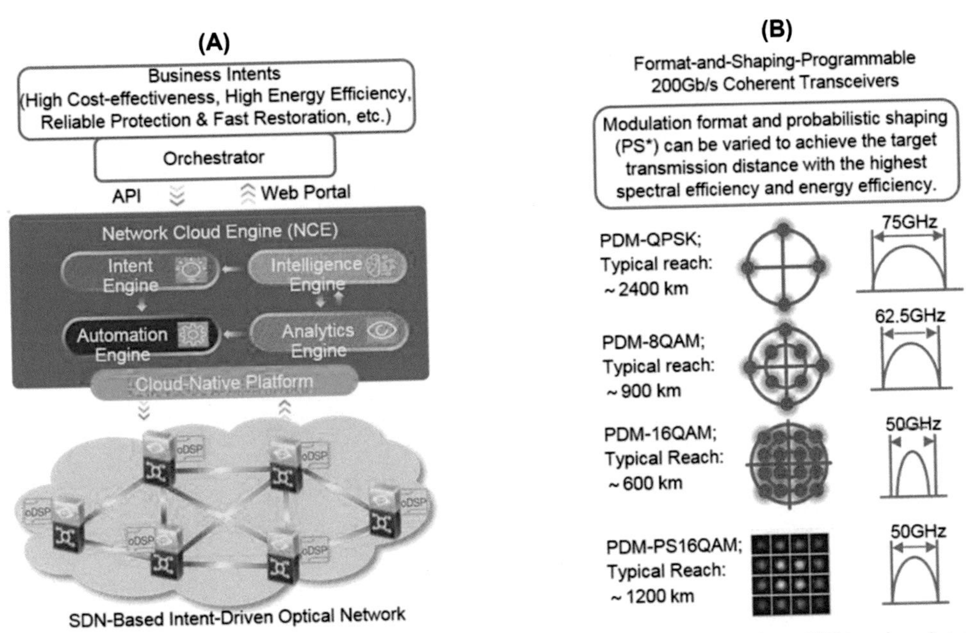

Figure 8.21 (A) Schematic of an intelligent intent-driven optical network with an NCE and an intelligent optical layer; (B) Illustration of format-and-shaping-programmable coherent transceivers. (After Ref. [117]).

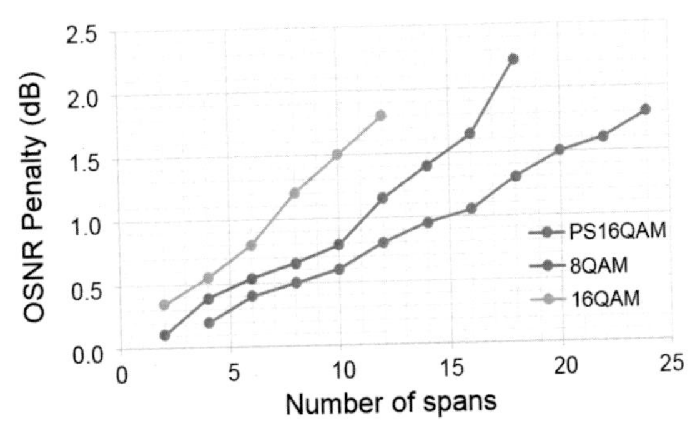

Figure 8.22 Measured transmission performance comparison among 200 G PDM-PS16QAM, PDM-8QAM, and PDM-16QAM in a lab environment. (After Ref. [117]).

PDM-16QAM and PDM-PS-16QAM, respectively, which represent 1 and 2.5 dB improvements over the previous PDM-16QAM result reported in Ref. [120]. At the FEC threshold, PDM-16QAM requires a received OSNR of 20 dB after 571-km transmission, while PS-16QAM requires a received OSNR of 17 dB after 1142-km transmission, confirming that PS-16QAM offers a doubled reach as compared to 16QAM. The measured average OSNR values after 571 and 1142-km transmissions

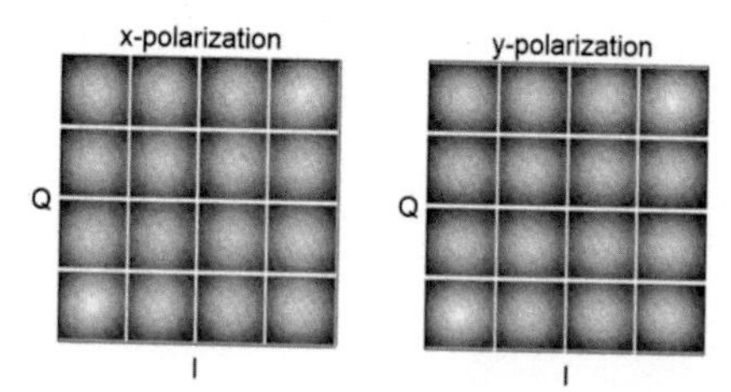

Figure 8.23 Recovered 200-Gb/s PDM-16QAM constellations after 571-km transmission (without loop-back) in the field trial. (After Ref. [117]).

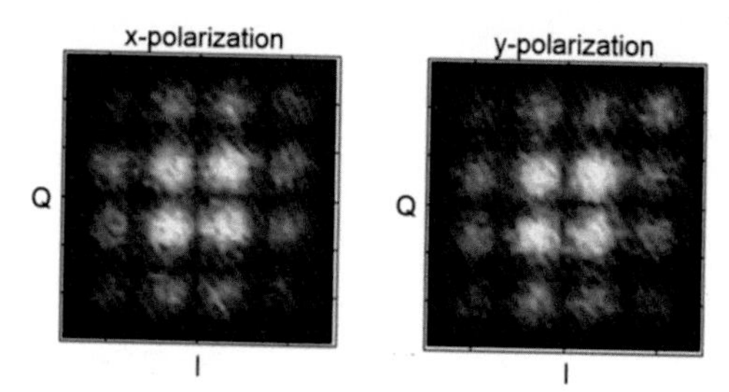

Figure 8.24 Recovered 200-Gb/s PDM-PS-16QAM constellations after 1142-km transmission (with loop-back) in the field trial. (After Ref. [117]).

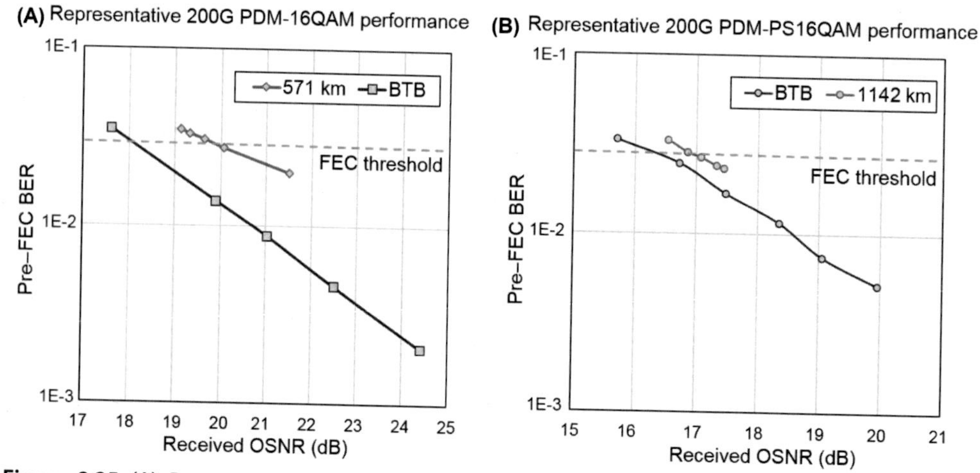

Figure 8.25 (A) Representative 200 G PDM-16QAM performance after the 571-km field test; (B) Representative 200 G PDM-PS-16QAM performance after the 1142-km field test. (After Ref. [117]).

are 22.3 and 19.3 dB, respectively. Thus there is an average OSNR margin of 2.3 dB. Ethernet tester report showed error-free over 24 h with over 2×10^{15} bytes received [117].

PDM-PS-16QAM real-time coherent transceivers have also been used for 200Gb/second upgrade of legacy metro/regional WDM networks [99]. While the unshaped 16-QAM allowed error-free transmission of up to 1500 km, the PS-16QAM extended the reach to 2000 km with additional OSNR margin. The effective OSNR gain of PS-16QAM over unshaped 16-QAM was found to be about 2 dB [99]. More recently, real-time demonstration of a PDM-PS-16QAM transceiver at 400-Gb/s net data rate using 7-nm application-specific integrated circuit (ASIC) has been reported [121]. Modulated at 69 Gbaud, the 400-Gb/second PDM-PS-16QAM coherent transceiver requires an OSNR of 20.1 dB at the FEC threshold of 2.4×10^{-2} [121]. Remarkably, the OSNR performance is about 4 dB better than those reported for early silicon photonics-based 400ZR transceivers [96]. Moreover, real-time demonstration of PDM-PS-64QAM at 800-Gb/second net data rate using 7-nm ASIC has also been reported [57].

In these real-time demonstrations, the measured effective shaping gains were typically larger than the theoretical shaping gains, indicating that PS is more tolerant to transmission impairments (such as nonlinear distortions) and implementation imperfections than its unshaped counterpart. These demonstrations affirm the benefits of PS-enabled coherent optical transceivers in high-capacity long-haul optical networks.

8.6 Improved fibers and EDFAs for high-capacity long-haul transmission

New optical fibers that provide lower loss coefficient (α), higher dispersion, and/or smaller nonlinear coefficient (γ), for example, by increasing its effective area (A_{eff}), can increase SE and transmission distance. A useful method to compare the transmission performances of different fiber types is to use a figure of merit (FOM) for a given fiber that determines its transmission distance advantage (in dB) over a reference fiber, which is typically set to be the standard International Telecommunication Union (ITU) G.652.D optical fiber. With the aforementioned analytical results, we can approximately define the FOM [5] as

$$\mathrm{FOM} = \frac{2}{3}(\alpha_{\mathrm{ref}}L - \alpha L) + \frac{1}{3}\cdot 10\log\left[\left(\frac{A_{\mathrm{eff}}}{A_{\mathrm{eff,ref}}}\right)^2\left(\frac{L_{\mathrm{eff,ref}}}{L_{\mathrm{eff}}}\right)\left(\frac{D}{D_{\mathrm{ref}}}\right)\right] \tag{8.36}$$

For example, we can set the reference fiber as a G.652.D fiber that has the following parameters, $L = 100$ km, $\alpha_{\mathrm{ref}}L = 20$ dB, $A_{\mathrm{eff,ref}} = 85\ \mu\mathrm{m}^2$, $L_{\mathrm{eff,ref}} = 21.5$ km, $D_{\mathrm{ref}} = 17$ ps/nm/km.

Recently, a new class of fiber suitable for long-haul transmission has been standardized by the ITU as G.654.E. This class of fibers are commercially available, such as OFS's TeraWave fiber and Corning's TXF fiber [122]. The new ITU G.654.E-compliant fiber typically has the following parameters [122,123], $L = 100$ km, $\alpha L = 17$ dB, $A_{\mathrm{eff}} = 125\ \mu\mathrm{m}^2$, $L_{\mathrm{eff}} = 25$ km, $D = 21$ ps/nm/km.

Based on Eq. (8.36), the FOM of the new G.654.E fiber is about 3 dB with respect to the reference fiber. This means that the new fiber may provide twice the reach as compared to the reference fiber. If the transmission distance is set to be the same as the reference fiber link, a 3-dB increase in FOM can be used to increase the SE by about 2 b/second/Hz [5]. A more rigorous definition of FOM with the consideration of splicing loss differences is provided by Hasegawa et al. [124].

With the combined use of G.654.E-compliant fiber, $C+L$ band EDFAs, and advanced oDSP such as PS, a single-fiber capacity of 100.5 Tb/second was recently demonstrated over a 600-km fiber link in the laboratory [125]. More specifically, WDM transmission of 195 50-GHz-spaced channels with a net WDM SE of 10.3 b/s/Hz and a total EDFA amplification bandwidth of 9.75 THz was demonstrated, achieving a total single-fiber capacity of 100.5 Tb/second. The transmission fiber link consisted of six spans of 100-km G.654E-compliant fiber. Fig. 8.26 shows the measured optical spectrum of the 195 wavelength channels after 600-km optical fiber transmission. This demonstration reveals the power of the combined use of better fiber, wider EDFA amplification bandwidth, and advanced oDSP for achieving high single-fiber capacity.

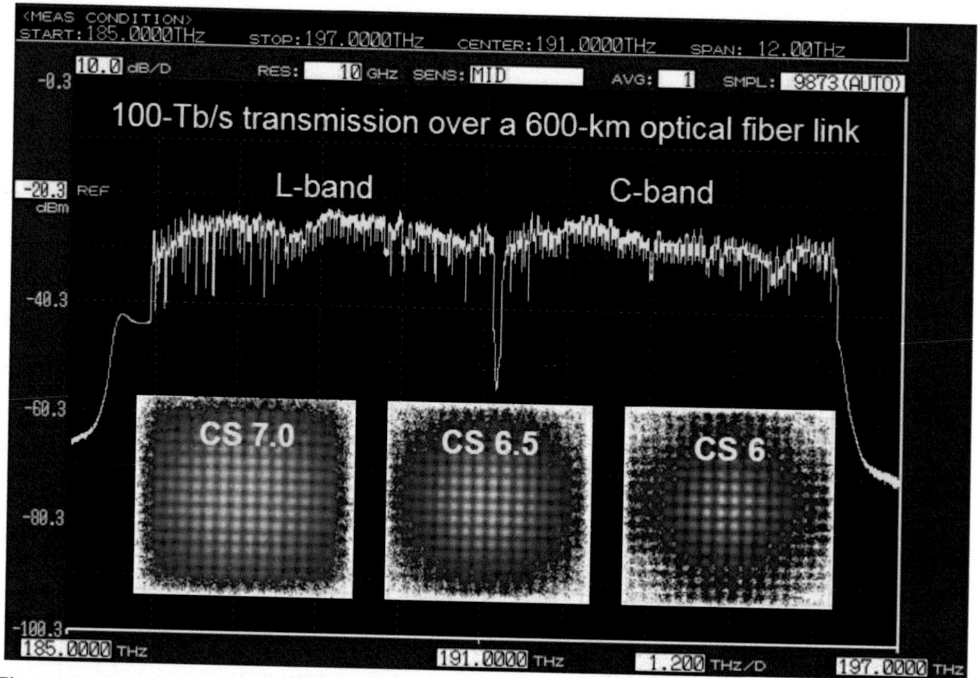

Figure 8.26 Experimentally measured optical spectrum of 195 50-GHz-spaced wavelength channels with a net WDM SE of 10.3 b/second/Hz and a total EDFA amplification bandwidth of 9.75 THz, achieving a total single-fiber capacity of 100.5 Tb/second. Insets: constellation diagrams with different CS entropy values. (After [125]).

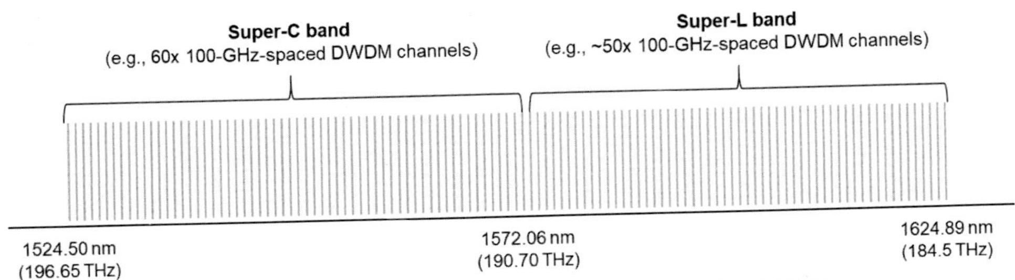

Figure 8.27 The widened wavelength window enabled by the super C + L EDFAs.

More recently, the combined use of the super *C* and super *L* bands has widened the EDFA amplification window to 11 THz [126]. Fig. 8.27 shows the widened wavelength window supported by the super *C* + *L* EDFAs. In the super *C* band, there is a 6-THz amplification window that can be used to support 60 100-GHz-spaced DWDM channels, or 120 50-GHz-spaced DWDM channels. In the super *L* band, there is a 5-THz amplification window that can be used to support 50 100-GHz-spaced DWDM channels, or 100 50-GHz-spaced DWDM channels. With the use of 800-Gb/second wavelength channels on a 100-GHz grid, a record-setting total single-fiber transmission capacity of 88 Tb/second has been achieved for data center interconnect applications [126].

8.7 Undersea optical transmission

Undersea or submarine optical fiber transmission is a key part of the global optical network that links all the continents together. As an example, the SEA-ME-WE-3 submarine telecommunications fiber cable system links multiple regions in South (Australia), East Asia, Middle East, and Western Europe together [127]. It was regarded as the longest in the world with a total length of 39,000 km and 39 landing points in over 30 countries, as illustrated in Fig. 8.28. Completed in late 2000, it was led by France Telecom and China Telecom, and was administered by Singtel, a telecommunications operator owned by the Government of Singapore. It was commissioned in March 2000. In 2015, the SEA-ME-WE-3 had been upgraded to carry 100-Gb/second coherent wavelength channels with capacity of 4.6 Tb/second [127]. Later systems offered higher capacity, such as 38 Tb/second over 3 fiber pairs in such as SEA-ME-WE-5 [128].

Owing to the ever-increasing global communication traffic for new services such as 5G and cloud services, the capacity demands for undersea fiber cable systems are also fast growing. More and more undersea fiber cable systems with increased per-fiber capacity have been built. As of 2021, the title of "the highest-capacity submarine cable in the world" was given to MAREA transatlantic telecommunications fiber cable

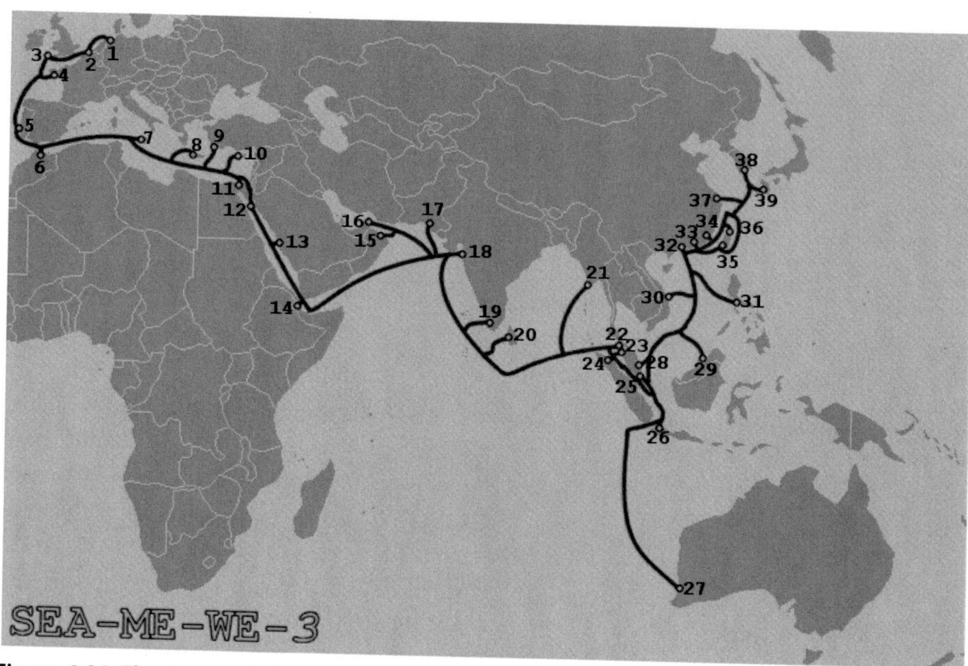

Figure 8.28 The layout of the SEA-ME-WE-3 submarine telecommunications fiber cable that links multiple regions in South (Australia), East Asia, Middle East, and Western Europe with a total length of 39,000 km and 39 landing points. *(Courtesy J.P. Lon, CC BY 2.5, https://commons.wikimedia.org/w/index.php?curid = 2308375).*

system, which was owned and funded by Microsoft and Facebook but constructed and operated by Telxius, a subsidiary of the Spanish telecom company Telefónica [129]. MAREA is a 6600 km-long transatlantic communication cable connecting Virginia Beach, Virginia, in the United States, with Bilbao, Spain. The MAREA system was initially expected to have a transmission capacity of 160 Tb/second, by using 16 fibers each supporting 10 Tb/second [129]. Remarkably, a research team demonstrated 26.2 Tb/second capacity over a single MAREA fiber with a record SE of 6.21 b/second/Hz by using real-time PDM-16-QAM superchannels [130]. With these technical advances, MAREA alone is expected to provide over 400 Tb/second of total bidirectional communication capacity between the United States and Europe.

The superior transmission performance of modern undersea cable systems such as MAREA is attributed to technical advances in both coherent transceivers and fiber links. MAREA fiber link consists of 117 amplified fiber spans with an average length of 56.5 km. The fiber has a large effective area of 150 μm^2 (to reduce the nonlinear impairments) and a low loss of 0.156 dB/km. The EDFAs have a flat amplification window of about 4.5 THz. Using the FOM formula given in Eq. (8.36), we can find that the MAREA fiber has a FOM that is ~8.5 dB better than the conventional G.652 fiber link with 100-km spans. Taking into consideration the shorter fiber span (by ~2.5 dB), the MAREA fiber link is expected to support a reach that is ~6 dB longer than (or ~4 times as long as) that supported by the conventional G.652 fiber link with 100-km spans. Since PDM-16QAM has been shown to reach 1500 km in the conventional G.652 fiber link with 100-km spans [99], it can then reach ~6000 km with the MAREA fiber link. In addition, the EDFAs used in undersea applications typically have a low noise figure (NF), making it feasible for PDM-16QAM to reach 6600 km in the MAREA link.

When even longer transmission distance needs to be supported, for example, for transpacific links of over 13,000 km, lower-level modulation such as PDM-8QAM and PDM-QPSK can be used. Indeed, PDM-8QAM was shown to allow the loop-back in the MAREA experiment to reach 13,210 km [130]. When more capacity is needed, a wider EDFA bandwidth can be used. In the Pacific Light Cable Network (PLCN), which connects Hong Kong, Taiwan, the Philippines and the United States, *C* and *L* bands were used [131]. It can be expected that with further enhancements such as PS, more capacity can be offered in new-generation undersea fiber cable systems such as MAREA and PLCN.

In modern undersea cable systems, more fiber pairs are packaged inside each cable to increase the overall transmission capacity of each cable. This is often referred to as space-division multiplexing (SDM) in undersea transmission. Since the total electrical power provided to an undersea cable is limited, it is meaningful to maximize the total cable capacity with the power constraint. Recently, there had been intensive studies to explore SDM to maximize the undersea system capacity under the power constraint [132–141]. A relevant performance indicator is the system power efficiency (PE) defined as [141]

$$PE = \frac{C}{P} = \frac{2B\log_2(1 + \mathrm{SNR})}{P} \tag{8.37}$$

where C is the Shannon capacity, P is the total electrical power provided to the cable system, B is the amplification bandwidth, SNR is the signal-to-noise ratio, and the factor "2" is for PDM. As the electronic power is mostly spent in the EDFAs to amplify the signals, P is proportional to the output power of each EDFA and is related to SNR as

$$P = p(1 + \mathrm{SNR}) \tag{8.38}$$

where p is a constant. Substituting Eq. (8.38) into Eq. (8.27), we have

$$PE = \frac{2B\log_2(1 + \mathrm{SNR})}{p(1 + \mathrm{SNR})} \tag{8.39}$$

There exists an optimal SNR that gives the maximal PE

$$SNR_{opt.} = 2^{1/(\ln 2)} - 1 = 1.72 \text{ (or 2.35 dB)} \tag{8.40}$$

At this optimal SNR, the optimal SE is then

$$SE_{opt.} = 2\log_2(1 + 1.72) = 2.89\ (b/s/\text{Hz}) \tag{8.41}$$

The above optimization process is quite general, as it does not depend on link parameters such as span length, total reach, and amplifier NF. This result is also in a reasonable agreement with experiments [141]. Thus future undersea cable systems may leverage this finding to use more fiber pairs per cable to further increase the total cable capacity for a given electrical power limitation. With the use of more fiber pairs per cable, higher integration of multifiber EDFA could be desirable, opening new opportunities for technical innovations [141]. In addition, the cost per bit also matters in undersea fiber communications. It was found that while the cost per bit reduces with the increase in the capacity, the cost saving saturates at a large capacity [141]. Therefore continued technical innovations are needed to advance undersea fiber communications to the next level.

8.8 Summary and concluding remarks

The field of high-capacity long-haul optical fiber transmission has witnessed great technical advances in the 5G era, such as flexible PDM-n-QAM modulation, digital coherent detection, digital subcarrier multiplexing, fiber nonlinearity mitigation and compensation, capacity-approaching FEC, probabilistic constellation shaping, improved optical fibers with less nonlinearity and lower loss, and wide-band EDFAs to name a few. With these advances, per-fiber capacity has been increased to 88 Tb/second in commercial data center interconnect (DCI) applications on the one hand [126], and over 26 Tb/second in ultra-long-haul transatlantic undersea links [130] on the other hand. Via flexible modulation, coding and shaping, different transmission data rates, capacities, and distances can be realized to optimally address the diverse applications of high-capacity and long-haul transmission in metro, regional, national, continental, intercontinental, and trans-oceanic optical networks.

Table 8.2 summarizes the key capabilities of commercial high-capacity long-haul optical fiber transmission systems in the early 2020s [57,99,117,121,126,130,142–150].

Going forward, we can expect new innovations to be made to further approach the nonlinear Shannon limit. Better ASIC technologies will help reduce the power consumption per bit further. Higher speed modulation beyond 100 Gbaud will help reduce the cost per bit. Tighter integration between photonics and electronics would help reduce the overall power consumption, size and cost. The superchannel concept

Table 8.2 Key capabilities of commercial high-capacity long-haul optical fiber transmission systems in the early 2020s.

Data rate per wavelength channel	100/200/400/600/800 Gb/s
Modulation format	PDM-BPSK/QPSK/8QAM/16QAM/64QAM
Modulation speed	30–100 Gbaud
FEC	Capacity-approaching with 15–33% overhead
Shaping	PS with flexible entropy loadings
WDM channel spacing	50/75/100/125 GHz
EDFA bandwidth	4.8 THz (extended C), 6 THz (super C), and 11 THz (super C + L)
Per-fiber capacity	10–88 Tb/s
System reach	200–15,000 km

PDM, polarization-division multiplexed; *BPSK*, binary-phase-shift keying; *QAM*, quadrature-amplitude modulation; *FEC*, forward-error correction; *WDM*, wavelength-division multiplexing; *EDFA*, Erbium-doped fiber amplifier.

can be used to achieve multiTb/second throughput per transceiver module, leverage large-scale photonic integration, and increase WDM SE. In the following chapter, we will discuss superchannel transmission, flexible-grid WDM, reconfigurable optical add/drop, and optical cross-connect.

References

[1] Shannon CE. A mathematical theory of communication. Bell Syst Tech J 1948;27(3):379–423.
[2] Bosco G, Curri V, Carena A, Poggiolini P, Forghieri F. On the performance of Nyquist-WDM terabit superchannels based on PM-BPSK, PM-QPSK, PM-8QAM or PM-16QAM subcarriers. J Light Technol 2011;29(1):53–61.
[3] Poggiolini P, Bosco G, Carena A, Curri V, Jiang Y, Forghieri F. The GN model of fiber non-linear propagation and its applications. J Light Technol 2014;32(4):694–721.
[4] Chandrasekhar S, Liu X. Advances in Tb/s superchannels. In: Kaminov IP, Li T, Willner A, editors, Chapter 3 of optical fiber telecommunications VIB; 2013.
[5] Liu X. Evolution of fiber-optic transmission and networking toward the 5G era. iScience 2019;22:489–506.
[6] Liu X. et al. 3x485-Gb/s WDM transmission over 4800 km of ULAF and 12x100-GHz WSSs using CO-OFDM and single coherent detection with 80-GS/s ADCs. In: Optical fiber communication conference (OFC), paper JThA37; 2011.
[7] Mitra PP, Stark JB. Nonlinear limits to the information capacity of optical fiber communications. Nature 2001;411(6841):1027–30.
[8] Essiambre R-J, Foschini GJ, Kramer G, Winzer PJ. Capacity limits of information transport in fiber-optic networks. Phys Rev Lett 2008;101 Paper 163901.
[9] Essiambre RJ, Kramer G, Winzer PJ, Foschini GJ, Goebel B. Capacity limits of optical fiber networks. J Light Technol 2010;28(4):662–701.
[10] Chen X, Shieh W. Closed-form expressions for nonlinear transmission performance of densely spaced coherent optical OFDM systems. Opt Express 2010;18:19039–54.
[11] Rafique D, Ellis AD. Impact of signal-ASE four-wave mixing on the effectiveness of digital back-propagation in 112 Gb/s PM-QPSK systems. Opt Express 2011;19(4):3449–54.

[12] Dar R, Feder M, Mecozzi A, Shtaif M. Properties of nonlinear noise in long, dispersion-uncompensated fiber links. Opt Express 2013;21:25685−99.
[13] Gnauck AH, Raybon G, Chandrasekhar S, Leuthold J, Doerr C, Stulz L. et al. 2.5 Tb/s (64 x 42.7 Gb/s) transmission over 40 x 100 km NZDSF using RZ-DPSK format and all-Raman-amplified spans. In: Optical fiber communication conference (OFC), Anaheim, CA, post-deadline paper FC2; 2002.
[14] Xu C, Liu X, Wei X. Differential phase-shift keying for high spectral efficiency optical transmissions. IEEE J Sel Top Quant Electron 2004;10(2):281−93.
[15] Gnauck AH, Winzer PJ. Optical phase-shift-keyed transmission. J Light Technol 2005;23 (1):115−30.
[16] Taylor MG. Coherent detection method using DSP for demodulation of signal and subsequent equalization of propagation impairments. IEEE Photon Technol Lett 2004;16(2):674−6.
[17] Noe R. PLL-free synchronous QPSK polarization multiplex/diversity receiver concept with digital I&Q baseband processing. IEEE Photo Technol Lett 2005;17(4):887−9.
[18] Ly-Gagnon D-S, Tsukamoto S, Katoh K, Kikuchi K. Coherent detection of optical quadrature phase shift keying signals with carrier phase estimation. J Light Technol 2006;24(1):12−21.
[19] Leven A, Kaneda N, Klein A, Koc U-V, Chen Y-K. Real-time implementation of 4.4 Gbit/s QPSK intradyne receiver using field programmable gate array. Electron Lett 2006;42(24):1421−2.
[20] Savory S. Digital filters for coherent optical receivers,". Opt Express 2008;16:804.
[21] Ip E, Lau APT, Barros DJF, Kahn JM. Coherent detection in optical fiber systems. Opt Express 2008;16(2):753−91.
[22] Fludger CRS, et al. Coherent equalization and POLMUX-RZDQPSK for robust 100-GE transmission. J Light Technol 2008;26(1):64−72.
[23] Sun H, Wu K-T, Roberts K. Real-time measurements of a 40 Gb/s coherent system. Opt Express 2008;16(2):873−9.
[24] Liu X, Chandrasekhar S, Leven A. Self-coherent optical transport systems. In: Kaminov IP, Li T, Willner A, editors, Chapter 4 of Optical fiber telecommunications VB; 2008 February.
[25] Winzer PJ. High-spectral-efficiency optical modulation formats. J Light Technol 2012;30:3824−35.
[26] Laperle C, O'Sullivan M. Advances in high-speed DACs, ADCs, and DSP for optical coherent transceivers. J Light Technol 2014;32:629−43.
[27] Haykin S. Signal processing: where physics and mathematics meet. IEEE Signal Process Mag 2001;18(4):6−7.
[28] Spinnler B, Hauske FN, Kuschnerov M. Adaptive equalizer complexity in coherent optical receivers. In: European conference on optical communications (ECOC), Belgium, paper We.2.E.4; 2008.
[29] Kudo R, Kobayashi T, Ishihara K, Takatori Y, Sano A, Miyamoto Y. Coherent optical single carrier transmission using overlap frequency domain equalization for long-haul optical systems. J Light Technol 2009;27(16):3721−8.
[30] Godard DN. Self-recovering equalization and carrier tracking in two-dimensional data communication systems. IEEE Trans Commun 1980;28:1867−75 vol. COM.
[31] Treichler JR, Agee BG. A new approach to multipath correction of constant modulus signals. IEEE Tran Acoust, Speech Signal Process 1983;31:459−72 vol ASSP.
[32] Treichler JR, Larimore MG. New processing techniques based on the constant modulus adaptive algorithm. IEEE Tran Acoust, Speech Signal Process 1985;33:420−31 vol ASSP.
[33] Johnson Jr. R, Schniter P, Endres TJ, Behm JD, Brown DR, Casas RA. Blind equalization using the constant modulus criterion: a review. Proc IEEE 1998;86:1927−50.
[34] Chi C-Y, Chen C-Y, Chen C-H, Feng C-C. Batch processing algorithms for blind equalization using higher-order statistics. IEEE Signal Process Mag 2003;20:25−49.
[35] Proakis JG. *Digital communications* (4th ed.). New York: McGraw Hill; 2000.
[36] Viterbi AJ, Viterbi AM. Nonlinear esitimation of PSK-modulated charier phase with application to burst digital transmission. IEEE Trans Inf Theory 1983;29:543−51 vol. IT.
[37] Zhang S, Kam P-Y, Yu C, Chen J. Decision-aided carrier phase estimation for coherent optical communications. J Light Technol 2010;28:1597−607.

[38] Zhang H, Cai Y, Foursa DG, Pilipetskii AN. Cycle slip mitigation in POLMUX-QPSK modulation. In: Optical fiber communication conference and exposition (OFC/NFOEC), Los Angeles, CA, paper OMJ7; 2011.

[39] Fludger CRS, Nuss D, Kupfer T. Cycle-slips in 100G DP-QPSK transmission systems. In: Optical fiber communication conference and exposition (OFC/NFOEC), Los Angeles, CA, paper OTu2G.1; 2012.

[40] Magarini M, Barletta L, Spalvieri A, Vacondio F, Pfau T, Pepe M, et al. Pilot symbols-aided carrier-phase recovery for 100-G PM-QPSK digital coherent receivers. IEEE Photon Technol Lett 2012;24(9):739−41.

[41] Gao Y, Lau APT, Lu C. Cycle-slip resilient carrier phase estimation for polarization multiplexed 16-QAM systems. In: Opto-electronics and communications conference (OECC), Busan; 2012, p. 4B2−4.

[42] Cheng H, Li Y, Zhang F, Wu J, Lu J, Zhang G, et al. Pilot-symbols-aided cycle slip mitigation for DP-16QAM optical communication systems. Opt Express 2013;21:22166−72.

[43] Chen X, Chandrasekhar S, Raybon G, Olsson S, Cho J, Adamiecki A, et al. Generation and intradyne detection of single-wavelength 1.61-Tb/s using an all-electronic digital band interleaved transmitter. In: Optical fiber communication conference (OFC), paper Th4C.1; 2018.

[44] Nakamura M, et al. 192-Gbaud signal generation using ultra-broadband optical frontend module integrated with bandwidth multiplexing function. In: Optical fiber communication conference (OFC), PDP Th4B.4; 2019.

[45] Buchali F, et al. 1.52 Tb/s single carrier transmission supported by a 128 GSa/s SiGe DAC. In: Optical fiber communication conference (OFC), Th4C.2, 2020.

[46] Pittalà F, et al. 220 GBaud signal generation enabled by a two-channel 256 GSa/s arbitrary waveform generator and advanced DSP. In: European conference on optical communications (ECOC), Brussels, Belgium, PDP1.4; 2020.

[47] Shieh W, Ho K-P. Equalization-enhanced phase noise for coherent-detection systems using electronic digital signal processing. Opt Express 2008;16:15718−27.

[48] Ho K-P. Subband equaliser for chromatic dispersion of optical fibre. Electron Lett 2009;45(24):1224−6.

[49] Nazarathy M, Tolmachev A. Subbanded DSP architectures based on underdecimated filter banks for coherent OFDM receivers: overview and recent advances,". IEEE Signal Process Mag 2014;31(2):70−81.

[50] Liu X, Chandrasekhar S, Zhu B, Winzer P, Gnauck A, Peckham D. 448-Gb/s reduced-guard-interval CO-OFDM transmission over 2000 km of ultra-large-area fiber and five 80-GHz-Grid ROADMs. J Light Technol 2010;29(4):483−90.

[51] Tolmachev A, Nazarathy M. Filter-bank based efficient transmission of reduced-guard-interval OFDM. Opt Express 2011;19:B370−84.

[52] Tang Y, Shieh W, Krongold BS. DFT-spread OFDM for fiber nonlinearity mitigation. IEEE Photon Technol Lett 2010;22(16):1250−2.

[53] Chen X, Li A, Gao G, Shieh W. Experimental demonstration of improved fiber nonlinearity tolerance for unique-word DFT-spread OFDM systems. Opt Express 2011;19:26198−207.

[54] Li C, Yang Q, Jiang T, He Z, Luo M, Li C, et al. Investigation of Coherent Optical Multiband DFT-S OFDM in Long Haul Transmission in IEEE Photon Technol Lett 2012;24(19):1704−7.

[55] Liu X, Winzer P, Chandrasekhar S, Randel S, Corteselli S. Multiband DFT-spread-OFDM equalizer with overlap-and-add dispersion compensation for low-overhead and low-complexity channel equalization. In Optical fiber communication conference (OFC), paper OW3B.2; 2013.

[56] Diniz PSR, da Silva EAB, Netto SL. Digital signal processing: system analysis and design. Cambridge Univ. Press; 2002.

[57] Sun H, et al. 800G DSP ASIC design using probabilistic shaping and digital sub-carrier multiplexing. J Light Technol 2020;38(17):4744−56 1 Sept.1.

[58] Liu X, Wei X, Slusher RE, McKinstrie CJ. Improving transmission performance in differential phase-shift-keyed systems by use of lumped nonlinear phase-shift compensation. Opt Lett 2002;27:1616−18.

[59] Ho K-P, Kahn JM. Electronic compensation technique to mitigate nonlinear phase noise. J Light Technol 2004;22:779−83.

[60] Roberts K, Li C, Strawczynski L, O'Sullivan M. Electronic precompensation of optical nonlinearity. IEEE Photon Technol Lett 2006;18:403–5.
[61] Johannisson P, Sjödin M, Karlsson M, Tipsuwannakul E, Andrekson PA. Cancellation of nonlinear phase distortion in self-homodyne coherent systems. IEEE Photon Technol Lett 2010;22:802–4.
[62] Ip E, Kahn JM. Compensation of dispersion and nonlinear impairments using digital back propagation. J Light Technol 2008;26:3416–25.
[63] Mateo EF, Zhu L, Li G. Impact of XPM and FWM on the digital implementation of impairment compensation for WDM transmission using backward propagation. Opt Express 2008;16:16124–37.
[64] Yariv A, Fekete D, Pepper DM. Compensation for channel dispersion by nonlinear optical phase conjugation. Opt Lett 1979;4:52–4.
[65] Pepper DM, Yariv A. Compensation for phase distortions in nonlinear media by phase conjugation. Opt Lett 1980;5:59–60.
[66] Fisher RA, Suydam BR, Yevick D. Optical phase conjugation for time-domain undoing of dispersive self-phase-modulation effects. Opt Lett 1983;8:611–13.
[67] Gnauck AH, Jopson RM, Derosier RM. 10-Gb/s 360-km transmission over dispersive fiber using midsystem spectral inversion. IEEE Photon Technol Lett 1993;5:663–6.
[68] Watanabe S, Chikama T, Ishikawa G, Terahara T, Kuwahara H. Compensation of pulse shape distortion due to chromatic dispersion and Kerr effect by optical phase conjugation. IEEE Photon Technol Lett 1993;5:1241–3.
[69] Chen X, Liu X, Chandrasekhar S, Zhu B, Tkach RW. Experimental demonstration of fiber nonlinearity mitigation using digital phase conjugation. In: Optical fiber communication conference (OFC), paper OTh3C.1; 2012.
[70] Liu X, Chraplyvy AR, Winzer PJ, Tkach RW, Chandrasekhar S. Phase-conjugated twin waves for communication beyond the Kerr nonlinearity limit. Nat Photon 2013;7:560–8.
[71] Agrawal GP. Nonlinear fiber optics. Academic Press; 2007.
[72] Wai PKA, Menyuk CR, Chen HH. Stability of solitons in randomly varying birefringent fibers. Opt Lett 1991;16:1231–3.
[73] Mecozzi A, Clausen CB, Shtaif M, Park S-G, Gnauck AH. Cancellation of timing and amplitude jitter in symmetric links using highly dispersed pulses. IEEE Photon Technol Lett 2001;13:445–7.
[74] Louchet H, Hodzic A, Petermann K, Robinson A, Epworth R. Simple criterion for the characterization of nonlinear impairments in dispersion-managed optical transmission systems. IEEE Photon Technol Lett 2005;17:2089–91.
[75] Wei X, Liu X. Analysis of intrachannel four-wave mixing in differential-phase-shift-keyed transmission with large dispersion. Opt Lett 2003;28:2300–2.
[76] Wei X. Power-weighted dispersion distribution function for characterizing nonlinear properties of long-haul optical transmission links. Opt Lett 2006;31:2544–6.
[77] Carena A, Curri V, Poggiolini P, Jiang Y, Forghieri F. Dispersion pre-compensation in PM-QPSK systems. In: Presented at the European conference on optical communication, Cannes, France, Sep. 2014, paper P.5.24.
[78] Poggiolini P, Nespola A, Jiang Y, Bosco G, Carena A, Bertignono L, et al. Analytical and experimental results on system maximum reach increase through symbol rate optimization. J Light Technol 2016;34(8):1872–85.
[79] Shieh W, Tang Y. Ultrahigh-speed signal transmission over nonlinear and dispersive fiber optic channel: the multicarrier advantage. IEEE Photon J 2010;2(3):276–83.
[80] Behrens C, Killey RI, Savory SJ, Chen M, Bayvel P. Nonlinear transmission performance of higher-order modulation formats. IEEE Photon Technol Lett 2011;23(6):377–9.
[81] Du LB, Lowery AJ. Optimizing the subcarrier granularity of coherent optical communications systems. Opt Exp 2011;19(9):8079–84.
[82] Shieh W, Chen X, Li A, Gao G, Al A. Amin What is the optimal symbol rate for long-haul transmission? In: 2011 Asia communications and photonics conference and exhibition (ACP), paper 83090L.

[83] Bononi A, Rossi N, Serena P. Performance dependence on channel baud-rate of coherent single-carrier WDM systems. In: European conference on optical communications (ECOC), paper Th.1.D.5; 2013.

[84] Qiu M, Zhuge Q, Chagnon M, Gao Y, Xu X, Morsy-Osman M, et al. Digital subcarrier multiplexing for fiber nonlinearity mitigation in coherent optical communication systems. Opt Express 2014;22:18770−7.

[85] Li X, Chen X, Goldfarb G, Mateo E, Kim I, Yaman F, et al. Electronic post-compensation of WDM transmission impairments using coherent detection and digital signal processing. Opt Express 2008;16(2):880−8.

[86] Tao Z, Dou L, Yan W, Li L, Hoshida T, Rasmussen JC. Multiplier-free intrachannel nonlinearity compensating algorithm operating at symbol rate. J Light Technol 2011;29(17):2570−6.

[87] Gao Y, Cartledge JC, Karar AS, Yam SS-H, O'Sullivan M, Laperle C, et al. Reducing the complexity of perturbation based nonlinearity pre-compensation using symmetric EDC and pulse shaping. Opt Express 2014;22:1209−19.

[88] Liu X, Chandrasekhar S, Winzer PJ. Phase-conjugated twin waves and fiber nonlinearity compensation. In: OECC'14, invited paper Th11B1; 2014.

[89] Ghazisaeidi A, Essiambre R-J. Calculation of coefficients of perturbative nonlinear pre-compensation for Nyquist pulses. In: European conference on optical communications (ECOC), paper We.1.3.3; 2014.

[90] Li Z, Peng W-R, Zhu F, Bai Y. Optimum quantization of perturbation coefficients for perturbative fiber nonlinearity mitigation. In: European conference on optical communications (ECOC), paper We.1.3.4; 2014.

[91] Peng W-R, Li Z, Zhu F, Bai Y. Extending perturbative nonlinearity mitigation to PDM-16QAM. In European conference on optical communications (ECOC), paper We.3.3.4; 2014.

[92] Liu X. Twin-wave-based optical transmission with enhanced linear and nonlinear performances. J Light Technol 2015;33(5):1037−43.

[93] Poggiolini P. Modeling of non-linear propagation in uncompensated coherent systems. In Optical fiber communication conference (OFC), tutorial paper OTh3G.1; 2013.

[94] Mizuochi T. Recent progress in forward error correction and its interplay with transmission impairments. IEEE J Sel Top Quant Electron 2006;12(4):544−54.

[95] Optical Internetworking Forum (OIF). Implementation agreement 400ZR [Online]. March 2020. Available from: https://www.oiforum.com/wp-content/uploads/OIF-400ZR-01.0_reduced2.pdf

[96] Nagarajan R, Lyubomirsky I. Low-complexity DSP for inter-data center optical fiber communications. In: 2020 European conference on optical communications (ECOC), tutorial paper SC04; 2020.

[97] Hirbawi Y, Zhu Q, Caia J-M. Continuation & results of FEC proposals evaluation for ITU G.709.3 200−400G 450km black link. In: Contribution to the ITU G.709.3, CD11-M10; 2019 May.

[98] Roese J, et al. Proposal to specify OFEC for FlexO-LR 450 km application. In: Contribution to the ITU study group 15, Q11, SG15-C1345R1; 2019 July.

[99] Loussouarn Y, Pincemin E. Probabilistic-shaping DP-16QAM CFP-DCO transceiver for 200G upgrade of legacy metro/regional WDM infrastructure. In Optical fiber communication conference (OFC), paper M2D.2; 2020.

[100] Smith BP, Farhood A, Hunt A, Kschischang FR, Lodge J. Staircase Codes: FEC for 100 Gb/s OTN in J Light Technol 2012;30(1):110−17.

[101] Gallager RG. Information theory and reliable communication. New York: Wiley; 1968.

[102] Kschischang FR, Pasupathy S. Optimal nonuniform signaling for gaussian channels. IEEE Trans Inf Theory 1993;39(3):913−29.

[103] Raphaeli D, Gurevitz A. Constellation shaping for pragmatic turbo-coded modulation with high spectral efficiency. IEEE Trans Commun 2004;52(3):341−5.

[104] Djordjevic IB, Batshon HG, Xu L, Wang T. Coded polarization multiplexed iterative polar modulation (PM-IPM) for beyond 400 Gb/s serialopticaltransmission. In: Optical fiber communication conference (OFC), paper OMK2; 2010.

[105] Liu X, Chandrasekhar S, Lotz TH, Winzer PJ, Haunstein H, Randel S, et al. Generation and FEC-decoding of a 231.5-Gb/s PDM-OFDM signal with 256-iterative-polar-modulation achieving 11.15-b/s/Hz intrachannel spectral efficiency and 800-km reach. In: Optical fiber communication conference (OFC), post-deadline paper PDP5B; 2012.
[106] Smith B, Kschischang F. A pragmatic coded modulation scheme for high-spectral-efficiency fiber-optic communications. J Light Technol 2012;30(13):2047–53.
[107] Lotz T, Liu X, Chandrasekhar S, Winzer P, Haunstein H, Randel S, et al. Coded PDM-OFDM transmission with shaped 256-iterative-polar-modulation achieving 11.15-b/s/Hz intrachannel spectral efficiency and 800 km reach. J Light Technol 2013;31(4):538–45.
[108] Dar R, Feder M, Mecozzi A, Shtaif M. On shaping gain in the nonlinear fiber-optic channel. In: Proceedings of IEEE international symposium on information Theory; 2014 June, p. 2794–2798.
[109] Yankov M, Zibar D, Larsen K, Christensen L, Forchhammer S. Constellation shaping for fiber-optic channels with QAM and high spectral efficiency. IEEE Photon. Technol. Lett. 2014;26(23):2407–10.
[110] Böcherer G. Probabilistic signal shaping for bit-metric decoding. In: 2014 IEEE international symposium on information theory, Honolulu, HI, USA; 2014, p. 431–435.
[111] Fehenberger T, Böcherer G, Alvarado A, Hanik N. LDPC coded modulation with probabilistic shaping for optical fiber systems. In Optical fiber communication conference (OFC), paper Th2A.23; 2015.
[112] Buchali F, Böcherer G, Idler W, Schmalen L, Schulte P, Steiner F. Experimental demonstration of capacity increase and rate-adaptation by probabilisticallyshaped64-QAM. In: Presented at the European conference on optical communication, Valencia, Spain, Paper PDP3.4; 2015.
[113] Böcherer G, Steiner F, Schulte P. Bandwidth efficient and rate matched low-density parity-check coded modulation. IEEE Trans Commun 2015;63(12):4651–65.
[114] Buchali F, Steiner F, Böcherer G, Schmalen L, Schulte P, Idler W. Rate adaptation and reach increase by probabilistically shaped 64-QAM: an experimental demonstration. J Light Technol 2016;34:1599–609.
[115] Schulte P, Böcherer G. Constant composition distribution matching. IEEE Trans Inf Theory 2016;62(1):430–4.
[116] Cho J, et al. Trans-atlantic field trial using high spectral efficiency probabilistically shaped 64-QAM and single-carrier real-time 250-Gb/s 16-QAM. J Light Technol 2018;36:103–13.
[117] Li J, Zhang A, Zhang C, Huo X, Yang Q, Wang J, et al. Field trial of probabilistic-shaping-programmable real-time 200-Gb/s coherent transceivers in an intelligent core optical network. In: Asia communications and photonics conference (ACP), PDP Su2C.1; 2018.
[118] Böcherer G, Schulte P, Steiner F. Probabilistic shaping and forward error correction for fiber-optic communication systems. J Light Technol 2019;37:230–44 January 15.
[119] Cho J, Winzer PJ. Probabilistic constellation shaping for optical fiber communications,". J Light Technol 2019;37(6):1590–607.
[120] Buchali F, et al. Optimization of time-division hybrid-modulation and its application to rate adaptive 200Gb transmission. In: European conference on optical communications (ECOC), paper Tu.4.3.1; 2014.
[121] Castrillón A, Xu H, Morero D, Taddei A, Asinari M, Fan SH, et al. First real-time demonstration of probabilistic shaping 400G transmission enabling high-performance pluggable module applications. In: 2020 IEEE photonics conference (IPC), Vancouver, BC, Canada, paper MG1.3; 2020 September.
[122] See for example, https://fiber-optic-catalog.ofsoptics.com/Asset/TeraWave-Fiber-161-web.pdf; and https://www.corning.com/media/worldwide/coc/documents/Fiber/WP8105_1.17.pdf.
[123] Shen S, Wang G, Wang H, He Y, Wang S, Zhang C, et al. G.654.E fibre deployment in terrestrial transport system. In: Optical fiber communication conference (OFC), paper M3G.4; 2017.
[124] Hasegawa T, Yamamoto Y, Hirano M. Optimal fiber design for large capacity long haul coherent transmission [Invited]. Opt Express 2017;25(2):706–12.

[125] Yu Y, Jin L, Xiao Z, Yu F, Lu Y, Liu L, et al. 100.5Tb/s MLC-CS-256QAM transmission over 600-km single mode fiber with C + L Band EDFA. In: Asia communications and photonics conference (ACP), PDP Su2C.3; 2018.
[126] See for example, https://www.cio.com/article/3607195/huawei-optixtrans-dc908-ranked-dci-leader-again-by-globaldata.html.
[127] See for example, https://en.wikipedia.org/wiki/SEA-ME-WE_3.
[128] See for example, https://en.wikipedia.org/wiki/SEA-ME-WE_5.
[129] See for example, https://en.wikipedia.org/wiki/MAREA.
[130] Grubb S, Mertz P, Kumpera A, Dardis L, Rahn J, O'Connor J, et al. Real-time 16QAM transatlantic record spectral efficiency of 6.21 b/s/Hz enabling 26.2 Tbps capacity. In: Optical fiber communication conference (OFC), paper M2E.6; 2019.
[131] See for example, https://www.submarinenetworks.com/en/systems/trans-pacific/plcn.
[132] Winzer PJ. Energy-efficient optical transport capacity scaling through spatial multiplexing. IEEE Photon Technol Lett 2011;23(13):851−3.
[133] Pilipetskii A, Foursa D, Bolshtyansky M, Mohs G, Bergano NS. Optical designs for greater power efficiency. In: Proceedings SubOptic 2016, Dubai 2016, paper TH1A.5; 2016.
[134] Sinkin OV, Turukhin AV, Patterson WW, Bolshtyansky MA, Foursa DG, Pilipetskii AN. Maximum optical power efficiency in SDM-based optical communication systems. IEEE Photon Technol Lett 2017;29(13):1075−7.
[135] Domingues OD, Mello DAA, da Silva R, Arık SÖ, Kahn JM. Achievable rates of space-division multiplexed submarine links subject to nonlinearities and power feed constraints. J Light Technol 2017;35(18):4004−10.
[136] Sinkin OV, Turukhin AV, Sun Y, Batshon HG, Mazurczyk MV, Davidson CR, et al. SDM for power-efficient under sea transmission. J Light Technol 2018;36(2):361−71.
[137] Dar R, Winzer PJ, Chraplyvy AR, Zsigmond S, Huang K-Y, Fevrier H, et al. Cost-optimized submarine cables using massive spatial parallelism. J Light Technol 2018;36(18):3855−65.
[138] Downie JD. Maximum capacities in submarine cables with fixed power constraints for C-band, C + L-band, and multicore fiber systems. J Light Technol 2018;36(18):4025−32.
[139] Turukhin A, Paskov M, Mazurczyk MV, Patterson WW, Batshon HG, Sinkin OV, et al. Demonstration of potential 130.8 Tb/s capacity in power-efficient SDM transmission over 12,700 km using hybrid microassembly based amplifier platform. In Optical fiber communication conference (OFC), paper M2I.4; 2019.
[140] Bolshtyansky MA, Sinkin OV, Paskov M, Hu Y, Cantono M, Jovanovski L, et al. Single-mode fiber SDM submarine systems. J Light Technol 2020;38(6):1296−304.
[141] Pilipetskii A, Bolshtyansky M, Foursa D, Sinkin O. SDM power-efficient ultra high-capacity submarine long haul transmission systems (tutorial). In: Optical fiber communication conference (OFC) 2020, paper M3G.5; 2020.
[142] Okamoto S, Yonenaga K, Horikoshi K, Yoshida M, Miyamoto Y, Tomizawa M, et al. 400Gbit/s/ch field demonstration of modulation format adaptation based on pilot-aided OSNR estimation using real-time DSP. IEICE Trans Commun 2017;E100-B(10):1726−33.
[143] Smith K, Zhou YR, Gilson M, Pan W, Huang W, Shen L, et al. Field trial of real-time 400GbE superchannel using configurable modulation formats. IEEE Photon Technol Lett 2018;30(23):2044−7.
[144] Kawahara H, Saito K, Nakagawa M, Kubo T, Seki T, Kawasaki T, et al. Real-time demonstration of 600Gbps/carrier WDM transmission and highly-survivable adaptive restoration on field installed fiber. In: OECC/PSC2019, TuB2−2, Fukuoka, Japan; 2019.
[145] Hamaoka F, Sasai T, Saito K, Kobayashi T, Matsushita A, Nakamura M, et al. Dual-carrier 1-Tb/s transmission over field-deployed large-core pure-silica-core fiber link using real-time transponder. In: OECC/PSC2019, PDP1, Fukuoka, Japan; 2019.
[146] Li J, Zhang A, Zhang C, Huo X, Yang Y, Zhang J, et al. Field trial of real time 200G and 400G flex-rate transmission using 69 Gbaud signal. In: European conference on optical communications (ECOC), paper Th.1.A.2; 2019.

[147] Yu Y, Wang G, Shen S, Wang S, Zhao C, Gui T, et al. Real-time demonstration of C-band 19.2Tb/s field trial by constellation shaping 16QAM in 2000-km G.654.E terrestrial link. in European conference on optical communications (ECOC), P71; 2019.
[148] Zhang H, Zhu B, Pfau T, Aydinlik M, Nadarajah N, Park S, et al. Real-time transmission of single-carrier 400Gb/s and 600Gb/s 64QAM over 200km-Span link. In: European conference on optical communications (ECOC), paperTu.2.D.6; 2019.
[149] Zhou YR, Smith K, Duff S, Wang H, Pan W, Hackett P, et al.. Field trial demonstration of real-time 1.2Tb/s (2 × 600Gb/s) optical channel over a live G.652 fiber link achieving net spectral efficiency of 8bit/s/Hz. In: Asia communications and photonics conference/international conference on information photonics and optical communications 2020 (ACP/IPOC), paper T2B.3; 2020.
[150] Hamaoka F, et al. Dual-carrier 1-Tb/s transmission over field-deployed G.654.E fiber link using real-time transponder. IEICE Trans Commun 2020;E103–B(11):1183–9 VOL.

CHAPTER 9

Superchannel transmission and flexible-grid wavelength routing

9.1 Introduction on superchannel

To satisfy the ever-increasing capacity demand in optical fiber communications, the data rate carried by each wavelength channel in wavelength-division multiplexing (WDM) systems has been increasing exponentially [1–3]. The evolution of serial interface rate and system capacity of optical networks is shown in Fig. 9.1 [3]. The dramatic system capacity increase that occurred in the 1990s was due to the introduction of WDM that allows multiple wavelength channels to be carried on the same fiber link. Subsequent to the year 2000, the WDM system capacity has increased at a pace similar to that of the serial interface rate. In the late-2000s, digital coherent detection enabled coherent modulation and polarization-division multiplexing (PDM) and speeded up the increase in serial interface rate. 100-Gb/s per-channel data rates had been realized in commercial systems since the early-2010s [3]. Subsequently, 200/400/600/800-Gb/s per-channel data rates had been realized prior to 2021, as discussed in the previous chapter.

With the introduction of optical superchannels, which avoid the electronic bottleneck via optical parallelism, optical transmission with per-channel data rates reaching and even exceeding Terabits/sec (Tb/s) has been experimentally demonstrated [4–7]. The term "superchannel" was first coined by Chandrasekhar et al. [4] for multiple single-carrier-modulated signals arranged under the orthogonal frequency-division multiplexing (OFDM) conditions [8–10]. The superchannel concept was later generalized to any collection of optical signals that are:

- modulated and multiplexed together with a high spectral efficiency (SE) at a same originating site
- transmitted and routed together over a same optical link
- received at a same destination site.

To achieve high-SE multiplexing, "Nyquist-WDM," "quasi-Nyquist-WDM," and "super-Nyquist-WDM" [11–18] have been introduced as promising alternatives of optical OFDM (O-OFDM). Since the first report of 1.2-Tb/s superchannel transmission in 2009 [4], many Tb/s-class superchannel transmission demonstrations have been reported, such as 19.2-Tb/s comb-based superchannel transmission over the C-band

Optical Communications in the 5G Era
DOI: https://doi.org/10.1016/B978-0-12-821627-9.00005-X

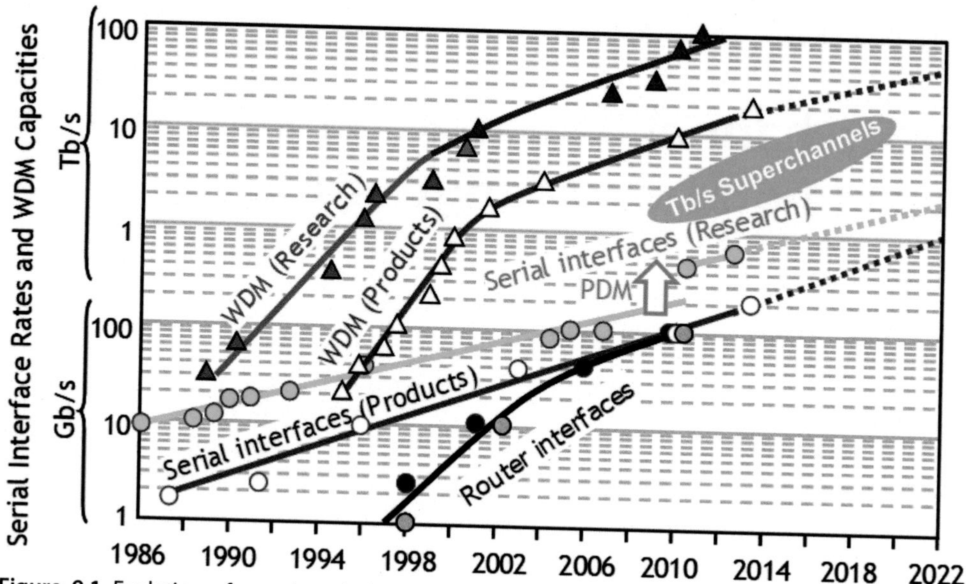

Figure 9.1 Evolution of wavelength-division multiplexing capacity, serial interface rate, and Tb/s-class superchannel data rate.

in 2019 [19], as indicated in Fig. 9.1. There are five key benefits of using superchannels in WDM systems as follows [20]:

1. Higher per-channel data rate, to meet the demand of ever-increasing serial interface rates, by avoiding the electronic bottleneck imposed by opto-electronic converters, electro-optical converters, digital-to-analog converters (DAC), and analog-to-digital converters (ADC)
2. Higher SE in WDM transmission, especially in transparent optical networks based on reconfigurable optical add/drop multiplexers (ROADMs), by reducing the percentage of wasted optical spectrum between channels that are on different routes
3. Better leverage of large-scale photonic integrated circuits (PICs) and application-specific integrated circuits (ASICs) to process multiple wavelength channels within each superchannel together
4. The feasibility to perform joint processing of superchannel constituents to further improve the transmission performance
5. Reduced the number of optical-routing paths/ports in optical cross-connect (OXC), by aggregating more capacity in each optically routed superchannel.

Tb/s-class superchannels have also led to a rethinking of spectral bandwidth allocation beyond the current fixed-grid architectures, leading to the so-called flexible-grid architecture that allows for a more efficient utilization of the optical spectrum. In this chapter, we review recent progress on Tb/s-class superchannel transmission and wavelength routing. This chapter is organized as follows. Section 9.2 describes the superchannel principle and multiplexing schemes. Section 9.3 presents superchannel transceiver designs including the joint signal processing of superchannel constituents. Section 9.4 addresses the transmission performance of superchannels based on joint nonlinearity compensation (NLC). Section 9.5 discusses high-capacity wavelength routing with Tb/s-class superchannels. Finally, Section 9.6 concludes this chapter with an outlook of future perspectives on superchannel-enabled high-capacity optical transmission and wavelength routing.

9.2 Superchannel principle and multiplexing schemes

Over the last several decades, the field of WDM transmission has evolved from sparsely populated channels to very densely populated channels, as in the case of superchannel transmission. It is therefore instructive to classify WDM systems on the basis of the channel bandwidth allocation (or channel spacing), Δf, relative to the modulation symbol rate of the channel B. In Table 9.1, different classes of WDM systems are defined [20]. One can clearly see that the recent progress in high-SE systems using advanced modulation formats with coherent detection has opened new regimes, identified as "quasi-Nyquist" WDM (for $1 \leq \Delta f/B \leq 1.2$), "Nyquist-WDM" (for $\Delta f/B = 1$) and "super-Nyquist" WDM (for $\Delta f/B < 1$), respectively. Here, the term "Nyquist" is borrowed from the well-known Nyquist–Shannon sampling theorem, named after Nyquist [21] and Shannon [22], which essentially states that if a function $x(t)$ contains no frequencies higher than $B/2$ Hertz, then it is completely determined by giving its ordinates at a series of points spaced $1/B$ seconds apart. This is a fundamental result in the field of information theory, particularly in telecommunications and signal processing. More specifically, Nyquist-WDM here means that the optical spectral bandwidth of an optical signal modulated at a symbol rate of B can be limited to B without losing its fidelity, and such signals can be packed at a frequency spacing equal to B in the ideal case, or $\Delta f/B = 1$.

Table 9.1 Definitions of various classes of wavelength-division multiplexing.

Definition	Condition	Example
Coarse WDM	$\Delta f/B > 20$	10 Gb/s on a 20-nm grid
WDM	$5 < \Delta f/B \leq 20$	10 Gb/s on a 100-GHz grid
Dense wavelength division multiplexing (DWDM)	$1.2 < \Delta f/B \leq 5$	28-Gbaud PDM-QPSK on 50 GHz
"Quasi-Nyquist" WDM	$1 < \Delta f/B \leq 1.2$	96-Gbaud PDM-Quadrature amplitude modulation (QAM) on 100 GHz
"Nyquist" WDM	$\Delta f/B = 1$	25-Gbaud PDM-QPSK on 25 GHz
"superNyquist" WDM	$\Delta f/B < 1$	28-Gbaud PDM-QPSK on 25 GHz

Packing wavelength channels closer than the Nyquist frequency, termed as "super-Nyquist" WDM or "faster-than-Nyquist" WDM [23], will cause crosstalk, which can be compensated, albeit with added digital signal processing (DSP) complexity and reduced optical signal-to-noise (OSNR) performance. The definition makes no assumptions on how a channel is modulated or any physical impairment associated with placing channels close together, such as crosstalk from overlapping spectral content. As an example, optical prefiltering (PreF) has been employed to mitigate crosstalk in the demonstration of "quasi-Nyquist" WDM and "super-Nyquist" WDM. Alternatively, electronic PreF has also been employed for such demonstrations.

A special case of "Nyquist" WDM is the one that additionally satisfies the O-OFDM conditions, as described below, allowing for crosstalk-free reception of symbol-rate–spaced channels without using optical or electrical PreF [4,8,24–27]. The O-OFDM conditions that must be fulfilled for multiplexing multiple modulated carriers to form a superchannel [8] can be enumerated as follows:

- The carrier spacing must equal the symbol rate with sufficient accuracy (inversely proportional to the duration of each processing block at the receiver). This implies that the carriers on which the modulation is imprinted need to be frequency-locked.
- The modulated symbols on the carriers need to be time aligned at the point of demultiplexing.
- Typically, the frequency-domain response of the modulated symbols is a *sinc* function. This implies that sufficient bandwidth is needed at the transmitter and the receiver to modulate each subcarrier. At the receiver, there must also be sufficient oversampling speed to capture most of the sinc function for each of the modulated subcarriers.

The transmitter and receiver bandwidth requirements for fulfilling the O-OFDM conditions can be relaxed by using OFDM to modulate each signal [28], which effectively reduces the bandwidth of each orthogonal optical subcarrier. In this case, one has N_e electronic subcarriers for each of the N_o optical subcarriers. The bandwidth requirements can also be relaxed by introducing a guard interval (GI) between adjacent modulated symbols, albeit at the expense of reduced WDM SE [29]. While O-OFDM requires that the orthogonality conditions be met by the modulated signals that construct a superchannel, quasi-Nyquist-WDM relaxes the multiplexing requirement by allocating a guard band (GB) between adjacent modulated signals, which leads to a slight reduction in the system SE.

Table 9.2 summarizes various demonstrated multiplexing schemes for superchannel formation [4,19,27,30–36]. Key performance indicators include intrachannel SE (ISE) and the ISE-distance product (ISEDP) [37]. Using OFDM-based seamless multiplexing, Chandrasekhar et al. demonstrated the transmission of a 1.2-Tb/s superchannel, consisting of 24 PDM quadrature phase-shift keying (QPSK) signals, over 7200 km of

Table 9.2 Various multiplexing schemes for superchannel formation.

Modulation format	Superchannel data rate	Formation	ISE (b/s/Hz)	Reach (km)	ISEDP (km · b/s/Hz)
Seamless CO-OFDM with frequency-locked optical carriers					
PDM-QPSK [4]	1.2 Tb/s	24 × 50 Gb/s	3.74	7200	26,928
PDM-16QAM [27]	1.5 Tb/s	15 × 100 Gb/s	7.00	1200	8400
OFDM-16QAM [30]	0.485 Tb/s	10 × 48.5 Gb/s	6.20	4800	29,760
Guard-banded CO-OFDM with frequency-unlocked carriers					
OFDM-16QAM [31]	1.864 Tb/s	8 × 233 Gb/s	5.75	5600	32,000
Quasi-Nyquist-WDM with frequency-unlocked carriers					
PDM-32/64QAM [32]	0.504 Tb/s	5 × 100.8 Gb/s	8	1200	9600
PDM-16QAM [33]	1.28 Tb/s	2 × 640 Gb/s	5.0	3200	16,000
Quasi-Nyquist-WDM with frequency-locked optical carriers					
PDM-64QAM [34]	6 Tb/s	25 × 240 Gb/s	~11.3	160	~1800
PDM-256QAM [19]	19.2 Tb/s	50 × 384 Gb/s	12	40	480
Super-Nyquist-WDM with frequency-unlocked optical carriers					
PDM-QPSK [35]	8.8 Tb/s	20 × 440 Gb/s	4.11	3600	~15,000
Super-Nyquist-WDM with frequency-locked optical carriers					
PDM-QPSK [36]	1.232 Tb/s	11 × 112 Gb/s	4.19	640	~2700

ultra-large area fiber (ULAF) [4]. With the use of 16-QAM, Huang et al. demonstrated the transmission of a 1.5-Tb/s superchannel over 1200 km of a standard single-mode fiber (SSMF) [27]. With the use of reduced-guard-interval OFDM, which essentially performs full chromatic dispersion compensation prior to OFDM processing and hence allows to reduce the guard-interval to accommodate mostly polarization effects with much lower channel memory, Liu et al. demonstrated the transmission of three 485-Gb/s superchannels at an ISEDP as high as 30,000 km · b/s/Hz [30].

To avoid the use of accurate frequency-locked optical carriers, a small frequency guard-band between adjacent carriers within a superchannel can be used to allow one to trade a small fraction (e.g., <10%) of link capacity for simplicity, scalability, and performance. With the use of this "guard-banded OFDM," Liu et al. demonstrated the transmission of a 1.864-Tb/s superchannel, consisting of eight PDM-OFDM-16QAM signals, over 5600 km of ULAF, achieving a record ISEDP of over 32,000 km · b/s/Hz for Tb/s-class superchannel transmission with a net SE of 5.75 b/s/Hz [31].

Using Nyquist-WDM and hybrid 32- and 64-QAM modulation, Zhou et al. demonstrated the transmission of five 400-Gb/s superchannels on a 50-GHz grid over

1200-km ULAF with a WDM SE as high as 8 b/s/Hz [32]. With the use of 80-Gbaud modulation and detection, Raybon et al. demonstrated the transmission of a 1.28-Tb/s superchannel using only two optical carriers and two pairs of transmitter/receiver front-ends [33]. There, close-to-Nyquist spectral shaping was achieved by optical filters instead of digital pulse shaping. Fig. 9.2 shows the spectra of three representative Tb-class superchannels [4,31,33]. The flexibility in choosing the number of carriers and the modulation speed per carrier is evident.

More recently, Lundberg et al. demonstrated the use of a single optical frequency comb source to replace the 25 independent lasers needed for 25 wavelength channels within a superchannel and achieved a raw superchannel data rate of 6 Tb/s (25 × 240 Gb/s) [34]. In addition, another optical frequency comb source is used at the receiver side to serve as the optical local oscillator (OLO) lasers needed to receive the 25 wavelength channels. The use of the optical frequency comb source not only saves the needed lasers but also enables frequency locking of the optical carriers, which reduces the bandwidth of the guard band needed in the quasi-Nyquist-WDM system and further increases the system SE. Moreover, the broadband phase coherence of frequency combs could allow for a better joint processing of superchannel constituents such as low-complexity phase recovery [34] and high-performance nonlinearity compensation [38], which will be described in more depth in the following sections.

Using optical frequency combs, Mazur et al. demonstrated a record superchannel capacity of 19.2 Tb/s raw data rate by aggregating 50 24-Gbaud PDM-256QAM signals on a 25-GHz frequency grid [19]. After 40-km SSMF transmission with three 19.2-Tb/s superchannels, the average generalized mutual information (GMI) was 13.1 bits per PDM-256QAM symbol, which results in a system SE as high as 12 b/s/Hz after deducting the overhead for pilots and guard bands [19]. Remarkably, the entire C-band of 4 THz was filled with three 19.2 Tb/s superchannels that only requires three tunable external-cavity lasers (ECLs) to seed the three frequency comb

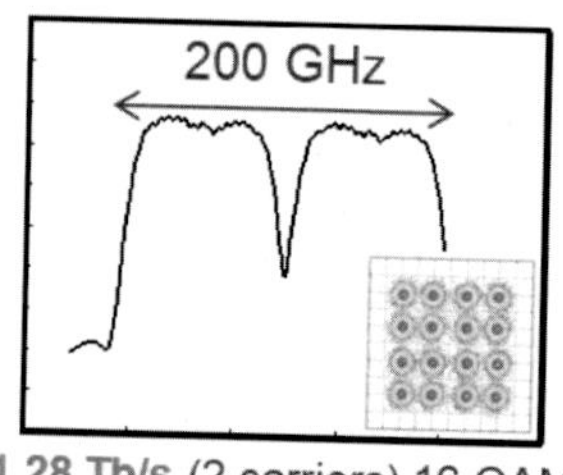

1.28 Tb/s (2 carriers) 16-QAM
5.0 bit/s/Hz
3200 km transmission
[Raybon et al., IPC'12]

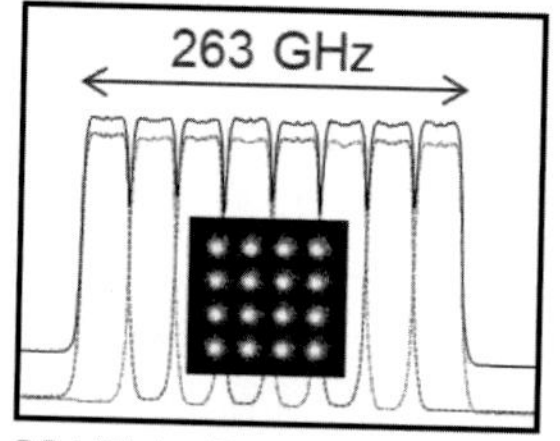

1.864 Tb/s (8 carriers) 16-QAM
5.75 bit/s/Hz
5600 km transmission
[Liu et al., ECOC'12]

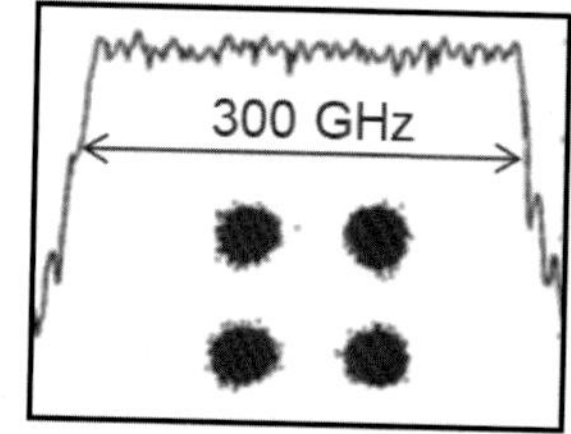

1.2 Tb/s (24 carriers) QPSK
3.74 bit/s/Hz
7200 km transmission
[Chandrasekhar et al., ECOC'09]

Figure 9.2 Representative Tb-class superchannel spectra.

generators at the transmitter side. At the receiver side, a pilot carrier generated by each transmitter-side comb generator was used to seed a receiver-side comb generator after injection-locking a standard distributed feedback laser [19]. This not only saves three ECLs at the receiver side but also automatically ensures frequency locking between the superchannel optical carriers and their corresponding OLOs. This demonstration shows the potential for using comb-based superchannels to reduce system complexity and improve system performance [19].

For super-Nyquist-WDM transmission, signal generation can be based on optical or electrical spectrum PreF, while signal detection can be based on maximum a posteriori or maximum likelihood sequence estimation (MLSE) [35,36,39–42]. Transmission of 20 super-Nyquist-filtered 110-Gbaud PDM-QPSK wavelength channels based on frequency-unlocked optical carriers has been demonstrated over a 100-GHz frequency grid, achieving an aggregated superchannel data rate of 8.8 Tb/s, an ISE of 4.11 b/s/Hz [after removing a 7% overhead for forward-error correction (FEC)], and a transmission distance of 3600 km [35]. With frequency-locked optical carriers, a 1.232-Tb/s super-Nyquist-WDM superchannel consisting of 11 28-Gbaud PDM-QPSK wavelength channels on a 25-GHz frequency grid was transmitted over 640 km with an ISE of 4.19 b/s/Hz [36]. These demonstrations affirm the promise of using receiver-side DSP such as MLSE to realize fast-than-Nyquist WDM transmission to achieve a high system SE.

9.3 Superchannel transceiver designs

9.3.1 Overall superchannel transceiver architecture

Fig. 9.3 shows the schematic of a Tb/s-class superchannel transponder embedded in a WDM optical network [37]. The full optical field information of an optical signal can be decomposed into four orthogonal real-valued components, namely, the inphase (I) and quadrature (Q) components of the two orthogonal polarization states (x and y) supported by an SSMF. At the transmitter, a PDM I/Q modulator is commonly used to imprint these four high-speed electrical waveforms onto an optical carrier, emerging from a semiconductor laser oscillating at around 193 THz and stabilized to within about ± 1 GHz. At the receiver, a polarization-diversity 90-degree hybrid and an OLO are used to decompose the received signal into four orthogonal components, which can then be processed in a digital signal processor to recover the original signal. To avoid the bandwidth bottleneck of electronic and opto-electronic components in the transmitter and receiver, optical parallelism is utilized to generate and detect a superchannel whose aggregate bandwidth by definition exceeds those of the individual transponder components.

It is evident from Fig. 9.3 that the generation and detection of superchannels can greatly benefit from large-scale integration of both optical and electronic components,

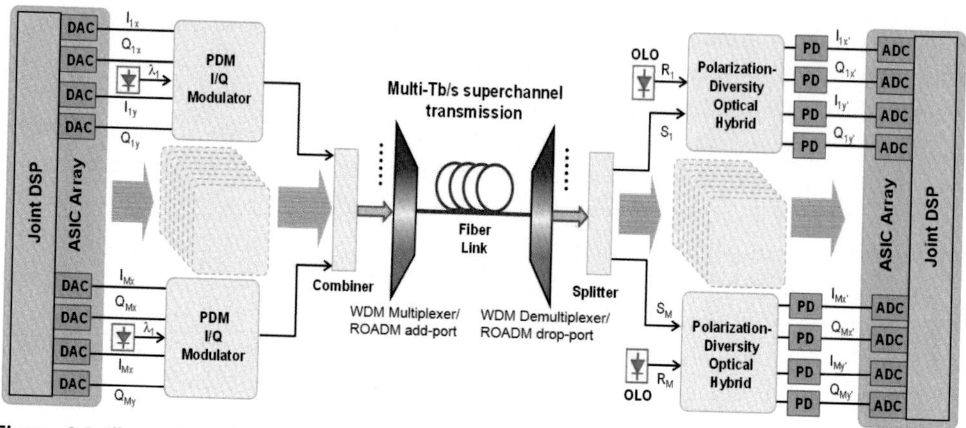

Figure 9.3 Illustration of Tb/s-class superchannel transmitter and receiver in a wavelength-division multiplexing system. *After Liu X, Chandrasekhar S, Winzer PJ. Digital signal processing techniques enabling multi-Tb/s superchannel transmission: an overview of recent advances in DSP-enabled superchannels. IEEE Signal Process Mag 2014;31(2):16–24 [37].*

leading to potential savings in cost, size, and power. With all the constituents of a superchannel being available at the transmitter and at the receiver, joint digital signal processor may be leveraged to improve the transmission performance and/or reduce DSP complexity, as will be discussed in more depth later.

9.3.2 Superchannel transceiver DSP

With access to the full optical field information at both transmitter and receiver, modern coherent optical communications have benefited from many powerful DSP techniques originally developed for wireless communication. Fig. 9.4 schematically shows the DSP architecture of a superchannel consisting of two optical carriers [37]. Key signal processing steps include FEC encoding, constellation mapping, fiber NLC, upsampling, electronic dispersion precompensation, Nyquist PreF, and equalization of a nonideal transmitter frequency response through a static transmit equalizer. The four resulting electrical signals representing I and Q components of the baseband signal in both x and y polarizations are then modulated onto an optical carrier. Note that some of the transmitter-side DSP modules shown here, such as NLC, may bring valuable performance gains in long-haul optical transmission, but their DSP complexities need to be taken into consideration. In some cases, these modules can be built to be optionally available depending on the requirements on performance and power consumption.

Fig. 9.4 also shows the schematic architecture of the receiver DSP for the superchannel [37]. After coherent opto-electronic conversion, the DSP is presented with four electrical baseband signals representing I and Q components in x and y polarization. Key processing steps include electronic dispersion compensation (EDC), timing

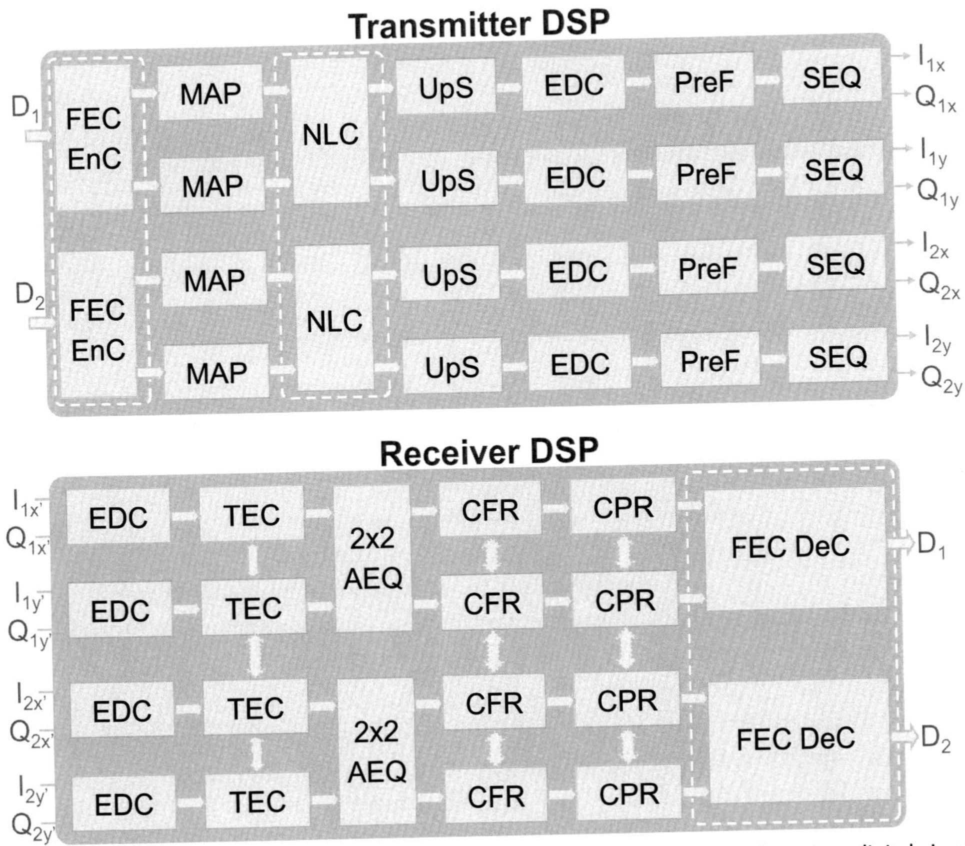

Figure 9.4 Schematics of the transmitter digital signal processor (upper) and receiver digital signal processor (lower) for generating and detecting two optical carriers of a superchannel. *After Liu X, Chandrasekhar S, Winzer PJ. Digital signal processing techniques enabling multi-Tb/s superchannel transmission: an overview of recent advances in DSP-enabled superchannels. IEEE Signal Process Mag 2014;31(2):16–24 [37].*

error correction, adaptive 2-by-2 multiple-input-multiple-output equalization and polarization source separation compensating for polarization rotation and PMD, carrier frequency recovery (CFR), carrier phase recovery, and FEC decoding based on both soft-decision and/or hard-decision. Typically, transmitter-side digital signal processor has the advantages of (1) absence of optical noise and (2) efficient implementation operating on one sample per symbol. On the other hand, receiver-side digital signal processor has the advantage of being able to adaptively and quickly react to dynamic channel changes. Note in this context that receiver-to-transmitter feedback is often impossible in optical systems: For a 1000-km link, signal round-trip times are on the order of 10 Ms, whereas channel dynamics can be in the 10-kHz range or above, owing to acoustically and mechanically induced vibrations of transmission fiber segments. As optical channels are usually well defined in terms their frequency response, blind channel equalization and phase recovery has been extensively used in optical coherent detection [43]. For fast channel equalization and/or sophisticated modulation formats, training symbols can be inserted in the data symbol stream to aid channel equalization [44]. To enable the efficient implementation of high-performance FEC codes, it is desirable to recover the phase of the modulated symbols (as opposed to just the differential phase between symbols), which can be readily realized by pilot-assisted phase estimation [45]. For superchannel transmission, some of the processes such as CFR, FEC, and NLC may be performed jointly over multiple signals within the superchannel to improve the overall superchannel performance, which will be discussed further in later sections.

9.3.3 Flexible superchannel modulation and entropy loading

With transmitter-side digital signal processor and collaborative receiver-side digital signal processor, the modulation format can be software defined to enable the optimization of system performance depending on the link conditions. Fig. 9.5 illustrates some square-QAM signal constellations generated in Tb/s-class superchannels by software-defined flexible transmitters [20]. In addition to conventional square QAM, advanced modulation formats based on constellation shaping [46—48] can be utilized to increase signal's tolerance to both linear noise and nonlinear fiber transmission impairments.

As the signal modulation speed continues to increase, the signal-to-noise ratio (SNR) within each modulated signal becomes nonflat in the frequency domain. Typically, the SNR is lower at frequencies farther away from the signal center frequency because of the following effects:

- Bandwidth limitations of the transceiver components that cause a frequency roll-off at high frequencies
- Bandwidth limitations of wavelength-routing filters, such as ROADMs, in the optical transmission link
- Nonlinear distortions due to modulation that degrade higher frequency more
- *I*/*Q* mismatch during modulation and coherent detection that also further degrades higher frequency

To address the issue of nonflat SNR in the frequency domain, multicarrier modulation or digital subcarrier multiplexing with flexible entropy loading [49] can be applied to transmit different carriers at different entropy or SE to match their corresponding SNRs such that all the carriers have the same bit error ratio (BER), thereby achieving the optimal performance of the entire multicarrier superchannel. With the use of the newly developed probabilistic shaping (PS) technique [46—48], fine-granularity entropy loading can be achieved to approach the theoretically optimal water-filling performance [49]. As described in Chapter 8, High-Capacity Long-Haul

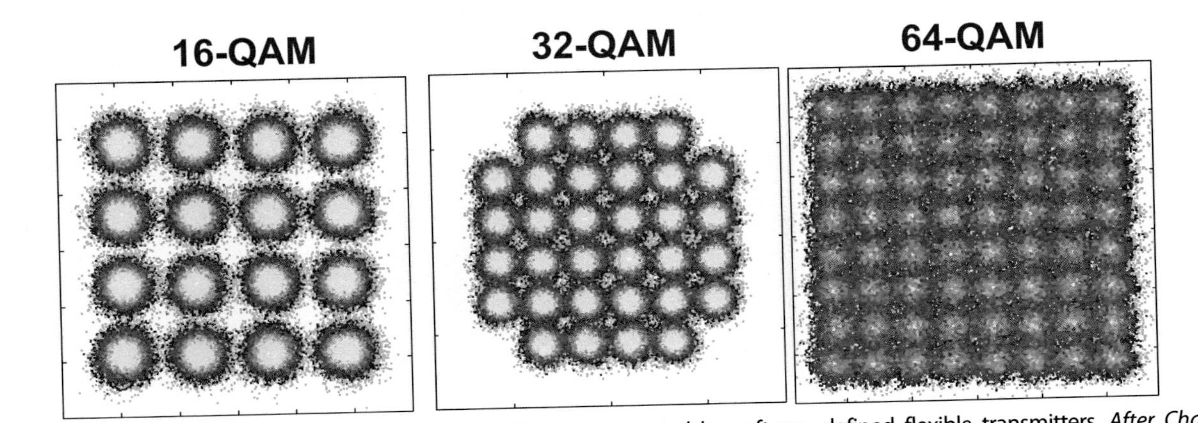

Figure 9.5 Exemplary Tb/s-class superchannel signal constellations generated by software-defined flexible transmitters. *After Chandrasekhar S, Liu X. Advances in Tb/s superchannels. In: Kaminov IP, Li T, Willner A, editors. Chapter 3 of Optical fiber telecommunications VIB; May 2013 [20].*

Optical Fiber Transmission, PS conveniently allows the same square-QAM constellation and FEC to be used, so a universal transmitter and receiver digital signal processor architecture can be used.

Fig. 9.6 illustrates the use of flexible entropy loading in a superchannel so that different frequency bands within the superchannel are loaded with different entropies (or SEs) based on their corresponding SNRs to approach the ideal water-filling performance. In this illustrative example, the superchannel is divided into six bands. Assuming symmetric frequency roll-off, three entropy values, (A) 4.3 bits, (B) 5.7 bits, and (C) 5.0 bits per QAM-symbol, are used for the two outermost bands (1 and 6), the two innermost bands (3 and 4), and the two intermediate bands (2 and 5). The SNR scales with the entropy per QAM symbol roughly as 6 dB per bit, so this entropy loading example can cover an SNR variation range of about 8.4 dB.

Recently, capacity optimization via flexible entropy loading has been demonstrated in a real-time 1.6-Tb/s superchannel that consists of two 800-Gb/s optical signals each having eight PDM-PS-64QAM subcarriers with different entropies [50]. This real-time demonstration further shows the benefit of combining superchannel formation

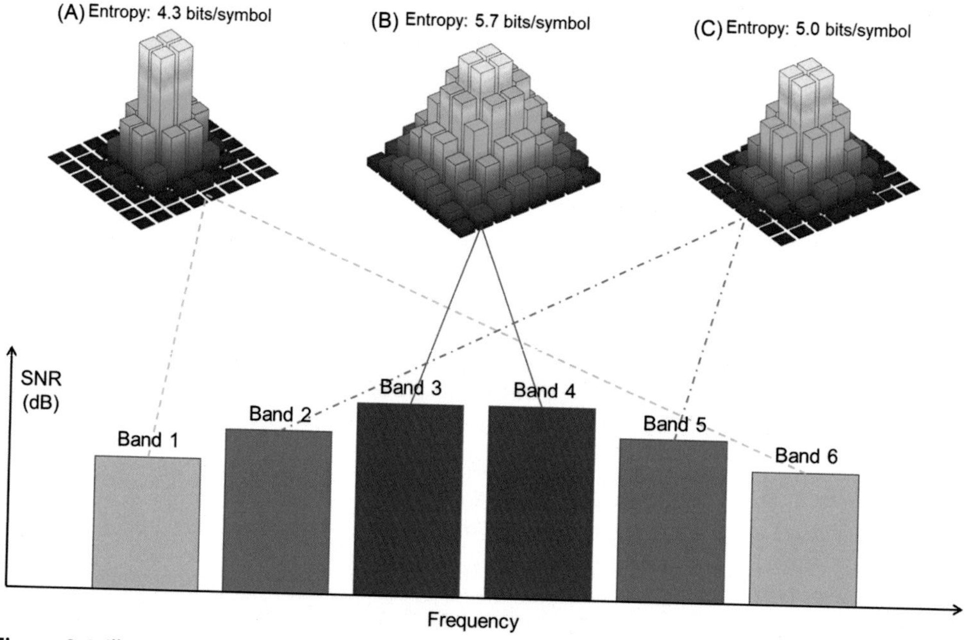

Figure 9.6 Illustration of the use of flexible entropy loading in a superchannel where different frequency bands within the superchannel are loaded with different entropies (or spectral efficiencies) to match their corresponding signal-to-noise ratios in order to approach the ideal water-filling performance.

with PS-based flexible entropy loading to approach the ideal water-filling performance in high symbol-rate transmission.

9.3.4 Joint processing of superchannel constituents

As the constituents of a superchannel share the same optical transmission path, they experience some common transmission properties. Therefore the monitoring of some transmission properties of the superchannel can be simplified by monitoring only one of the constituents, or more accurate and faster measurements can be obtained by averaging certain channel estimates across superchannel constituents [20]. The joint processing of superchannel constituents can be applied for the following linear functions:

- Joint carrier recovery
- Joint crosstalk cancellation
- Joint FEC processing
- Joint processing for polarization-dependent loss (PDL) mitigation

Joint carrier recovery can be readily applied to a superchannel where its constituents are frequency locked, as in the case of coherent optical OFDM [4]. In the context of space-division multiplexing (SDM) systems, joint carrier frequency recovery in spatial superchannel transmission has been experimentally demonstrated [51]. In a recent demonstration, joint carrier recovery was achieved in a frequency comb-based transmission system [34]. This demonstration utilizes the broadband phase coherence of optical frequency combs. The results show that joint carrier recovery reduces the needed digital signal processor complexity when multiple carriers share one carrier recovery digital signal processor. Alternatively, joint carrier recovery can be used to speed up the carrier recovery (by simultaneously processing multiple carriers) and increase the tolerance to rapid phase fluctuations due to linear and nonlinear effects [34]. With advances in compact chip-scale optical frequency comb sources, for example, based on soliton microcombs and integrated photonics [52], joint signal processing in comb-based superchannels may enable future high-performance, energy-efficient optical transceivers.

As the signal fields of all superchannel constituents are available at the receiver, joint DSP can be applied to mitigate coherent crosstalk between adjacent signals [53]. In typical Nyquist-WDM transmission systems, optical signals are tightly packed at a frequency spacing that is near or equal to the signal modulation baud rate, potentially inducing strong interchannel interference (ICI) [53]. One way to mitigate the impact of the ICI is to apply aggressive optical filtering for each signal at the transmitter side and compensate for resulting intersymbol interference via receiver-side equalization, as discussed previously. With joint signal processing at superchannel receiver, all the interfering channels can be jointly detected and the ICI can be digitally compensated [53].

Joint FEC processing over all the superchannel constituents can provide additional performance improvements. FEC encoding over two PDM-QPSK channels spaced at 200 GHz was implemented to reduce the transmission penalties due to PDL, which is optical frequency−dependent, as compared to that without joint FEC processing [54]. In a recent WDM transmission experiment, short-wavelength channels were observed to be worse performing than long-wavelength channels, and joint FEC processing using a long-wavelength channel and a short-wavelength channel has also found to be beneficial [55]. In effect, joint FEC processing allows the superchannel receiver to average the raw BER performances of multiple superchannel constituents and thus achieves an overall superchannel performance that is more stable and better than the worst-case performance of each superchannel constituent. The joint processing of superchannel constituents can also be applied for nonlinearity mitigation, which will be discussed in the following section.

9.4 Joint nonlinearity compensation

9.4.1 Joint nonlinearity compensation with frequency-locked optical comb

Fiber nonlinearity is a major transmission impairment in optical fiber communications. Because of fiber nonlinearity, an optical channel suffers from power-dependent amplitude and phase distortions within the channel and from other copropagating WDM channels. Coupled with chromatic dispersion, the nonlinear distortions on a given symbol may result from the nonlinear interactions with many other symbols. Mathematically, fiber transmission in two polarizations is governed by the coupled nonlinear Schrödinger equation (NLS) [56], which, under the common assumption that the nonlinear interaction length is much greater than the length scale of random polarization rotations, can be reduced to the Manakov equation as [57]

$$\left[\frac{\partial}{\partial z}+\frac{\alpha(z)-g(z)}{2}+i\frac{\beta_2(z)}{2}\frac{\partial^2}{\partial t^2}\right]E_{x,y}(z,t)=i\frac{8}{9}\gamma\left(|E_x(z,t)|^2+|E_y(z,t)|^2\right)E_{x,y}(z,t) \tag{9.1}$$

where E_x (E_y) is the signal E-field along the x (y) polarization, i is the imaginary unit, z, α, g, β_2, and γ are respectively the propagation distance, the loss coefficient, the gain coefficient, the group-velocity dispersion (or second-order dispersion) coefficient, and the fiber nonlinear Kerr coefficient along a transmission link. NLC based on digital backpropagation (DBP) to invert the NLS at the receiver [58,59] has been introduced to mitigate signal-to-signal nonlinear distortions from intra-channel nonlinear effects and interchannel nonlinear effects. Interchannel cross-phase modulation (XPM) compensation using DBP has been experimentally demonstrated [60]. It was found that interchannel XPM is insensitive to the relative phase between the interacting channels, so the phase-locking among the interacting signals is not needed for interchannel XPM compensation. On the other hand, it was found that stable frequency locking among the optical carriers of interacting channels makes the interchannel nonlinear interactions more deterministic and helps improve the overall performance of multichannel NLC [38]. Stable frequency locking among optical carriers can be realized by using an optical comb source with multiple frequency-locked optical carriers.

With the availability of the signal fields of all the constituents of a superchannel, joint DBP can be applied to mitigate both intra- and interchannel nonlinear effects. Particularly, a spectrally sliced receiver can be used to measure a superchannel spectrum slice-by-slice using a phase-locked and equally spaced optical frequency comb, followed by stitching the slices digitally together to reconstruct the full-band electrical field of the received superchannel [61]. We will discuss the potential gain of such multichannel NLC and the implication to joint NLC in superchannel in the following.

9.4.2 Experimental demonstration of joint NLC

Most of the deployed long-haul fiber transmission systems are dispersion-managed (DM), which incur more severe nonlinearity impairments as compared to dispersion-unmanaged (DUM) systems [62,63]. For DM systems, multichannel dispersion-folded DBP was shown to provide better NLC performance than single-channel NLC and lower digital signal processor complexity than traditional DBP [63]. Fig. 9.7 shows the experimental setup of joint NLC for a quasiNyquist-WDM superchannel, consisting of eight 37.5-GHz-spaced 128-Gb/s PDM-QPSK signals, by a spectrally sliced coherent receiver and dispersion-folded DBP after transmission over a 2560-km DM fiber link [63].

To compare the performance with different aggregate WDM transmitter bandwidths, the Nyquist-WDM transmitter was first composed of three and then of eight 100-kHz ECLs centered near 1550 nm with a 37.5-GHz spacing. The outputs from the eight ECLs lasers were divided into even and odd groups and separately modulated by two PDM I/Q modulators at 32 Gbaud. The drive signals for each modulator were from four DACs operating at two samples per symbol or 64 GS/s. To obtain close channel spacing, a root raised-cosine digital filter with a roll-off factor of 0.1 was applied to confine the optical spectrum of each signal. After the modulation, the even group was delayed for signal decorrelation and recombined with the odd group. The transmission link was a re-circulating loop consisting of four spans of 80-km TrueWave reduced slope (TWRS) fiber, each followed by a dispersion-compensating fiber (DCF). The dispersion and Kerr nonlinear coefficient of the TWRS fiber were 4.66 ps/nm and 1.79 W/km, respectively. The residual dispersion per span was about

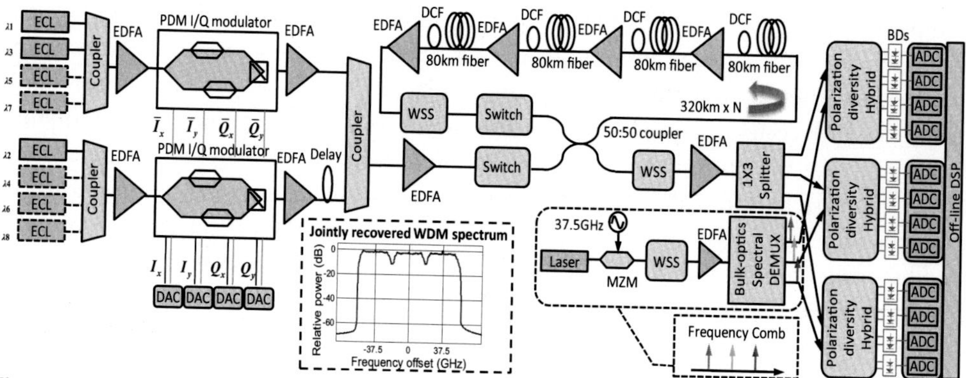

Figure 9.7 Experimental setup of joint nonlinearity compensation for a quasiNyquist-wavelength-division multiplexing superchannel by a spectrally sliced coherent receiver and dispersion-folded digital backpropagation after transmission over a dispersion-managed fiber link. *After Xia C, Liu X, Chandrasekhar S, Fontaine NK, Zhu L, Li G. Multi-channel nonlinearity compensation of PDM-QPSK signals in dispersion-managed transmission using dispersion-folded digital backward propagation. Opt Express 2014;22:5859–66 [63].*

36 ps/nm on average. Inline erbium-doped fiber amplifiers (EDFAs) were added after each span to compensate for the span loss. A wavelength-selective switch (WSS) was used to reject the out-of-band noise. The signals were propagated over 2560 km (32×80 km) and only three center channels were detected by a spectrally sliced receiver based on a phase-locked and equally spaced optical frequency comb [61]. The optical carriers in the frequency comb were precisely spaced at 37.5 GHz so that each carrier could serve as an OLO for coherent detection of the corresponding spectrum slice. Because the phase and amplitude were locked among the frequency comb lines, the received spectrum slices could be used to seamlessly reconstruct the full field of the superchannel, and joint DBP was applied to realize multichannel NLC. The frequency comb was generated by a Mach–Zehnder modulator that was driven by a 37.5-GHz sinusoidal RF carrier to modulate an optical carrier. The frequency comb was then equalized by a WSS and finally separated by a bulk-optics–based wavelength demultiplexer. To detect the spectrum slices, the WDM signals were split into three copies, each detected by a digital coherent receiver consisting of a polarization-diversity optical hybrid followed by four balanced detectors (BDs). The BD outputs were recorded by 40-GS/s ADCs, which also acted as optical filters of 40-GHz cut-off bandwidth to slice the spectrum. Finally, the signal spectrum of the three channels was reconstructed, as shown in the inset of Fig. 9.7. The received signals were processed using a dispersion folded DBP algorithm and evaluated by the Q^2-factor, which is directly calculated from the measured BER using $Q^2 = 20\log_{10}\left[\sqrt{2}\cdot \mathrm{erfcinv}(2\cdot \mathrm{BER})\right]$.

The comparison between the folded DBP and the conventional DBP are as follows. Fig. 9.8A shows the schematic of the conventional DBP that uses the split-step Fourier method to separately deal with dispersion and nonlinearity step-by-step in the reverse way of signal propagation along the fiber. The total step numbers thus determine the computational load. Folded DBP attempts to calculate the steps with the similar dispersion and nonlinearity in one, that is, fold the steps to save the large computation load. This becomes possible for a dispersion-managed system where the accumulated dispersion repeats often. Furthermore, it is found that the nonlinear distortions can be assumed the same as well at locations where the accumulated dispersion is identical. This is because the waveform is dominated by the accumulated dispersion since the total nonlinear shift of the entire link at optimum power is rather small, on the order of 1 radian. Fig. 9.8B shows the schematic of the dispersion folded DBP, where the fiber segments with the same range of accumulated dispersion are folded into one step of DBP instead of being computed separately in conventional DBP. First, a lumped dispersion compensator ($\hat{D}_{\mathrm{lump}}$) is applied to bring the received signal field E_{in} to the first folded DBP fiber segment E_1. Then dispersion ($\hat{D}_i$ of i-th step) and nonlinear compensator ($\hat{N}_i$) are performed step-by-step with folded segments. Note that $W_i = \sum_k \gamma \int P_{i,k}(z)\mathrm{d}z$ is a weighting factor that effectively takes into account of varied power levels $P_{i,k}(z)$ within each segment.

Fig. 9.9A shows the required steps for folded and conventional DBP to achieve maximum Q^2 at optimum signal launch power (P_{in}) when three channels were transmitted, detected, and processed. Folded DBP provides equal maximum gain as the conventional DBP while saving the computation load by a factor of ~3.4 since the computation per step for both DBPs is similar. Fig. 9.9B shows the performances of the two DBP schemes at a higher signal launch power (−2 dBm). Folded DBP saves

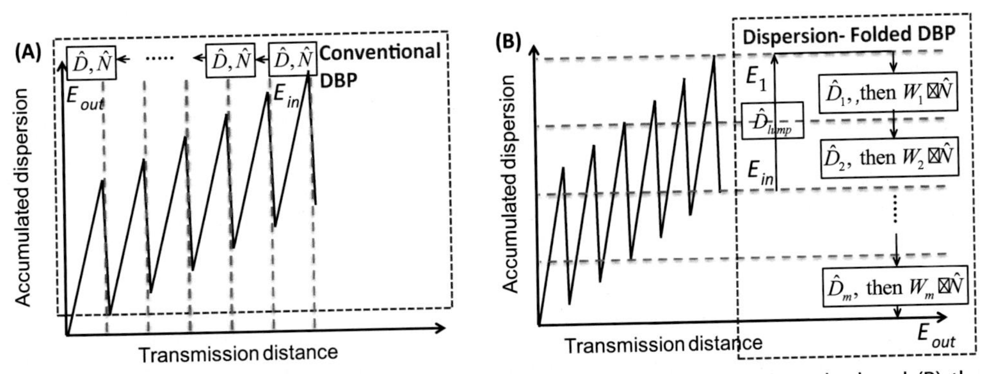

Figure 9.8 Illustrations of (A) the conventional digital backpropagation (DBP) method and (B) the dispersion folded DBP method. *After Xia C, Liu X, Chandrasekhar S, Fontaine NK, Zhu L, Li G. Multi-channel nonlinearity compensation of PDM-QPSK signals in dispersion-managed transmission using dispersion-folded digital backward propagation. Opt Express 2014;22:5859–66 [63].*

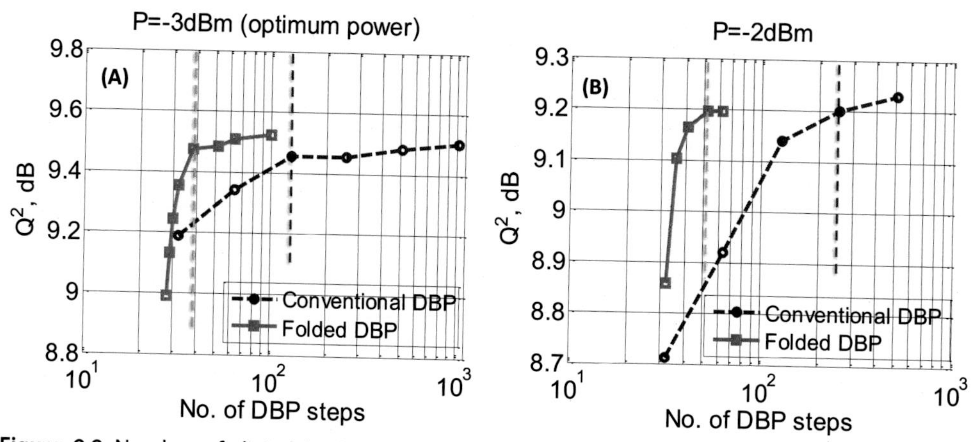

Figure 9.9 Number of digital backpropagation (DBP) steps needed in folded DBP compared to that needed in conventional DBP at (A) $P_{in} = -3$ dBm and (B) $P_{in} = -2$ dBm. *After Xia C, Liu X, Chandrasekhar S, Fontaine NK, Zhu L, Li G. Multi-channel nonlinearity compensation of PDM-QPSK signals in dispersion-managed transmission using dispersion-folded digital backward propagation. Opt Express 2014;22:5859–66 [63].*

the digital signal processor load by a factor of 5 for a same Q^2 factor of 9.2 dB, giving a hint that folded DBP could save more DSP resource when nonlinearity is larger. In addition, folded DBP algorithm has the potential to perform even better because more segments can be folded if residual dispersion per span is smaller or fiber dispersion is larger.

Figs. 9.10 and 9.11 compare the signal quality of the center channel after DBP for two different cases, three channels transmitted with a transmitted WDM superchannel bandwidth (B_{WDM}) of 112.5 GHz and eight channels transmitted with $B_{WDM} = 300$ GHz, respectively. For each case, three compensation methods were used: (1) EDC only; (2) intra-channel NLC using the folded DBP with the effective receiver NLC bandwidth (B_{RX}) being 37.5 GHz; (3) three-channel nonlinearity compensation with the knowledge of adjacent channels using folded DBP, that is, B_{RX} being 112.5 GHz. Method (2) was enabled by filtering the center channel digitally before compensation. With intra-channel NLC, the highest achievable Q^2 improvement compared to EDC only, that is, NLC gain, is 1.0 dB when three channels are transmitted ($B_{WDM} = 112.5$ GHz) and 0.6 dB when eight channels are transmitted ($B_{WDM} = 300$ GHz). Furthermore, the use of three-channel NLC with the knowledge of two adjacent channels ($B_{RX} = 112.5$ GHz) gives 2.0 and 1.1 dB NLC gains for these two cases, respectively. These results suggest that multichannel NLC provides substantial performance gain over intra-channel NLC by additionally mitigating interchannel nonlinear impairments, especially when B_{RX} is the same as B_{WDM}. When B_{WDM} becomes larger than B_{RX}, the interchannel nonlinear impairments resulting

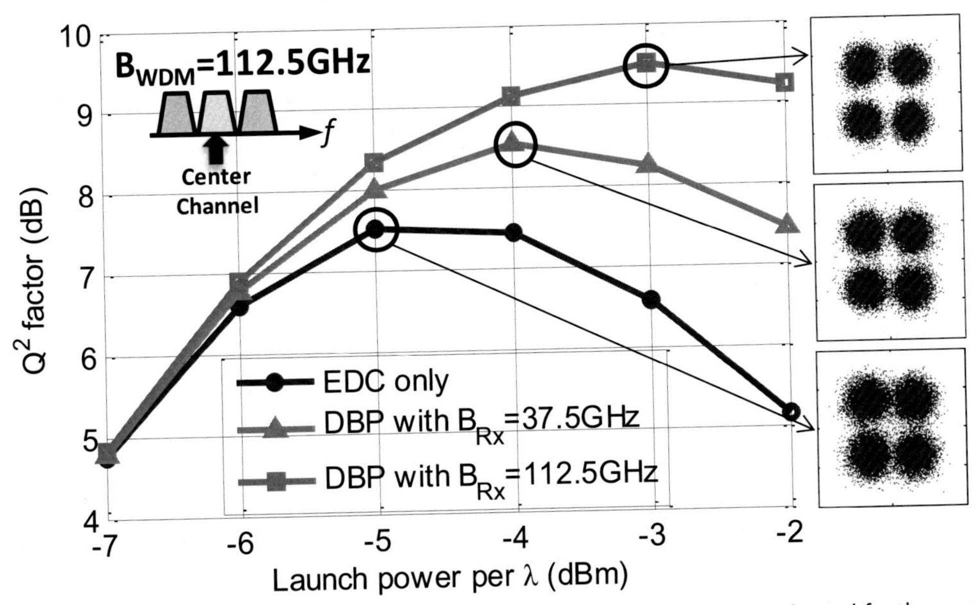

Figure 9.10 Measured signal quality as a function of signal launch power per channel for the case of three copropagating channels. *After Xia C, Liu X, Chandrasekhar S, Fontaine NK, Zhu L, Li G. Multi-channel nonlinearity compensation of PDM-QPSK signals in dispersion-managed transmission using dispersion-folded digital backward propagation. Opt Express 2014;22:5859–66 [63].*

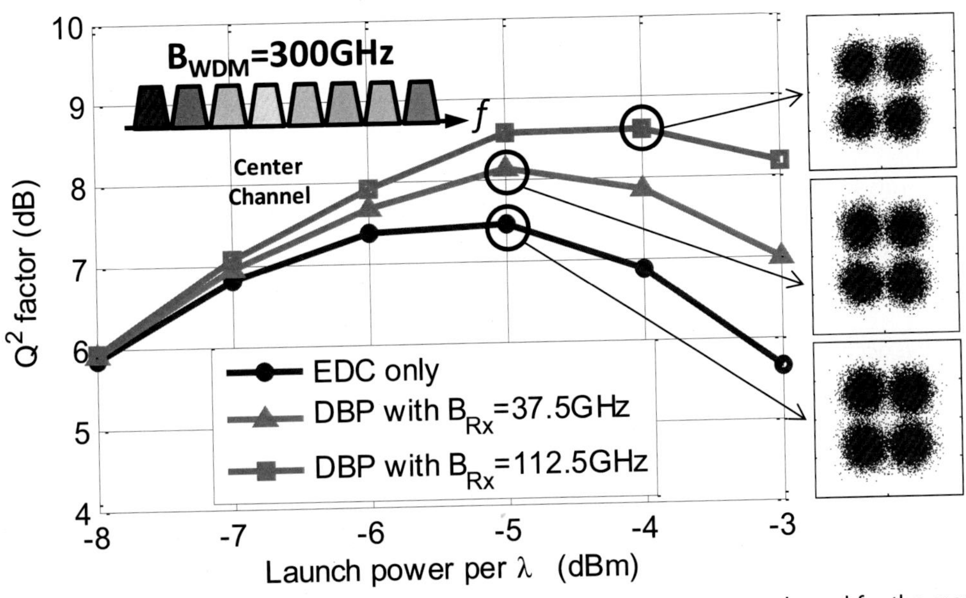

Figure 9.11 Measured signal quality as a function of signal launch power per channel for the case of eight copropagating channels. *After Xia C, Liu X, Chandrasekhar S, Fontaine NK, Zhu L, Li G. Multi-channel nonlinearity compensation of PDM-QPSK signals in dispersion-managed transmission using dispersion-folded digital backward propagation. Opt Express 2014;22:5859–66 [63].*

from the WDM channels that are outside the receiver detection bandwidth cannot be compensated digitally, which decreases the effectiveness of DBP. These experimental findings are in agreement with theoretical assessments such as those reported by Dar and Winzer [64].

9.4.3 Scaling rules for single- and multichannel NLC

The experimental demonstration described in the previous section shows the performances of multichannel NLC in two cases where three ($B_{WDM} = 112.5$ GHz) and eight ($B_{WDM} = 300$ GHz) WDM channels are transmitted over the same fiber link. However, practical WDM systems typically have much more WDM channels per system, for example, when the C-band is fully populated. To estimate the effectiveness of the intra- and multichannel NLC in a fully loaded WDM system, we can assess the nonlinear noise power (σ_{NL}^2) as a function of receiver bandwidth by using the Gaussian noise (GN) model [65], which treats the nonlinear signal distortions as additive GN. Although the GN model is originally proposed for DUM systems, it works reasonably well for long-distance dispersion-compensated transmission with high-speed modulation and sufficiently large residual dispersion per span (RDPS) [63]. Under the GN model, the power spectral density (PSD) of the nonlinear noise in a DUM system can be evaluated by one integral equation related to basic system parameters as below [65],

$$G_{NLI}(f) = \frac{16}{27}\gamma^2 L_{eff}^2 \cdot \int_{-\infty}^{+\infty}\int_{-\infty}^{+\infty} G_{WDM}(f_1)G_{WDM}(f_1)G_{WDM}(f_1)\cdot \rho(f_1,f_2,f)\cdot \chi(f_1,f_2,f)df_2 df_1 \tag{9.2}$$

where γ is the fiber nonlinearity coefficient and L_{eff} is the effective fiber span length, $G_{WDM}(f)$ denotes to the PSD of the transmitted WDM optical signal, $\rho(f_1,f_2,f)$ accounts for single-span nonlinear noise [65], and $\chi(f_1,f_2,f)$ is the phase-array factor

$$\chi(f_1,f_2,f) = \frac{\sin^2\left[2N_s\pi^2(f_1-f)(f_2-f)\beta_2 L_s\right]}{\sin^2\left[2\pi^2(f_1-f)(f_2-f)\beta_2 L_s\right]} \tag{9.3}$$

which accounts for the coherent interference of the nonlinear noises of N_S spans, each having a length of L_S and a dispersion coefficient of β_2. One can use Eq. (9.2) to integrate over a given frequency band and obtain the bandwidth dependence of the nonlinear noise. For the modeling of the DM case, $\beta_2 L_s = \text{RDPS}\cdot\lambda^2/2\pi c$ is substituted into the phase-array factor instead of using $\beta_2 L_s = |D|L_s\cdot\lambda^2/2\pi c$ as for DM case [63].

In what follows, the effective bandwidth B_{RX} is represented by n, the number of channels detected and processed. The total number of WDM channels is N. Fig. 9.12 plots the normalized σ_{NL}^2 versus B_{RX} for DUM and DM transmission over SSMF and TWRS fibers. The entire nonlinear noise power σ_{NL}^2 is normalized by the nonlinear

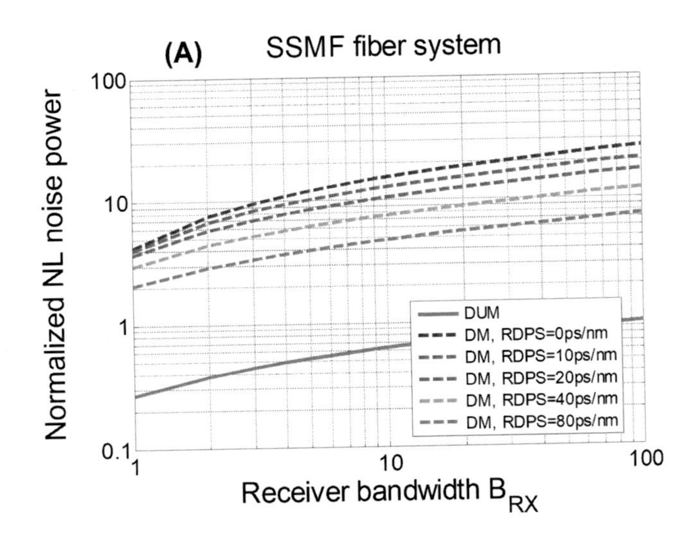

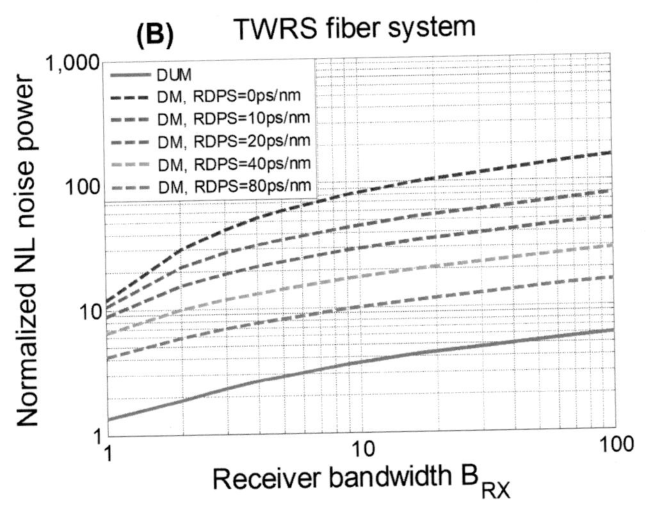

Figure 9.12 Normalized nonlinear noise power versus the effective receiver nonlinearity compensation bandwidth (in units of channel bandwidth) for both dispersion-unmanaged transmission and dispersion-managed transmission with different residual dispersion per span values in links with (A) standard single-mode fiber spans and (B) TWRS spans. *After Xia C, Liu X, Chandrasekhar S, Fontaine NK, Zhu L, Li G. Multi-channel nonlinearity compensation of PDM-QPSK signals in dispersion-managed transmission using dispersion-folded digital backward propagation. Opt Express 2014;22:5859–66 [63].*

noise power in the DUM SSMF transmission with 100 WDM channels. The nonlinear Kerr coefficients used for SSMF and TWRS fibers are 1.3 and 1.79 W/km respectively while the fiber dispersion parameters for SSMF and TWRS fibers are 16.5 and 4.65 ps/nm/km, respectively. As compared to TWRS, SSMF has a lower nonlinear coefficient and larger dispersion-induced walk-off effect that mitigates the nonlinear noise, so the nonlinear noise in SSMF transmission is lower than that in TWRS transmission for the DUM case and for each DM case with the same RDPS. In DM SSMF transmission, the nonlinear noise power reduces more at a larger RDPS, slowly approaching the performance of DUM SSMF transmission. For a TWRS system, the trend is similar, but the impact of RDPS is stronger because of smaller fiber dispersion.

It can be seen from Fig. 9.12 that the ratio between the nonlinear noise in a limited receiver bandwidth B_{RX} (e.g., $B_{RX}=3$) and the total nonlinear noise over the entire WDM spectrum is larger when the RDPS is larger, so larger multichannel NLC gain may be expected for larger RDPS. Assuming an NLC ratio of $a_{NLC}(B_{RX}=n)$ for the nonlinear noise inside a receiver bandwidth of n channels, the NLC gain can be represented as [63]

$$\Delta Q^2 = \frac{1}{3}\cdot 10\log_{10}\frac{\sigma_{NL}^2(B_{RX}=N)}{\sigma_{NL}^2(B_{RX}=N) - a_{NLC}(B_{RX}=n)\cdot\sigma_{NL}^2(B_{RX}=n)}. \tag{9.4}$$

In effect, the larger the portion of σ_{NL}^2 within the effective NLC bandwidth B_{RX}, the larger the NLC gain. Fig. 9.13 shows the normalized σ_{NL}^2 as a function of B_{RX} for three cases of $N=3$, 8, and 100 under the conditions of the previous DM experiment using TWRS spans with an RDPS of 36 ps/nm. The corresponding NLC gains

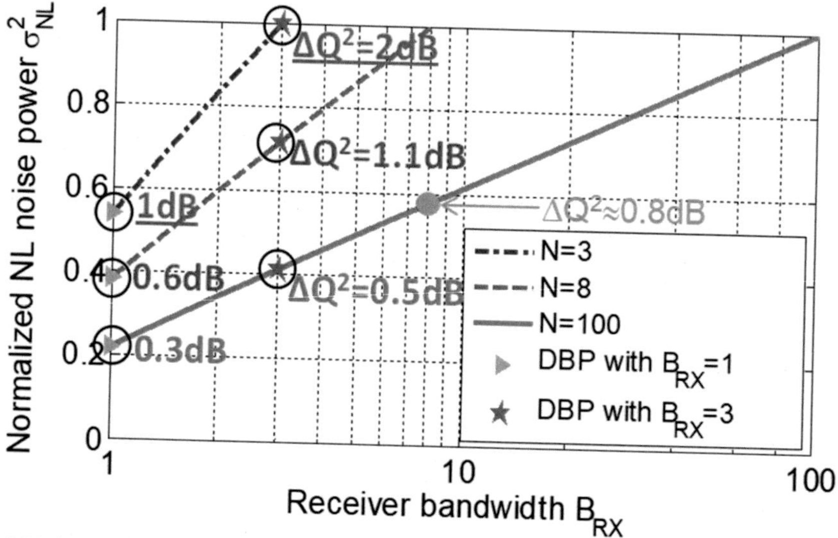

Figure 9.13 Normalized nonlinear noise power versus the effective receiver bandwidth for the cases of 3, 8, and 100 37.5-GHz-spaced wavelength-division multiplexing channels. The potential use of a superchannel receiver with 300-GHz bandwidth is indicated by the solid circle. *After Xia C, Liu X, Chandrasekhar S, Fontaine NK, Zhu L, Li G. Multi-channel nonlinearity compensation of PDM-QPSK signals in dispersion-managed transmission using dispersion-folded digital backward propagation. Opt Express 2014;22:5859–66 [63].*

calculated are marked on the figure by the triangular and star markers for single-channel ($B_{RX}=1$) and three-channel DBP ($B_{RX}=3$), respectively. Taking the single-channel NLC for example, the detailed calculation is as follows. First, single-channel NLC ratio $a_{NLC}(B_{RX}=1)$ is estimated to be 93% given the experimentally demonstrated 1 dB NLC gain for $N=3$, indicated by the top blue triangular markers. The NLC gain for $N=8$ and 100 can then be calculated as 0.64 and 0.33 dB, respectively. Similarly, for a three-channel NLC, $a_{NLC}(B_{RX}=3)$ is estimated to be 75%, given the experimentally demonstrated three-channel NLC gain for $N=3$. The respective estimated NLC gains of 0.64 and 1.1 dB using single-channel and three-channel NLC in an eight-channel WDM system are in good agreement with the experimental results of 0.69 and 1.0 dB, respectively [63].

In the fully loaded case with 100 WDM channels ($N=100$) spanning over 3.75 THz, the estimated NLC gains from the single- and three-channel NLC are reduced to 0.27 and 0.54 dB, respectively, as indicated in Fig. 9.13. If we assume an ideal NLC (instead of the experimentally measured 75% NLC ratio), the NLC gains from the single- and three-channel NLC would be increased to approximately 0.4 and 0.8 dB, respectively. It can be seen from Fig. 9.12 that the ratio between the nonlinear noise within the receiver bandwidth and the total nonlinear noise in the WDM system is larger for DUM transmission than for the DM transmission example used in Fig. 9.12. Thus the NLC gains in DUM transmission would be higher. The above results are in good agreement with the theoretical results reported by Dar and Winzer [64], which shows that the single- and three-channel NLC gains in DUM transmission with fully loaded WDM channels are typically limited to 0.5 and 1 dB, respectively.

To increase the NLC gain in a fully loaded WDM transmission, we can further increase the receiver bandwidth. Assuming the potential use of a superchannel receiver with a bandwidth of 300 GHz, that is, $B_{RX}=8$ in Fig. 9.13, the amount of nonlinear noise inside the receiver bandwidth would account for 58% of the total nonlinear noise, and the ideal NLC gain could be increased to about 1.3 dB. The NLC gain can be even higher in DUM transmission over high dispersion coefficient fibers such as SSMF, ULAF, and G.654-compliant fibers. This shows the potential benefit of using wider bandwidth superchannel receiver to achieve a higher NLC gain.

9.5 High-capacity wavelength routing with Tb/s-class superchannels

9.5.1 Flexible-grid WDM

In modern DWDM systems, transparent optical routing based on ROADMs is an essential technology that enables network flexibility and efficiency. In DWDM systems, a fixed channel grid defined by the International Telecommunications Union (ITU) was used traditionally. The channel spacing is typically 50 GHz. For high-symbol-rate and Tb/s-class superchannels, the needed optical bandwidths are not only more than 50 GHz but also variable. This calls for a new DWDM channel allocation scheme where the channel bandwidth is flexible or adjustable in order to support these high-symbol-rate channels and superchannels [20]. This new type of DWDM can be called flexible DWDM or elastic DWDM. ITU Telecommunication Standardization Sector (ITU-T) published a new edition of DWDM frequency grid recommendation (G.694.1) that includes the specifications for a flexible DWDM grid [66]. The frequency grid is generally anchored to 193.1 THz and supports a variety of channel spacings ranging from 12.5 to 100 GHz and wider. According to the ITU-T G.694.1 recommendation, the allowed flexible DWDM frequency slots have a nominal central frequency defined by

$$f_n = 193.1 + n \times 0.00625(\text{THz}) \quad (9.5)$$

where n is a positive or negative integer including 0 and 0.00625 THz (or 6.25 GHz) is the nominal central frequency granularity. The allowed frequency a slot bandwidth is defined by

$$B_m = 12.5 \times m(\text{GHz}) \quad (9.6)$$

where m is a positive integer and 12.5 GHz is the frequency slot bandwidth granularity. Any combination of frequency slots is allowed as long as no two frequency slots overlap [66]. The reason that the nominal central frequency granularity needs to be 6.25 GHz is to allow a slot that has a width of an even multiple of 12.5 GHz to be next to another slot with a width of an odd multiple of 12.5 GHz without a spectral gap or GB. The flexible DWDM architecture is also supported by the recent availability of flexible bandwidth ROADMs and WSS's [67].

The benefits of the flexible DWDM grid in optical networking have been studied extensively [68–72]. It was shown that the introduction of the elasticity and adaptation in the optical domain yields significant spectral savings and leads to increased network capacity and enhanced survivability in the systems with 400-Gb/s and 1-Tb/s channels [68]. The blocking performance of spectrum-efficient superchannels in dynamic flexible-grid networks was studied, and the results demonstrated that increased SE and flexible superchannel assignment translate into network efficiency gains [69]. The energy efficiency of flexible-grid DWDM networks has also been studied [71,72]. The flexible-grid DWDM is becoming a popular choice in modern optical networks.

The use of flexible-grid DWDM allows one to take advantage of the improved cost per bit from high-data-rate channels or superchannels and increase the overall DWDM system capacity by reducing the GB overhead. Table 9.3 shows four exemplary DWDM system designs based on the fixed 50-GHz DWDM grid and flexible 75/125/237.5-GHz DWDM grids. Fig. 9.14 illustrates these four channel plans to indicate their spectral utilizations.

It can be seen from Table 9.3 and Fig. 9.14 that the flexible DWDM grids can better match the needed bandwidths for high-data-rate channels/superchannels (with the consideration of the required GB to accommodate the optical filtering during

Table 9.3 Exemplary system designs based on fixed and flexible dense wavelength division multiplexing grids.

Data rate per channel or superchannel	Modulation symbol rate and format	Channel spacing[a] (GHz)	Number of channels in the system[b]	System SE (b/s/Hz)	System capacity
100 Gb/s (200 Gb/s)	32-Gbaud PDM-QPSK (16QAM)	50	120	2 (4)	12 Tb/s (24) Tb/s
400 Gb/s	64-Gbaud PDM-QPSK	75	80	5.3	32 Tb/s
800 Gb/s	110-Gbaud PDM-PS-64QAM	125	48	6.4[c]	38.4 Tb/s
1.6 Tb/s	2-carrier 110-Gbaud PDM-PS-64QAM superchannel	237.5	25	6.7[c,d]	40 Tb/s

[a]The channel spacing is assumed to include a GB of at least 11 GHz to allow for wavelength routing.
[b]A super C band with an available bandwidth of 6 THz is assumed.
[c]The system spectral efficiency is increased by advanced techniques such as PS and entropy loading.
[d]The system spectral efficiency is further increased by using superchannel with closely spaced optical carriers.

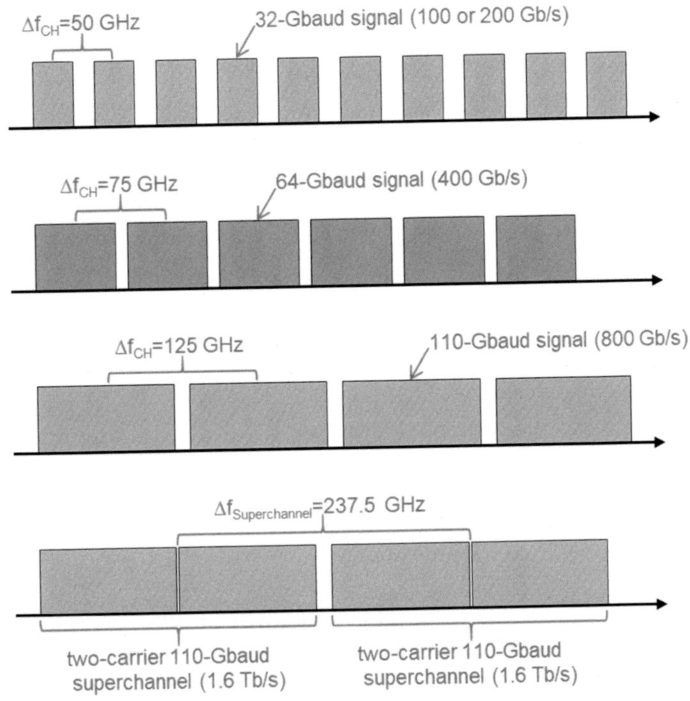

Figure 9.14 Illustration of the four channel plans presented in Table 9.3.

wavelength routing) and provide a higher system SE and capacity than the systems with the fixed 50-GHz DWDM grid. Particularly, the combined use of flexible grid and superchannel can offer the highest system SE and capacity. With the assumption of the use of the super C band with 6 THz bandwidth [73], the DWDM system SE and capacity are increased from 2 b/s/Hz and 12 Tb/s in the case of 100-Gb/s channel on the 50-GHz grid to 6.7 b/s/Hz and 40 Tb/s in the case of 1.6-Tb/s two-carrier superchannel on the 237.5-GHz grid, respectively. With the use of the super C + L band with a total bandwidth of 11 THz [74], 46 1.6-Tb/s two-carrier superchannels can be supported and the overall system capacity can be further increased to 73.6 Tb/s.

Note that in the illustrative DWDM system designs shown in Table 9.3, the system SE is set below 7 b/s/Hz to allow for long-haul transmission, for example, over 1000 km. A guard band of at least 11 GHz is also allocated between routed channels/superchannels to accommodate the successive optical filtering due to wavelength routing in ROADM/OXC nodes. With shorter transmission distance and less optical filtering, even higher system SE and capacity can be achieved, as shown in the demonstration of 88-Tb/s transmission over the super C + L band with 8-b/s/Hz SE for data center interconnect applications [74]. In addition, it is important to note that the use of large bandwidth superchannels may reduce the number of transceiver modules, save rack space and electrical power consumption, and reduce the number of the needed ROADM/OXC ports for wavelength routing. In the examples shown in Table 9.3, the number of routed DWDM channels is dramatically reduced from 120 in the case of 100-Gb/s on the 50-GHz grid to 25 in the case of 1.6-Tb/s superchannel on the 237.5-GHz grid. We will discuss in more depth the benefits of using superchannels for wavelength routing in the Section 9.5.3.

9.5.2 Low-latency wavelength routing

As discussed in Chapter 1, Introduction, 5G applications impose a stringent requirement on overall network latency. To effectively transport WDM channels with low latency, it is highly preferred to adopt transparent optical wavelength routing as much as possible to minimize the overall transmission latency, as illustrated in Fig. 9.15. Such low-latency optically routed networks have the following features [75,76]:

- *One-hop transmission between two sites*—Using the shortest path to directly connect the two sites, enabled by mesh connection in the optical network.
- *Least pass-through sites*—Using the route with the least number of pass-through sites to reduce processing latency, again enabled by mesh connection in the network.
- *Transparent wavelength pass-through*—Using transparent wavelength routing in OXC nodes to realize wavelength pass-through at the optical layer (L0) without going

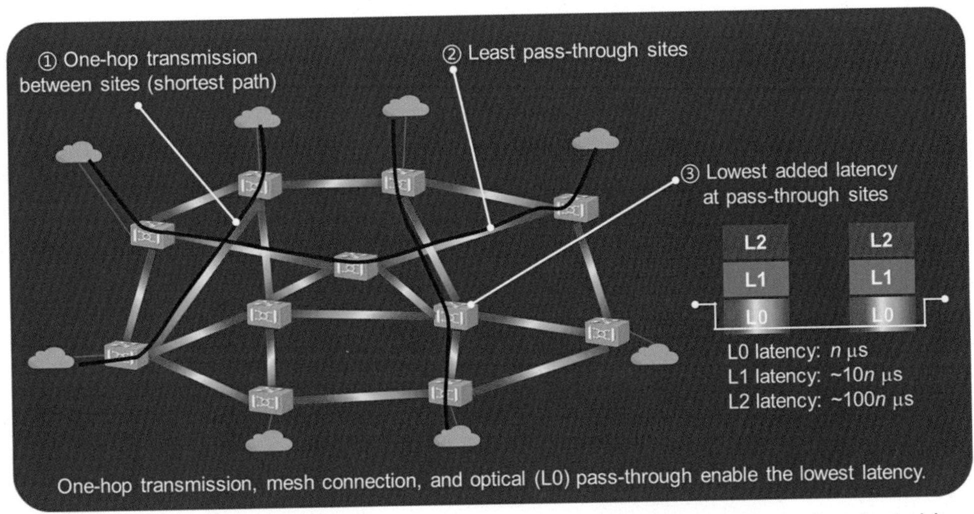

Figure 9.15 Illustration of the use of transparent optical wavelength routing in the physical layer (L0) to achieve low-latency transmission.

through optical-to-electrical (OE) conversion, electrical-to-optical (EO) conversion, and signal processing at higher layers such as the optical transport network (OTN) layer (L1) and the Ethernet layer (L2), thereby substantially reducing the latency at each pass-through site.

- *Energy-efficient routing*—Optical routing on the wavelength level avoids unnecessary OE/EO conversions and high layer signal processing, which are energy consuming. This in turn improves the energy efficiency of the optical network.

At the heart of the low-latency optically routed networks are multidegree (MD) OXCs. The optical communication industry has evolved from fixed-grid ROADM to flexible-grid-compatible MD-OXC. Fig. 9.16 illustrates a flexible-grid-compatible MD-OXC node supporting mesh connectivity from any direction to any direction, as well as adding/dropping any wavelength channel (or superchannel) to/from any direction. A compact optical backplane, which is free of manual fiber connections, is used to realize the full mesh connectivity. In addition, the MD-OXC supports the adding and dropping of wavelength channels at the OXC node. Assuming that an MD-OXC connects with M directions or degrees (such as north, east, south, and west directions) and allows the adding/dropping of N wavelength channels (or superchannels), we need a $M \times N$ add/drop module, which can be based on a multicast switch (MCS) or $M \times N$ WSS [77,78].

Fig. 9.17 illustrates an MCS-based $M \times N$ add/drop module for adding wavelength channels from N optical transmitters to M directions and dropping wavelength channels from M directions to N optical receivers. When adding a local wavelength channel to one of the M directions, the transmitter is connected to one of the add ports of

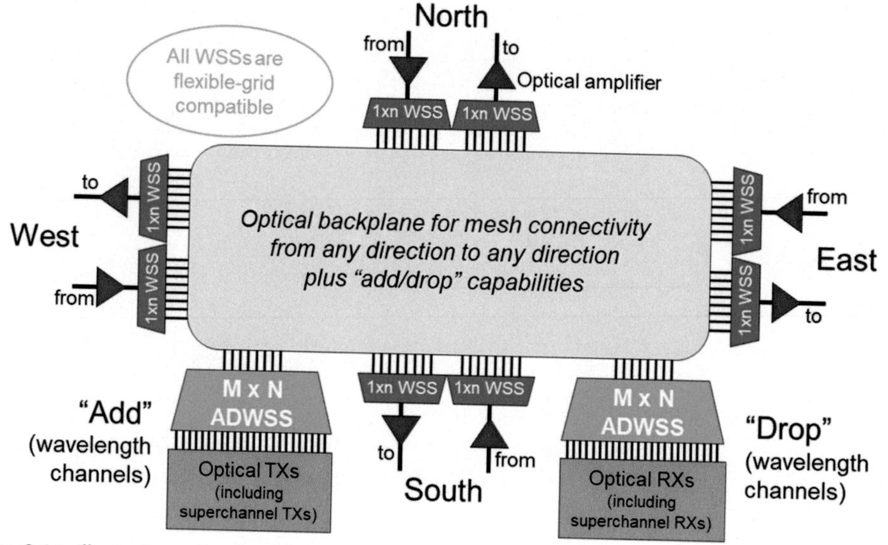

Figure 9.16 Illustration of a flexible-grid–compatible multidegree-optical cross-connect node supporting mesh connectivity from any direction to any direction, as well as adding/dropping any wavelength channel (or superchannel) to/from any direction.

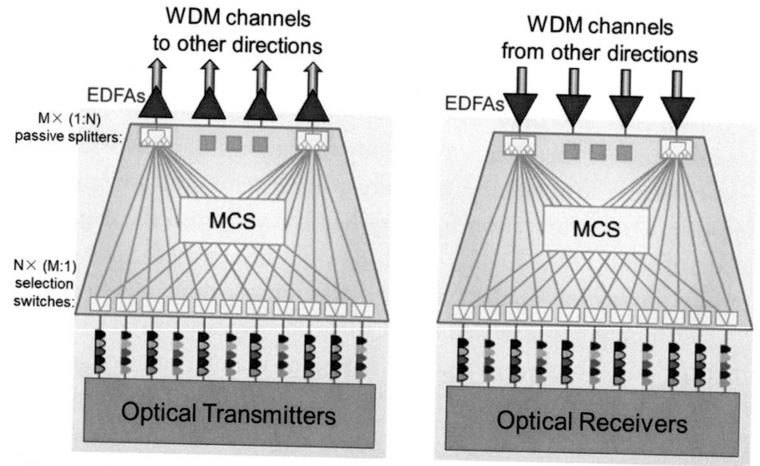

Figure 9.17 Illustration of an multicast switch-based $M \times N$ add/drop module for adding wavelength channels from N optical transmitters to M directions (left) and dropping wavelength channels from M directions to N optical receivers (right).

the $M \times N$ add/drop module. Through an M:1 selection switch, this channel is connected to the intended direction. Up to N local channels can be connected to each of the M directions via a 1:N passive splitter. The intrinsic loss of the 1:N passive splitter is $10 \cdot \log_{10}(N)$ dB, which is over 12 dB for $N \geq 16$. To compensate for the splitter loss, as well as other losses in the OXC path, an array of M EDFAs are used before the optical channels exit the OXC node. When dropping a local wavelength channel from one of the M directions, the reverse of the above propagation path is provided in drop portion of the add/drop module.

There are three key drawbacks of the MCS-based add/drop approach. First, because of the high loss of the 1:N passive splitters, 2M EDFAs are needed, which in turn leads to a higher energy consumption, higher cost, and larger form factor. Second, the total optical power of the WDM channels in a fiber depends on the aggregated bandwidth of the populated WDM channels, and thus the output power of each EDFA needs to vary with the formation of the WDM channels. There is an inherent "bandwidth bottleneck" that limits the maximum channel bandwidth and port count that an MCS can practically accommodate [79]. Third, the optical SNR (OSNR) of a given wavelength channel is degraded by the out-of-band optical noises from all the other wavelength channels on the same add path as the MCS-based add module does not filter out these out-of-band optical noises.

To address the above issues, "add/drop" module based on $M \times N$ WSS (ADWSS) has been introduced [77,78]. Fig. 9.18 illustrates the design of $M \times N$ ADWSS for adding wavelength channels from N optical transmitters to M directions and dropping wavelength channels from M directions to N optical receivers. The key difference from the MCS-based design is the replacement of the 1:N passive splitters with 1:N WSSs, which offer much lower intrinsic loss than the former. A contentionless twin 8×24 WSS has been developed, and the measured insertion loss for each port combination is between 3.3 and 6.8 dB [78]. With the low insertion loss, no EDFAs are

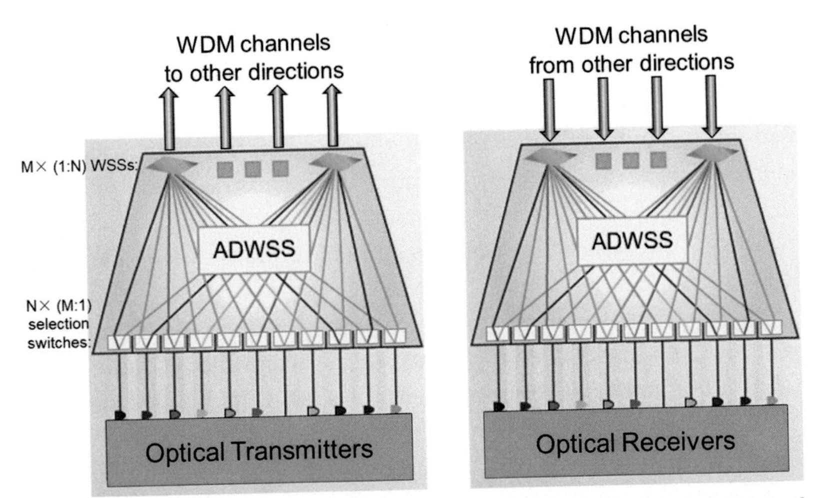

Figure 9.18 Illustration of an $M \times N$ add/drop wavelength-selective switch module for adding wavelength channels from N optical transmitters to M directions (left) and dropping wavelength channels from M directions to N optical receivers (right).

needed, which reduces the energy consumption, cost, and form factor. Moreover, the 1:N WSSs provide the wavelength filtering for each added wavelength channel and eliminate out-of-band noises from other wavelength channels. Furthermore, the 2M 1:N WSSs can be compactly implemented in a single device by sharing key components such as the grating and the beam-steering engine, to minimize the cost per module and the form factor [78]. As a result, the $M \times N$ ADWSS offers the following advantages over the MCS-based add/drop module:

- *Removal of EDFAs*—As the insertion loss of the ADWSS is independent of the port count and can be limited to below 7 dB, no EDFAs are needed, thus reducing the module power consumption, cost, and form factor, as well as removing the "bandwidth bottleneck" for scaling up the channel bandwidth and port count.
- *Improved OSNR for the added channels*—Owing to the wavelength filtering of the 1:N WSS in each channel adding path, the out-of-band noises from other wavelength channels are eliminated, thus improving the OSNR of all the added channels.
- *Improved receiver performance for the dropped channels*—Owing to the wavelength filtering of the 1:N WSS in each channel dropping path, only the intended channel is sent to its corresponding local receiver, thus improving the receiver performance and/or removing the need for additional filtering at the local receiver side.
- *High scalability*—Owing to the compact design and the absence of EDFAs, ADWSS is highly scalable for supporting more degrees and add/drop ports.

On the whole, the $M \times N$ ADWSS provides benefits in performance, energy consumption, cost, and form factor for next generation colorless, directionless, and contentionless wavelength routed networks [77–79].

9.5.3 Sustaining the capacity growth of OXC by superchannels

As discussed in Chapter 1, Introduction, the mobile data rate scales up roughly 30 times per generation of 10 years, indicating an increase of about 40% per year or doubling every 2 years. The mobile data rate's doubling every two years tracks well with the Moore's Law, which is reasonable considering that the mobile data are eventually processed by integrated circuits.

For optical networks, the network capacity has also been increasing at ~40% per year over the last 20 years [80]. As the switching capacity of OXC scales with the network capacity, the number of the routed channels per OXC node, together with the number of WSS ports, would grow exponentially with the network capacity growth if the channel data rate is fixed, which is unsustainable [79]. This scaling disparity can be mitigated by increasing the per-carrier interface rate and using the superchannel concept [80]. For example, the per-carrier optical interface rate was increased from ~100 Gb/s in 2010 to ~800 Gb/s in 2020 at a growth rate of ~25% per year, or

~0.5 dB lower than the network capacity growth rate. With the use of two-carrier superchannel, the aggregated channel data rate can be readily increased to 1.6 Tb/s [50]. Thus the number of routed channels per OXC node can grow at a gradual pace that is sustainable.

Fig. 9.19 shows the total number of routed wavelength channels/superchannels in an OXC node as a function of the channel data rate for OXC routing capacities of 100 Tb/s, 200 Tb/s, 500 Tb/s, and 1 Pb/s (or 10^{15} b/s). It is worth noting that Pb/s-capacity OXC has been implemented with 32 degrees of optical cross-connect grooming, fulfilling the requirements for mesh interconnection and large-capacity grooming in optical core networks [81]. The optical backplane containing around 1000 mesh connections was realized via an innovative fabrication process. When 200-Gb/s channels are used, a 200-Tb/s OXC needs to route 1000 wavelength channels. Assuming 25% of the channels being nominally added/dropped at the OXC node, we need 250 add/drop ports or 11 ADWSS modules each having 24 add/drop ports [79]. If 1.6-Tb/s superchannels are used instead, the number of the required 24-port ADWSS modules would be reduced from 11 to 2. Evidently, the use of high-data-rate superchannels significantly reduces the number of add/drop ports, as well as add/drop modules, required within an OXC node. In addition, the number of line-side WSS ports is also reduced as fewer add/drop modules need to be connected.

As the WDM system SE approaches the nonlinear Shannon limit, more optical bandwidth is needed to support the scaling of the optical network capacity. This can be achieved by increasing the optical amplification bandwidth per fiber, for example,

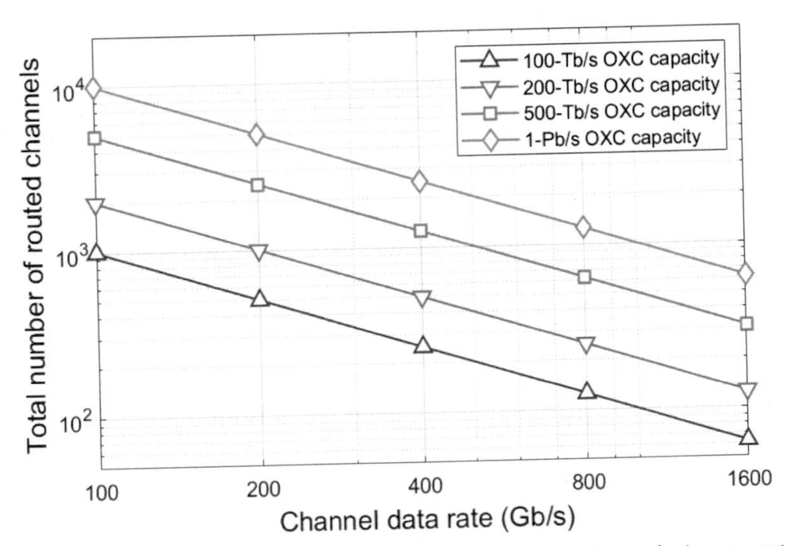

Figure 9.19 Total number of routed wavelength channels/superchannels in an optical cross-connect (OXC) node versus the channel data rate for different OXC routing capacities.

by using the super C band or super C + L band [73,74], and/or SDM with multiple fiber pairs per route [79,80]. With continued increase in per-carrier interface rate, aggregated superchannel data rate, optical amplification bandwidth, and spatial parallelism, the future scaling of OXC capacity can be supported with simple node configuration and reasonable cost.

9.6 Concluding remarks on superchannel transmission and routing

In high-capacity optical transmission, a superchannel with multiple modulated carriers offers several advantages over single-carrier channel, such as higher interface rate, better transmission performance (via joint signal processing to mitigate linear and nonlinear transmission impairments), and higher WDM system SE and capacity. In addition, superchannel naturally leverages the advances in large-scale PICs to reduce optical circuit complexity as compared to using discrete optical components [82–84]. Furthermore, superchannel architectures can advantageously leverage the advances in large-scale ASICs capable of high-speed parallel signal processing [85], especially when silicon photonics are used to achieve closer integration between optical and electrical components [84].

In modern optical networks, transparent wavelength routing via OXC reduces both latency and energy consumption, and the use of superchannel provides a much-needed help to support the future scaling of OXC routing/switching capacity with simple OXC node configuration and reasonable cost. This trend is similar to that in the evolution of data center switch for an EXC, where higher I/O interface rates are being provided via WDM- or SDM-based parallelism, as discussed in Chapter 7, Cloud Data Center Optics.

Going forward, as in the evolution trend of the data center EXC, a closer integration between the optical chip and the electrical chip is desired to reduce the cost and power consumption per bit while scaling up the switching capacity. Fig. 9.20 illustrates of the use of large-scale optical and electrical (O&E) integration for an OTN node with both OXC and EXC capabilities. The EXC used in OTN is being substantially enhanced to provide new capabilities such as network slicing with a fine granularity of 2 Mb/s, microsecond-level processing latency, and hitless bandwidth adjustment to better support 5G services and dedicated private line services [86].

With a focus on the optical transport and switching layer, both spectral and spatial superchannels are expected to benefit from future fiber-in-fiber-out coherent optical engines that will holistically cointegrate opto-electronic modulation and detection arrays for client and line traffic with CMOS DSP, as well as optical node architectures for switching and multiplexing of these superchannels [80]. With the enhanced O&E integration for both OXC and EXC functionalities, future high-capacity superchannel-assisted optical transport networks are expected to provide even more value to a diverse set of services in the 5G era.

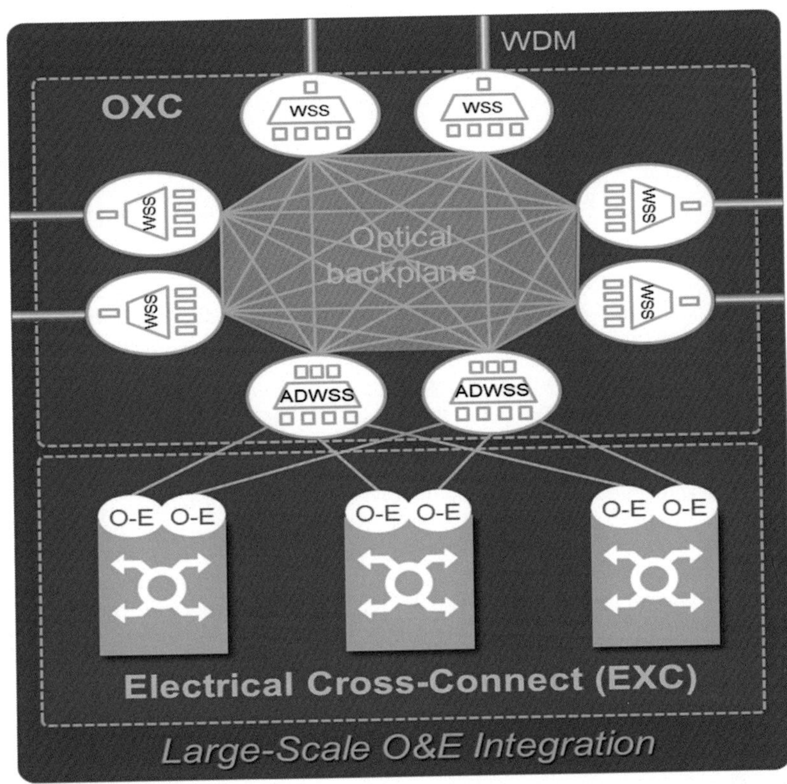

Figure 9.20 Illustration of large-scale optical and electrical (O&E) integration for an optical transport network node with both optical cross-connect and electrical cross-connect capabilities. O-E: Optical-to-electrical conversion.

References

[1] Chraplyvy AR. The coming capacity crunch. ECOC plenary talk; 2009.
[2] Tkach RW. Scaling optical communications for the next decade and beyond. Bell Labs Tech J 2010;14:3–10.
[3] Winzer PJ. Beyond 100G ethernet. IEEE Comm Mag 2010;48(7):26–30.
[4] Chandrasekhar S, Liu X, Zhu B, Peckham DW. Transmission of a 1.2-Tb/s 24-carrier no-guard-interval coherent OFDM superchannel over 7200-km of ultra-large-area fiber. In: ECOC'09, post-deadline paper PD2.6.
[5] Ma Y, Yang Q, Tang Y, Chen S, Shieh W. 1-Tb/s per channel coherent optical OFDM transmission with subwavelength bandwidth access. In: OFC'09, post-deadline paper PDPC1.
[6] Dischler R, Buchali F. Transmission of 1.2 Tb/s continuous waveband PDM-OFDM-FDM signal with spectral efficiency of 3.3 bit/s/Hz over 400 km of SSMF. In: Proceeding OFC 2009, Paper PDPC2; March 22–26, 2009.
[7] Zhu B, Liu X, Chandrasekhar S, Peckham DW, Lingle Jr. R. Ultra-long-haul transmission of 1.2-Tb/s multicarrier no-guardinterval CO-OFDM superchannel using ultra-large-area fiber. IEEE Photon Technol Lett 2010;22:826–8.
[8] Chandrasekhar S, Liu X. Experimental investigation on the performance of closely spaced multi-carrier PDM-QPSK with digital coherent detection. Opt Express 2009;17:12350–61.

[9] Chandrasekhar S, Liu X. Terabit superchannels for high spectral efficiency transmission. In: Proceeding ECOC'10, paper Tu.3.C.5; 2010.
[10] Sano A, Yamada E, Masuda H, Yamazaki E, Kobayashi T, Yoshida E, et al. No-guard-interval coherent optical OFDM for 100-Gb/s long-haul WDM transmission. J Light Technol 2009;27:3705–13.
[11] Gavioli G, et al. Investigation of the impact of ultra-narrow carrier spacing on the transmission of a 10-carrier 1 Tb/s superchannel. In: Proceeding OFC 2010, Paper OThD3; March 21–25, 2010.
[12] Gavioli G, Torrengo E, Bosco G, Carena A, Savory SJ, Forghieri F, et al. Ultra-narrow-spacing 10-channel 1.12 Tb/sD-WDM long-haul transmission over uncompensated SMF and NZDSF,". IEEE Photon Technol Lett 2010;22(19):1419–21.
[13] Torrengo E, et al. Transoceanic PM-QPSK terabit superchannel transmission experiments at baud-rate subcarrier spacing. In: Proceeding ECOC 2010, Paper We.7.C.2; September 19–23, 2010.
[14] Cai J-X, et al. Transmission of 96x100G pre-filtered PDM-RZQPSK channels with 300% spectral efficiency over 10,608 km and 400% spectral efficiency over 4,368 km. In: Proceeding OFC 2010, Post-deadline paper PDPB10; March 21–25, 2010.
[15] Bosco G, Carena A, Curri V, Poggiolini P, Forghieri F. Performance limits of Nyquist-WDM and CO-OFDM in high-speed PM-QPSK systems. IEEE Photon Technol Lett 2010;22:1129–31.
[16] Zhou X., et al. 1200 km tansmission of 50GHz spaced, 5x504-Gb/s PDM-32–64 hybrid QAM using electrical and optical spectral shaping. In: OFC 2012, paper OM2A.2.
[17] Kobayashi T., et al. 45.2Tb/s C-band WDM transmission over 240km using 538Gb/s PDM-64QAM single carrier FDM signal with digital pilot tone. In: ECOC 2011, PD Th.13.C.6.
[18] Randel S, Sierra A, Liu X, Chandrasekhar S, Winzer P. Study of multicarrier offset-QAM for spectrally efficient coherent optical communications. In: ECOC 2011, paper Th.11.A.
[19] Mazur M, Schröder J, Lorences-Riesgo A, Yoshida T, Karlsson M, Andrekson PA. 12 b/s/Hz spectral efficiency over the C-band based on comb-based superchannels. J Light Technol 2019;37:411–17.
[20] Chandrasekhar S, Liu X. Advances in Tb/s superchannels. In: Kaminov IP, Li T, Willner A, editors. Chapter 3 of Optical fiber telecommunications VIB; May 2013.
[21] Nyquist H. Certain topics in telegraph transmission theory. Trans Am Inst Electr Eng 1928;47:617–44.
[22] Shannon CE. Communication in the presence of noise. Proc Inst Radio Eng 1949;37(1):10–21 Reprint as classic paper in: Proc. IEEE, vol. 86, no. 2, (February 1998).
[23] Colavolpe G, Foggi T, Modenini A, Piemontese A. Faster-than-Nyquist and beyond: how to improve spectral efficiency by accepting interference. Opt Express 2011;19:26600–9.
[24] Ellis AD, Gunning FCG, Cuenot B, Healy TC, Pincemin E. Towards 1TbE using coherent WDM. In: Proceeding OECC/ACOFT 2008, Sydney, Australia, Paper WeA-1; 2008.
[25] Yu J, et al. Generation, transmission and coherent detection of 11.2 Tb/s (112x100Gb/s) single source optical OFDM superchannel. In: OFC'2011, PDPA6.
[26] Xia TJ, et al. 10,000-km Enhanced long-haul transmission of 1.15-Tb/s superchannel using SSMF only. In: OECC 2011, PD1.
[27] Huang Y-K, Ip E, Wang Z, Huang M-F, Shao Y, Wang T. Transmission of spectral efficient super-channels using all-optical OFDM and digital coherent receiver technologies. J Light Technol 2011;29(24):3838–44.
[28] Liu X, Chandrasekhar S, Zhu B, Winzer P, Gnauck A, Peckham D. 448-Gb/s reduced-guard-interval CO-OFDM transmission over 2000 km of ultra-large-area fiber and five 80-GHz-grid ROADMs. J Light Technol 2011;29(4):483–90.
[29] Hillerkuss D, et al. 26 Tbit s-1 line-rate super-channel transmission utilizing all-optical fast Fourier transform processing. Nat Photonics 2011;5:364–71.
[30] Liu X, et al. 3x 485-Gb/s WDM transmission over 4800 km of ULAF and 12x 100-GHz WSSs using CO-OFDM and single coherent detection with 80-GS/s ADCs. In: OFC'11, paper JThA037; 2011.
[31] Liu X, et al. 1.5-Tb/s guard-banded superchannel transmission over 56x 100-km (5600-km) ULAF using 30-Gbaud pilot-free OFDM-16QAM signals with 5.75-b/s/Hz net spectral efficiency. In: ECOC 2012, post-deadline paper Th.3.C.5; 2012.

[32] Zhou X, Nelson LE. 400G WDM transmission on the 50 GHz grid for future optical networks. J Light Technol 2012;30(24):3779–92.
[33] Raybon G, et al. 1-Tb/s dual-carrier 80-Gbaud PDM-16QAM WDM transmission at 5.2 b/s/Hz over 3200 km. In: IEEE photonics conference (IPC), paper PD1.2; 2012.
[34] Lundberg L, Mazur M, Mirani A, Foo B, Schröder J, Torres-Company V, et al. Phase-coherent lightwave communications with frequency combs. Nat Commun 2020;11:201.
[35] Zhang J, et al. Transmission of 20 × 440-Gb/s super-Nyquist-filtered signals over 3600 km based on single-carrier 110-GBaud PDM QPSK with 100-GHz grid. In: OFC 2014, paper Th5B.3; 2014.
[36] Li J, Sjödin M, Karlsson M, Andrekson PA. Building up low-complexity spectrally-efficient Terabit superchannels by receiver-side duobinary shaping. Opt Express 2012;20:10271–82.
[37] Liu X, Chandrasekhar S, Winzer PJ. Digital signal processing techniques enabling multi-Tb/s superchannel transmission: an overview of recent advances in DSP-enabled superchannels,". IEEE Signal Process Mag 2014;31(2):16–24.
[38] Temprana E, Myslivets E, Kuo BP-P, Liu L, Ataie V, Alic N, et al. Overcoming Kerr-induced capacity limit in optical fiber transmission. Science 2015;348:1445–8.
[39] Cai Y, Cai J-X, Davidson CR, Foursa D, Lucero A, Sinkin O, et al. High spectral efficiency long-haul transmission with pre filtering and maximum a posteriori probability detection. In: Presented at the European conference and exhibition on optical communication 2010, Torino, Italy, Paper We.7.C.4.
[40] Cai J-X, Davidson CR, Lucero AJ, Zhang H, Foursa DG, Sinkin OV, et al. 20 Tbit/s transmission over 6860 km with sub-Nyquist channel spacing. J Lightw Technol 2012;30(4):651–7.
[41] Zhang J, Yu J, Chi N, Dong Z, Yu J, Li X, et al. Multimodulus blind equalizations for coherent quadrature duobinary spectrum shaped PM-QPSK digital signal processing. J Lightw Technol 2013;31(7):1073–8.
[42] Jia Z, Cai Y, Chien H-C, Yu J. Performance comparison of spectrum-narrowing equalizations with maximum likelihood sequence estimation and soft-decision output. Opt Exp 2014;22:6047–59.
[43] Savory SJ. Digital coherent optical receivers: algorithms and subsystems. IEEE J Sel Top Quant Electron 2010;16(5):1164–79.
[44] Kikuchi K. Fundamentals of coherent optical fiber communications. J Light Technol 2016;34 (1):157–79.
[45] Zhuge Q, et al. Pilot-aided carrier phase recovery for M-QAM using superscalar parallelization based PLL. Opt Express 2012;20:19599–609.
[46] Buchali F, Steiner F, Böcherer G, Schmalen L, Schulte P, Idler W. Rate adaptation and reach increase by probabilistically shaped 64-QAM: an experimental demonstration. J Light Technol 2016;34:1599–609.
[47] Cho J, et al. Trans-atlantic field trial using high spectral efficiency probabilistically shaped 64-QAM and single-carrier real-time 250-Gb/s 16-QAM. J Light Technol 2018;36:103–13.
[48] Li J, Zhang A, Zhang C, Huo X, Yang Q, Wang J, et al. Field trial of probabilistic-shaping-programmable real-time 200-Gb/s coherent transceivers in an intelligent core optical network. In: Asia communications and photonics conference (ACP), PDP Su2C.1; 2018.
[49] Che D, Shieh W. Approaching the capacity of colored-SNR optical channels by multicarrier entropy loading. J Light Technol 2018;36:68–78.
[50] Sun H, et al. 800G DSP ASIC design using probabilistic shaping and digital sub-carrier multiplexing. J Light Technol 2020;38(17):4744–56.
[51] Feuer MD, et al. Joint digital signal processing receivers for spatial superchannels. IEEE Photonics Technol Lett 2012;24(21):1957–60.
[52] Kippenberg TJ, Gaeta AL, Lipson M, Gorodetsky ML. Dissipative Kerr solitons in optical microresonators. Science 2018;361:eaan8083.
[53] Liu C, et al. Super receiver design for superchannel coherent optical systems. In: Proceeding SPIE 8284, next-generation optical communication: components, sub-systems, and systems, 828405; 2012.
[54] Rahn J, et al. Transmission improvement through dual-carrier FEC gain sharing. In: OFC'13, paper OW1E.5; 2013.

[55] Zhang H, et al. 30.58 Tb/s transmission over 7,230 km using PDM half 4D-16QAM coded modulation with 6.1 b/s/Hz spectral efficiency. In: OFC'13, paper OTu2B.3; 2013.
[56] Agrawal GP. Nonlinear fiber optics. Academic Press; 2007.
[57] Wai PKA, Menyuk CR, Chen HH. Stability of solitons in randomly varying birefringent fibers. Opt Lett 1991;16:1231–3.
[58] Li X, Chen X, Goldfarb G, Mateo E, Kim I, Yaman F, et al. Electronic post-compensation of WDM transmission impai rments using coherent detection and digital signal processing. Opt Express 2008;16:880–8.
[59] Ip E, Kahn JM. Compensation of dispersion and nonlinear impairments using digital backpropagation. J Light Technol 2008;26:3416–25.
[60] Mateo EF, Yaman F, Li G. Efficient compensation of inter-channel nonlinear effects via digital backward propagation in WDM optical transmission. Opt Express 2010;18:15144–54.
[61] Fontaine K, Liu X, Chandrasekhar S, Ryf R, Randel S, Winzer P, et al. Fiber nonlinearity compensation by digital backpropagation of an entire 1.2-Tb/s superchannel using a full-field spectrally-sliced receiver. In: Proceeding ECOC'13, paper Mo.3.D.5; 2013.
[62] Xia C, Liu X, Chandrasekhar S, Fontaine NK, Zhu L, Li G. Multi-channel nonlinearity compensation of 128-Gb/s PDM-QPSK signals in dispersion-managed transmission using dispersion-folded digital backward propagation. In: Proceeding OFC'14, paper Tu3A.5; 2014.
[63] Xia C, Liu X, Chandrasekhar S, Fontaine NK, Zhu L, Li G. Multi-channel nonlinearity compensation of PDM-QPSK signals in dispersion-managed transmission using dispersion-folded digital backward propagation. Opt Express 2014;22:5859–66.
[64] Dar R, Winzer PJ. Nonlinear interference mitigation: methods and potential gain. J Light Technol 2017;35:903–30.
[65] Carena A, Curri V, Bosco G, Poggiolini P, Forghieri F. Modeling of the impact of nonlinear propagation effects in uncompensated optical coherent transmission links. J Light Technol 2012;30:1524–39.
[66] Recommendation ITU-T G.694.1. Spectral grids for WDM applications: DWDM frequency grid; October 2020.
[67] Frisken S, Poole S, Baxter G. Reconfiguration in transparent agile optical networks. Proc IEEE 2012;100(5):1056–64.
[68] Jinno M, Takara H, Sone Y. Elastic optical path networking: enhancing network capacity and disaster survivability toward 1 Tbps era. In: Proceedings of 2011 Optoelectronics and communications conference (OECC); 2011. pp. 401–04.
[69] Thiagarajan S, Frankel M, Boertjes D. Spectrum efficient super-channels in dynamic flexible grid networks—a blocking analysis. In: OFC/NFOEC 2011, paper OTuI6; 2011.
[70] Shieh W. OFDM for flexible high-speed optical networks. Light Technol J 2011;29(10):1560–77.
[71] Klekamp A, Gebhard U, Ilchmann F. Energy and cost efficiency of adaptive and mixed-line-rate IP over DWDM networks. Light Technol J 2012;30(2):215–21.
[72] Nag A, Wang T, Mukherjee B. On spectrum-efficient green optical backbone networks. In: Global telecommunications conference (GLOBECOM 2011), IEEE; 2011. pp. 1–5.
[73] See for example. Supersizing the C-Band. https://www.lightreading.com/optical-ip/400g-terabit/supersizing-the-c-band/a/d-id/751388.
[74] See for example, https://www.cio.com/article/3607195/huawei-optixtrans-dc908-ranked-dci-leader-again-by-globaldata.html.
[75] Liu X, Deng N. Emerging optical communication technologies for 5G. In: Willner A, editor. Chapter 17 of Optical fiber telecommunications VII; 2019.
[76] Liu X. Evolution of fiber-optic transmission and networking toward the 5G era. iScience 2019;22:489–506. Available from: https://www.cell.com/iscience/pdf/S2589-0042(19)30476-6.pdf.
[77] Zong L, Zhao H, Feng Z, Yan Y. Low-cost, degree-expandable and contention-free ROADM architecture based on M × N WSS. In: Optical fiber communication conference (OFC), paper M3E.3; 2016.

[78] Colbourne PD, McLaughlin S, Murley C, Gaudet S, Burke D. Contentionless twin 8x24 WSS with low insertion loss. In: Optical fiber communication conference (OFC), postdeadline paper Th4A.1; 2018.
[79] Lumentum White Paper. Efficiently supporting aggressive network capacity growth in next-generation ROADM networks. Available from: https://resource.lumentum.com/s3fs-public/technical-library-items/next-genroadm-wp-oc-ae.pdf.
[80] Winzer PJ, Neilson DT, Chraplyvy AR. Fiber-optic transmission and networking: the previous 20 and the next 20 years [Invited]. Opt Express 2018;26:24190−239.
[81] See for example, https://www.fibre-systems.com/news/etisalat-and-huawei-successfully-trial-optical-cross-connect.
[82] Chandrasekhar S, Liu X. Enabling components for future high-speed coherent communication systems. In: 2011 Optical fiber communication conference, Tutorial talk OMU5; 2011.
[83] Bennett G. Superchannels to the rescue!. Lightwave 2012;29(2).
[84] Koch T. III-V and silicon photonic integrated circuit technologies. In: OFC/NFOEC 2012, Tutorial talk OTh4D; 2012.
[85] Dedic I. High-speed CMOS DSP and data converters. In: Optical fiber communication conference, paper OTuN1; 2011.
[86] See for example, https://www.lightwaveonline.com/fttx/pon-systems/article/14168447/huawei-expands-intelligent-optix-network-line-with-liquid-otn-optical-transport-airpon-platforms.

CHAPTER 10

50-Gb/s passive optical network (50G-PON)

10.1 Motivation for 50G-passive optical network–based next-generation optical access

Time division multiplexing (TDM) passive optical network (PON) is the main fiber-optic telecommunications technology for delivering broadband access to end users. TDM-PON is based on a point-to-multipoint topology and has the following three elements:

- An optical line terminal (OLT) typically located at the central office
- Multiple optical network units (ONUs) located near the end users
- A passive optical distribution network (ODN), which consists of a single feeder fiber, a passive optical splitter, and multiple drop fibers that are connected to the ONUs.

The downstream signal from the OLT is broadcasted to all the ONUs, and encryption is used to prevent eavesdropping. The upstream signals from the ONUs are transmitted to the OLT based on time division multiple access, as illustrated in Fig. 10.1. TDM-PON offers the following key benefits:

- Low cost, by sharing the OLT and the feeder fiber with multiple end users
- Passive ODN, which requires no power suppliers or active devices throughout the distribution network, thereby simplifying the deployment, and minimizing the maintenance cost
- Statistical multiplexing gain by aggregating traffics for multiple end users.

Since 1995, the full service access network (FSAN) working group [1], consisting of major telecommunications service providers and system vendors, has worked on the fiber-to-the-home (FTTH) architectures based on TDM-PON. The international telecommunications union telecommunication standardization sector (ITU-T) and the Institute of Electrical and Electronics Engineers (IEEE) have standardized multiple generations of TDM-PON systems, as enumerated in Table 10.1. In 1998, ITU-T standardized the broadband PON (BPON) in the ITU-T G.983 Recommendation [2], providing 622 Mb/s downstream speed and 155 Mb/s upstream speed. In 2003, ITU-T standardized the gigabit-capable PON (GPON) in the ITU-T G.984 Recommendation [3], providing 2.5 Gb/s downstream speed and 1.25 Gb/s upstream speed. GPON encapsulation method enabled very efficient packaging of user traffic with frame segmentation. In parallel, IEEE standardized Ethernet PON (EPON) in the IEEE 802.3ah standard in 2004, offering

Optical Communications in the 5G Era
DOI: https://doi.org/10.1016/B978-0-12-821627-9.00006-1

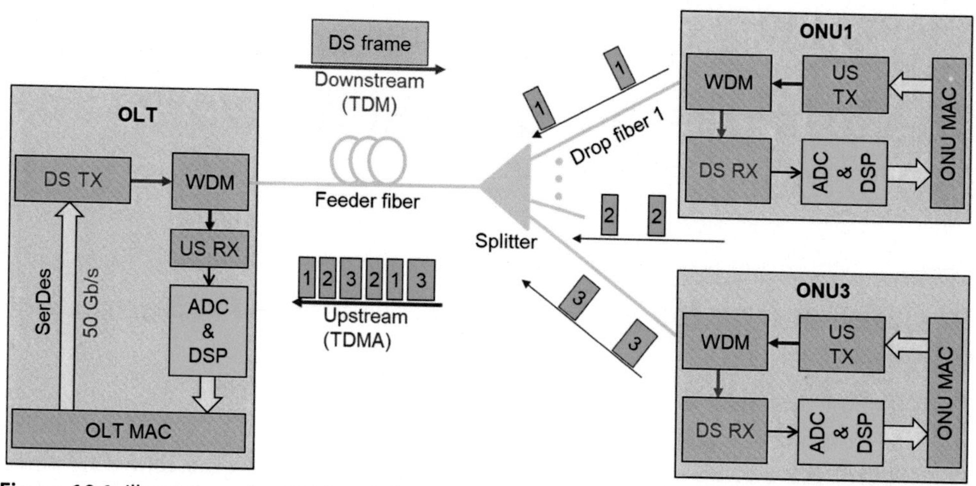

Figure 10.1 Illustration of a 50-Gb/s TDM-PON system with the use of DSP for receiver-side channel equalization. *ADC*, Analog-to-digital converter; *DS*, downstream; *DSP*, digital signal processing; *PON*, passive optical network; *RX*, receiver; *TDX*, time division multiplexing; *TX*, transmitter; *US*, upstream.

Table 10.1 Evolution of time division multiplexing-passive optical network systems and standards.

System	R_{DS} (b/s)	R_{US} (b/s)	λ_{DS} (nm)	λ_{US} (nm)	Standard	Approval year
BPON	622 M	155 M	1490	1310	ITU G.983	1998
GPON	2.5G	1.25G	1490	1310	ITU G.984	2003
EPON	1G	1G	1490	1310	IEEE 802.3ah	2004
10G-EPON	10G	1 or 10G	1577	1270	IEEE 802.3av	2009
XG-PON	10G	2.5G	1577	1270	ITU G.987	2010
XGS-PON	10G	10G	1577	1270	ITU G.9807	2016
50G-PON	50G	12.5, 25 or 50G	1342	1270, or 1300	ITU G.9804.1, G.9804.2/3	2019, 2021

λ_{DS}, Downstream center wavelength; R_{DS}, nominal downstream speed; R_{US}, nominal upstream speed; λ_{US}, upstream center wavelength.

symmetric downstream and upstream speeds at 1 Gb/s. GPON enjoyed a widespread deployment between 2010 and 2020.

To meet the demands in FTTH and fiber-to-the-building applications, IEEE standardized 10-Gb/s Ethernet PON (10G-EPON) in the IEEE 802.3av standard in 2009, offering 10 Gb/s downstream speed and 1 (or 10) Gb/s upstream speed. In the meantime, ITU-T standardized 10-Gigabit-capable PON (XG-PON) in the ITU-T G.987 Recommendation in 2010 [4], providing 10 Gb/s downstream speed and 2.5 Gb/s upstream speed. In 2016 10-Gigabit-capable symmetric PON (XGS-PON) was standardized in the ITU-T G.9807 Recommendation [5], providing symmetric downstream and upstream speeds at 10 Gb/s. According to a recent Ovum forecast [6], the annual deployment volume of 10G-PON will overtake that of GPON in the early 2020s.

In anticipation of the ever-increasing capacity demand in broadband access, especially in the 5G era, ITU-T established the higher speed PON (HSP) project to define the next-generation PON in 2018 [7]. 50-Gb/s PON (50G-PON) has been selected as a primary technology in the new G.9804 standard project series. The G.9804 series includes the following recommendations:

- ITU-T G.9804.1, describing the HSP requirements of the overall system, the evolution and coexistence, and the supported services and interfaces [8]. This recommendation was consented in July 2019 and officially approved in November 2019.
- ITU-T G.9804.2, describing the common transmission convergence (TC) layer specifications of the HSP series, such as the TC layer architecture, physical adaptation layer, business adaptation layer, management process, and message definition [9]. This recommendation was consented in April 2021.
- ITU-T G.9804.3 describing the physical media–dependent (PMD) layer specifications based on 50-Gb/s TDM-PON, such as the transmitter power, optical path penalty (OPP), and receiver sensitivity [10]. This recommendation was consented in April 2021.
- ITU-T G.9804.4, describing the PMD layer specifications of time-and-wavelength-division multiplexing PON [11].

To better support 5G applications, the ITU HSP project group also makes amendments to support cooperative dynamic bandwidth allocation/assignment (CoDBA) by adding in the TC layer the latency and jitter requirements, as well as guidance on how to build a DBA engine with proper controls [12]. In parallel, the Open Radio Access Network (ORAN) alliance is working on the specification of the cooperative transport interface [13]. Moreover, ITU-T is working on the G.sup.66 Supplement for the 5G wireless front-haul requirements in a PON context and specifying the needed interfaces such as the F1 and F*x* interfaces in the ORAN terminology [14].

In addition to 5G applications, other bandwidth-demanding applications, such as ultra-high-definition video streaming, teleconferencing, telecommuting, and remote education, accelerated the need for the next-generation PON. Given that each PON generation offers a capacity increase of four times over the previous generation, 50G-PON with a net data rate of slightly over 40 Gb/s is a natural choice for the next-generation PON after XG(S)-PON. Moreover, 50-Gbaud modulation and detection are being extensively deployed in data center optics, as discussed in Chapter 7, Cloud Data Center Optics; thus the development of 50G-PON can benefit from the 50-Gbaud ecosystem.

Historically, a new generation of PON needs to support the same ODNs used by the previous generations of PON, which means that 50G-PON needs to achieve similar loss budgets as XG(S)-PON. Theoretically, the receiver sensitivity of a 50-Gb/s signal is five times (i.e., 7 dB) worse than a 10-Gb/s signal of the same modulation format, which is nonreturn-to-zero (NRZ) on-off-keying (OOK). In addition, the dispersion tolerance of a 50-Gb/s signal is 25 times smaller than that of a 10-Gb/s signal of the same modulation format. Thus new physical layer technologies and system designs are required to enable 50G-PON to be deployable in the same ODNs used by previous generations of PON. The key new physical layer technologies and system designs used in 50G-PON are as follows:

- Receiver-side digital signal processing (DSP).
- Low-density parity check (LDPC) forward error correction (FEC), for high coding gain.
- Low-chirp downstream transmitter, such as low-chirp electroabsorption modulated laser (EML), for reducing the dispersion penalty.
- Semiconductor optical amplifier (SOA), for increasing the downstream transmitter power.
- Locating the downstream and upstream wavelengths near the zero-dispersion window of the standard single-mode fiber (SSMF), as shown in Table 10.1, for containing the dispersion penalty while supporting coexistence with GPON or XG(S)-PON.

Receiver-side DSP is a key enabling technology in high-speed data center connections, as described in Chapter 7, Cloud Data Center Optics. The use of receiver-side DSP in PON has been actively studied [15–22], as illustrated in Fig. 10.1 and is found to offer the following benefits:

- Improved receiver sensitivity, by performing matched filtering to maximize the signal-to-noise ratio and by allowing for soft-decision (SD) LDPC with further increased coding gain.
- Increased dispersion tolerance, by performing channel equalization (EQ) to mitigate the dispersion-induced intersymbol interference (ISI).
- Allowing the use of relatively low-cost 25-GHz-class transmitter and receiver, by performing channel EQ to compensate for the bandwidth limitations of the transmitter and the receiver.
- Shortened burst-mode upstream signal recovery time, which helps reduce the system latency and the burst-mode overhead.
- Support of TDM coexistence of different upstream data rates, via burst-by-burst data rate adaptation.

With the above new technologies and system designs, the ITU 50G-PON is capable of supporting common optical path loss (OPL) classes such as the Nominal 1 (N1) and the C+ that support maximal OPLs of 29 and 32 dB, respectively [10]. It is worth noting that the IEEE had approved the IEEE 802.3ca 50G-EPON standard on June 4, 2020 [23,24]. The IEEE 50G-EPON achieves an aggregated raw capacity of 50 Gb/s by using two fixed 25-Gb/s wavelength channels per direction, causing higher hardware complexity and more difficult maintenance/operation than the ITU 50G-PON that only requires one wavelength channel per direction [25]. In addition, the IEEE 50G-EPON has three upstream wavelength options, 1270 ± 10 nm (UW0), 1300 ± 10 nm (UW1), and 1320 ± 2 nm (UW2), where the last two options are inside the EPON/GPON upstream wavelength band, thus preventing the WDM coexistence of two 25-Gb/s upstream channels with EPON/GPON. Furthermore, when two 25-Gb/s upstream wavelength channels UW1 and UW2 coexist with 10G-EPON or XG(S)-PON, the UW2 transmitter in each ONU requires accurate wavelength control of within ±2 nm by using a thermoelectric cooler (TEC), which adds cost and power consumption.

In the following, we will present key 50G-PON PMD layer aspects, such as its wavelength plan, DSP algorithms used for channel equalization, burst-mode upstream transmission performance, the use of SOA for increased transmitter power and reduced chirp, and the use of transmitter and dispersion eye closure (TDEC) for assessing the transmitter quality and the dispersion penalty in Section 10.2. Key 50G-PON TC layer aspects such as LDPC, bit-interleaving, synchronization state machine, and low-latency 5G X-haul–related aspects will be presented in Section 10.3. Finally, Section 10.4 will conclude this chapter with a discussion on future perspectives of high-speed PON.

10.2 50G-PON physical media–dependent layer aspects

10.2.1 Overall PMD layer specification

Table 10.2 shows the preliminary PMD layer specification being considered by the ITU-T 50G-PON standard. Traditionally, the transmission line rate of a new ITU-T

Table 10.2 Preliminary physical media–dependent layer specification being considered by the international telecommunications union telecommunication 50G-passive optical network standard.

Specification	Downstream	Upstream
Line rate to be supported	49.7664 Gb/s	49.7664 Gb/s 24.8832 Gb/s 12.4416 Gb/s
Wavelength[a]	1342 ± 2 nm	1270 ± 10 nm (Option 1) 1300 ± 10(2) nm (Option 2)
Dispersion range	0 to 77 ps/nm	− 127 to 18 ps/nm
Modulation format	NRZ OOK	NRZ OOK
Default FEC used	LDPC (17,280, 14,592)	LDPC (17,280, 14,592)
Reference bit error ratio (BER) level[b]	1.0E-2	1.0E-2
Typical OPL classes supported	N2 (14–19 dB) C+ (17–32 dB)	N2 (14–19 dB) C+ (17–32 dB)
Exemplary launch power in OMA minus TDEC	9 dBm (in OMA)	5 dBm (in OMA)
Exemplary receiver sensitivity in OMA at the reference BER level	− 23 dBm (in OMA) @49.7664 Gb/s	− 27 dBm (in OMA) @24.8832 Gb/s

[a]For Option 2, the channel bandwidth has wideband and narrowband suboptions of 20 and 4 nm, respectively.
[b]This reference BER level assumes hard-decision (HD) forward error correction (FEC) decoding and can be higher when SD-FEC decoding is used.

PON generation is always a multiple of that of the previous generation. Therefore the downstream line rate of 50G-PON is chosen to be exactly five times as large as XG(S)-PON downstream line rate, that is, 5 × 9.95328 (Gb/s) or 49.7664 Gb/s. The upstream line rates have three options, 49.7664, 24.8832, and 12.4416 Gb/s, respectively, corresponding to 100%, 50%, and 25% of the downstream line rate.

Regarding the 50G-PON wavelength plan, the downstream wavelength is chosen to be 1342 ± 2 nm, which is the shortest wavelength that allows a 10-nm spacing from the long-wavelength edge of the GPON reduced upstream wavelength band (1330 nm), to minimize the dispersion penalty while also allowing sufficient spacing to realize the separation of downstream and upstream channels with a low-cost thin-film filter. The upstream wavelength has two options, 1270 ± 10 nm (Option 1) and 1300 ± 10(2) nm (Option 2). Option 2 has two suboptions for the channel bandwidth, 20 nm (wideband) and 4 nm (narrowband). The narrowband upstream channel bandwidth needs a TEC to limit the laser temperature range and thus the laser wavelength range. Although the use of the TEC adds some cost, the laser power can be increased because of the temperature control. In addition, the dispersion penalty is lower for the narrowband suboption. Thus the narrowband suboption is well suited for supporting high link loss budgets, especially at 49.7664 Gb/s upstream line rate.

The upstream wavelength Option 1 allows the WDM coexistence of 50G-PON and GPON, as illustrated in Fig. 10.2A. This helps service providers who have deployed GPON but not XG(S)-PON to smoothly upgrade their optical access networks to

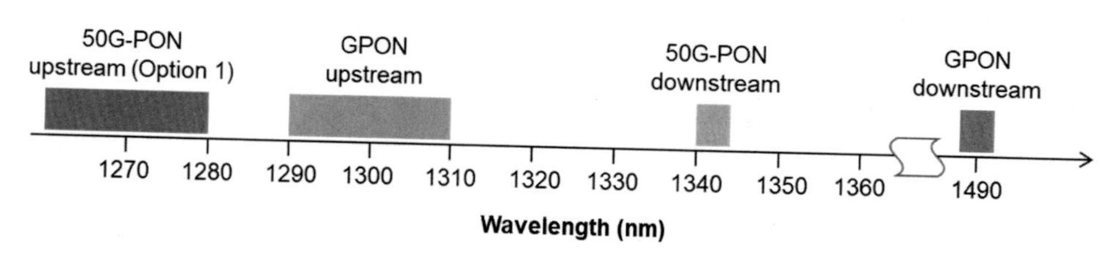

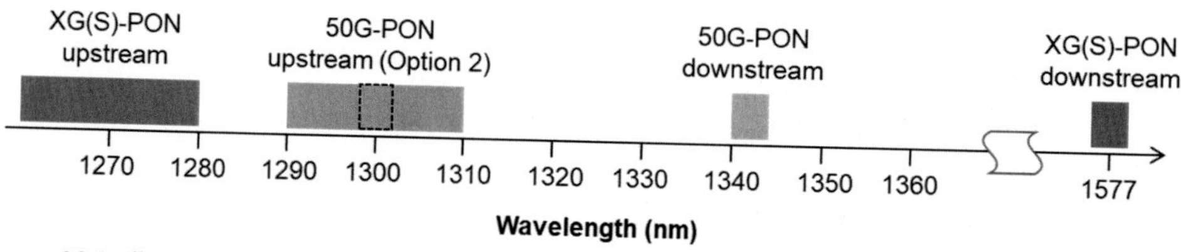

Figure 10.2 Illustration of wavelength-division multiplexing (WDM) coexistence of 50G-passive optical network (PON) with Gigabit-capable PON (GPON) (A) and 10-Gigabit-capable symmetric PON (XG(S)-PON) (B).

include the much-enhanced 50G-PON capabilities. On the other hand, the upstream wavelength Option 2 allows the WDM coexistence of 50G-PON and XG(S)-PON, as illustrated in Fig. 10.2B. This helps service providers who have deployed XG(S)-PON to add the 50G-PON capabilities by using the unoccupied or vacated GPON upstream wavelength band. Thus the ITU-T 50G-PON wavelength plan supports the WDM coexistence of 50G-PON with GPON or XG(S)-PON, which enables smooth system upgrade paths for many legacy PON systems. We will discuss other PMD layer aspects of 50G-PON in the following sections.

10.2.2 DSP for channel equalization in downstream transmission

The downstream wavelength of 50-PON is within 1342 ± 2 nm, and the dispersion penalty needs to be carefully evaluated. For the commonly used SSMF (G.652), the fiber dispersion coefficient (in ps/nm/km) can be expressed as

$$D(\lambda) \approx \frac{S_0 \lambda}{4}\left(1 - \frac{\lambda_0^4}{\lambda^4}\right) \tag{10.1}$$

where λ is the signal center wavelength (in nm), λ_0 is the zero-dispersion wavelength, and S_0 is the zero-dispersion slope. Because of fiber manufacturing uncertainties, we have $1300\ \text{nm} \leq \lambda_0 \leq 1324\ \text{nm}$ and $S_0 \leq 0.092\ \text{ps}/(\text{nm}^2 \cdot \text{km})$; therefore the maximum dispersion coefficient for $\lambda > 1324$ nm (in ps/nm/km) can be expressed as

$$D_{\max}(\lambda) \approx 0.023\lambda\left(1 - \frac{(1300\text{nm})^4}{\lambda^4}\right) \tag{10.2}$$

At the longest 50G-PON downstream wavelength of 1344 nm, we have $D_{\max}(1344\ \text{nm}) = 3.85$ ps/nm/km, resulting in a worst-case dispersion of 77 ps/nm after 20-km SSMF transmission.

In a recent study [19], the 1-dB dispersion tolerance of a chirp-free 25-Gb/s NRZ signal was found to be ~ 150 ps/nm at a FEC (BER) threshold of 10^{-3}. As dispersion tolerance scales as $1/R^2$, where R is the modulation symbol rate, the 1-dB dispersion tolerance of a chirp-free 50-Gb/s NRZ signal at 10^{-3} BER is expected to be ~ 37.5 ps/nm, which is only about half of the worst-case dispersion of 78 ps/nm after 20-km SSMF transmission. Thus the dispersion tolerance needs to be substantially improved.

To improve the dispersion tolerance, receiver-side (EQ) is a well-known technique. An experimental study was conducted on the dispersion tolerance of a 50-Gb/s NRZ OOK signal that was generated by a low-chirp EML and detected by a commercially available 25-GHz-class avalanche photodiode (APD) [19]. With receiver-side EQ, the receiver sensitivity penalty due to 85 ps/nm dispersion was reduced to less than 1 dB at the FEC BER threshold of between 1×10^{-2} and 2×10^{-2}. The results show the promise of using EML, receiver-side EQ, and advanced FEC to support 50G-PON downstream transmission over 20-km SSMF at the 1342-nm wavelength with over 32 dB loss budget.

Fig. 10.3 shows the experimental setup for the dispersion tolerance measurement of a 50-Gb/s NRZ signal [19]. An electric NRZ signal at 50 Gb/s by using an arbitrary waveform generator (AWG) was first generated. This electric NRZ signal was then used to drive a commercially available 40-GHz-bandwidth EML to generate a 50-Gb/s NRZ optical signal centered at 1550 nm. The EML output power was 5 dBm. The optical spectrum of the generated optical signal is shown as the inset in Fig. 10.3. The optical signal was then transmitted over a 5-km SSMF with a dispersion of 85 ps/nm. The eye diagrams of the 50-Gb/s optical signal before and after transmission over the 5-km SSMF are shown in Fig. 10.4. Clearly, there is strong ISI caused by the fiber dispersion, as well as the bandwidth limitations of the transmitter and the receiver. A variable optical attenuator (VOA) was used to vary the optical signal power for receiver sensitivity measurements. The optical signal was then detected by a commercially available APD with a 3-dB bandwidth of about 18 GHz. The detected signal was sampled by an 80-GSamples/s ADC, which was embedded in a real-time sampling scop. Then, receiver-side EQ based on feed-forward equalization (FFE), followed by a maximum-likelihood sequence estimation (MLSE) decoder or a Bahl-Cocke-Jelinek-Raviv (BJCR) decoder [26], was performed in the digital domain to recover the original signal, as showed in Fig. 10.5. The FFE had 41 taps. For hard

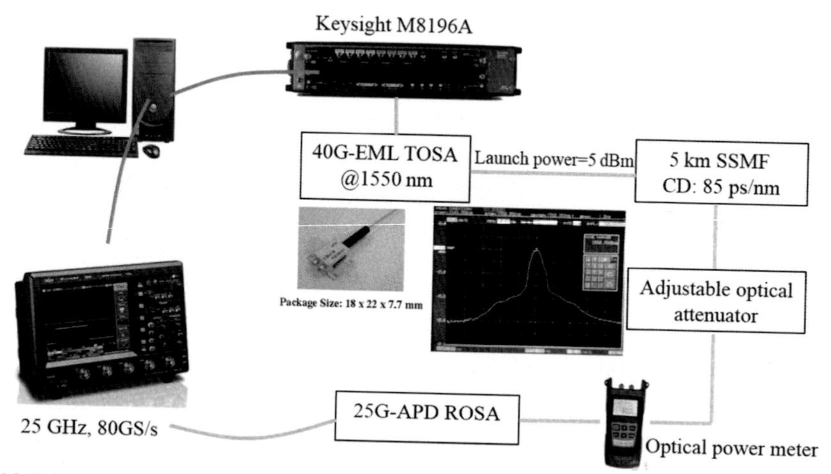

Figure 10.3 Experimental setup for the dispersion tolerance measurement of a 50-Gb/s nonreturn-to-zero (NRZ) signal. Inset: measured optical spectrum of the 50-Gb/s NRZ signal. *After Tao M, Zheng J, Dong X, Zhang K, Zhou L, Zeng H, et al. Improved dispersion tolerance for 50G-PON downstream transmission via receiver-side equalization. In: Optical fiber communication conference, paper M2B.3; 2019 [19].*

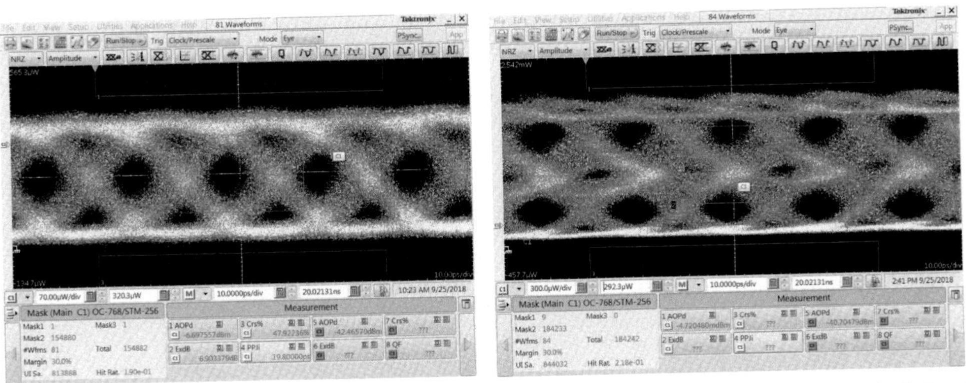

Figure 10.4 Experimentally measured eye diagrams of the 50-Gb/s nonreturn-to-zero optical signal before (left) and after (right) transmission with 85-ps/nm dispersion. *After Tao M, Zheng J, Dong X, Zhang K, Zhou L, Zeng H, et al. Improved dispersion tolerance for 50G-PON downstream transmission via receiver-side equalization. In: Optical fiber communication conference, paper M2B.3; 2019 [19].*

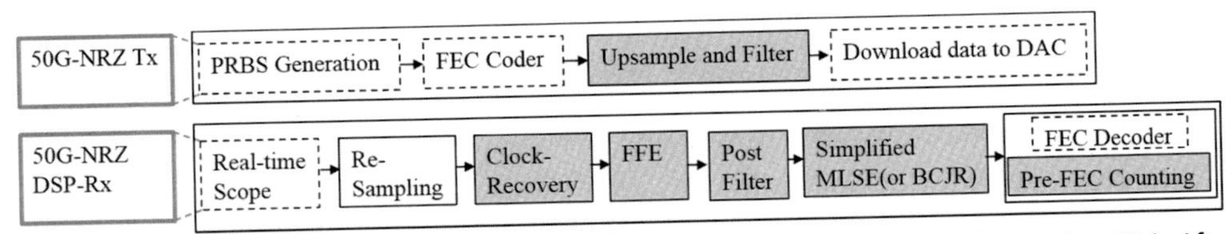

Figure 10.5 Digital signal processing blocks used in the transmitter (Tx) and the receiver (Rx). *After Tao M, Zheng J, Dong X, Zhang K, Zhou L, Zeng H, et al. Improved dispersion tolerance for 50G-PON downstream transmission via receiver-side equalization. In: Optical fiber communication conference, paper M2B.3; 2019 [19].*

decision (HD) FEC decoding, the conventional MLSE with a binary output was applied. For soft decision (SD) FEC decoding to be subsequently applied, a BCJR decoder with a soft (multilevel) output was used. The recovered signal bits were compared to the original signal bits to calculate pre-FEC BER for various received optical power levels.

Fig. 10.6 shows the experimentally measured pre-FEC BER performance of the 50-Gb/s NRZ signal as a function of receiver optical power with and without fiber dispersion. With HD-LDPC, a BER threshold of about 10^{-2} can be achieved. With SD-LDPC, the BER threshold can be increased to 2.4×10^{-2}. At BER thresholds of 10^{-3} and 2.4×10^{-2}, the dispersion-induced penalties are approximately 2 and 1 dB, respectively. Evidently, the combined use of receiver-side EQ and high-coding-gain FEC improves the dispersion tolerance. A receiver sensitivity of about -23.8 dBm is obtained for a dispersion of 85 ps/nm at the SD-LDPC BER threshold of 2.4×10^{-2}. Assuming a transmitter power of 9 dBm by using a booster SOA, a link power budget of over 32 dB can be achieved. As the back-to-back performance of the C-band EML

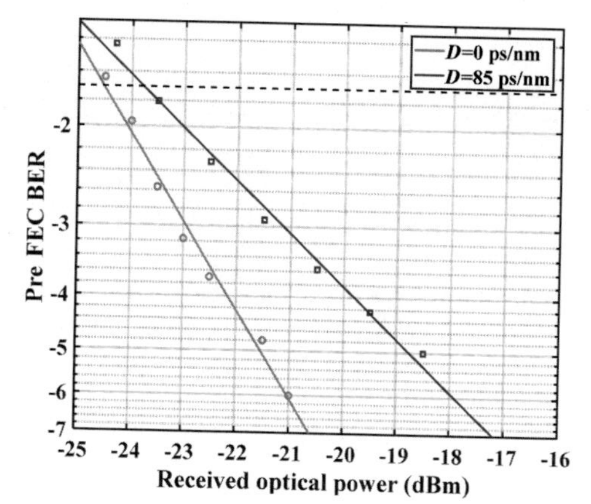

Figure 10.6 Experimentally measured pre-forward error correction (FEC) bit error ratio (BER) performance of the 50-Gb/s nonreturn-to-zero signal as a function s of received optical power for dispersion (D) equal to 0 and 85 ps/nm. *After Tao M, Zheng J, Dong X, Zhang K, Zhou L, Zeng H, et al. Improved dispersion tolerance for 50G-PON downstream transmission via receiver-side equalization. In: Optical fiber communication conference, paper M2B.3; 2019 [19].*

was not as good as that obtained by an O-band EML, further performance improvements could be obtained. This experimental demonstration shows the promise of using low-chirp EML, SOA, receiver-side EQ and advanced FEC to support 50G-PON downstream transmission over 20-km SSMF at the 1342-nm wavelength with a link loss budget of over 32 dB.

10.2.3 Burst-mode upstream transmission

For upstream transmission in the ITU-T 50G-PON standard (G.9804), three upstream line rates, nominally at 50, 25, and 12.5 Gb/s, are to be supported [8]. The first release of ITU-T G.9804.3 provides the PMD layer specifications for 25 and 12.5 Gb/s upstream line rates [10]. Prior to the approval of the ITU G.9804.3 standard, several experimental verifications were conducted on 25 Gb/s burst-mode transmission by using 25-GHz-class APD [27,28], 10-GHz-class APD/TIA/AGC receiver optical subassembly (ROSA) [29], and 25 Gb/s burst-mode ROSA based on PIN/TIA/AGC [30]. In Refs. [27] and [28], automatic gain control (AGC) was not implemented, and thus the dynamic range was limited to ~8 dB. In Ref. [29], a commercially available 10G APD/TIA/AGC ROSA was used to enable a large dynamic range of over 20 dB, but stress test with unequal burst powers was not conducted. In Ref. [30], it was found that the transient time response and burst-mode receiver performance depend on the power difference between two adjacent bursts. To determine the guard time (T_g) and preamble time (T_p) specifications for 50G-PON upstream transmission, it is necessary to study the performance of DSP-assisted burst-mode receiver with the consideration of unequal burst powers. In a recent report [21], the use of a 10 GHz-class burst-mode ROSA for multirate reception of 25, 12.5, and 10 Gb/s signals under the condition of unequal burst powers was demonstrated. With receiver-side DSP, fast burst synchronization and recovery were realized with a large dynamic range of 20 dB and a short preamble time (T_p) of 92 ns, showing the promise of this type of cost-effective ROSA for practical 50G-PON upstream applications.

Fig. 10.7 shows the experimental setup for multirate 25/12.5/10-Gb/s burst-mode upstream transmission with a commercial 10G burst-mode ROSA (MACOM-02238) based on APD/TIA/AGC. Inset (a) is a screen capture showing the transmitted optical signal spectrum. Inset (b) is a photo of the 10G burst-mode ROSA (on an evaluation board) used in the experiment. Periodic NRZ signal bursts were generated with a spacing of 40 μs, by using an AWG. Each burst had a duration of 9.95 μs. It contained a preamble that was 92 ns in duration, followed by payload data bits, which were taken from a pseudo random binary sequence (PRBS) 2^{20}-1 sequence. The preamble consisted of a known upstream physical synchronization block (PSBu) for synchronization purpose, similar to the PSBu used in XG(S)-PON. The upstream signal was amplified by a high-bandwidth RF amplifier, and the burst-enable signal was amplified by a low-bandwidth RF amplifier. These two RF signals were combined by a bias-T. The combined RF signal was then used to drive a directly modulated laser (DML) with a center wavelength of ~1270 nm to generate an upstream optical signal with a power of 5 dBm per burst. The extinction ratio (ER) of the transmitted optical signal is set at 7 dB by varying the amplitude of the upstream data. The signal was

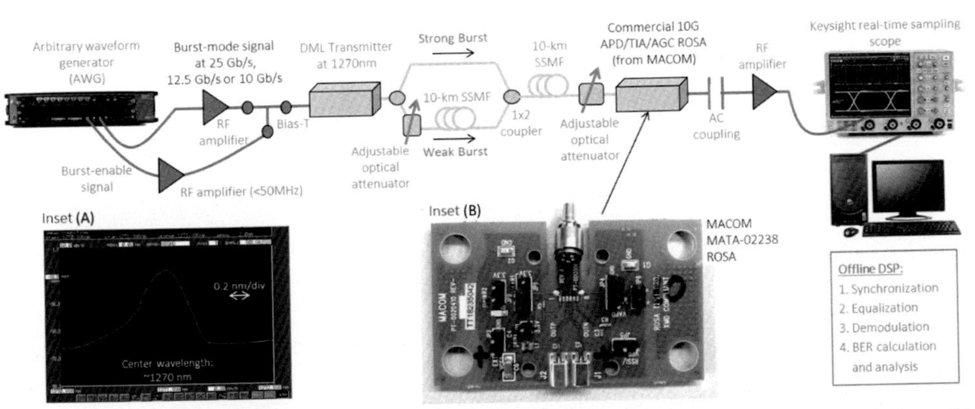

Figure 10.7 Experimental setup for multirate 25/12.5 Gb/s burst-mode upstream transmission with a commercial 10G burst-mode receiver optical subassembly (ROSA) based on APD/TIA/AGC (avalanche photodiode/transimpedance amplifier/automatic gain control). Inset (a): screen capture showing the transmitted optical signal spectrum. Inset (b): photo of the 10G burst-mode ROSA (on an evaluation board) used. *After Cheng N, Shen A, Luo Y, Zhang X, Cheng K, Steponick J, et al. Multi-rate 25/12.5/10-Gb/s burst-mode upstream transmission based on a 10G burst-mode ROSA with digital equalization achieving 20dB dynamic range and sub-100ns recovery time. In: 2020 European conference on optical communications (ECOC), paper Tu1J-7; 2020 [21].*

then split by a 1 × 2 coupler into two paths. The first path was directly connected to an input port of a second 1 × 2 coupler, while the second path was delayed by 50 μs by a 10-km SSMF, before being connected to the other input port of the second 1 × 2 coupler. The signal bursts passing through the first path and the second path are referred to as strong burst and weak burst, respectively. Because of the extra fiber loss and the use of an adjustable optical attenuator, the power of the weak burst was 10 dB lower than that of the strong burst, as shown in Fig. 10.8A. The transmitted optical signal waveform was measured by a DC-coupled opto-electronic (O/E) converter (from Keysight).

The combined burst signals were then launched into another 10-km SSMF. After fiber transmission, another adjustable optical attenuator with a built-in optical power monitor was used to vary the optical signal power for receiver sensitivity measurements. The optical signal was detected by the 10G burst-mode ROSA. As AC-coupled burst-mode transmission was shown to offer several advantages such as high performance and low power consumption [31,32], AC-coupled burst-mode reception was used in this experiment. The RF signal from the burst-mode ROSA was AC-coupled to an 80-GSamples/s ADC, which was embedded in a Keysight real-time sampling scope. The signal waveform received by the burst-mode ROSA is shown in Fig. 10.8B. Evidently, the AGC of the ROSA reduces the power difference between the strong burst and the weak burst from 10 dB to less than 1 dB. The guard time between these two bursts is only 50 ns, and it takes only 40 ns for the ROSA to produce NRZ signal with full swing, as shown in Fig. 10.8C. This configuration emulates real-world upstream transmission of multiple upstream bursts with different powers.

The received waveforms were well conditioned with a burst settling time of shorter than 50 ns. Offline DSP was performed on the sampled waveforms. Receiver-side burst synchronization based on the cross-correlation between the received signal and the known PSBu was conducted to find the beginning of each burst. At the same

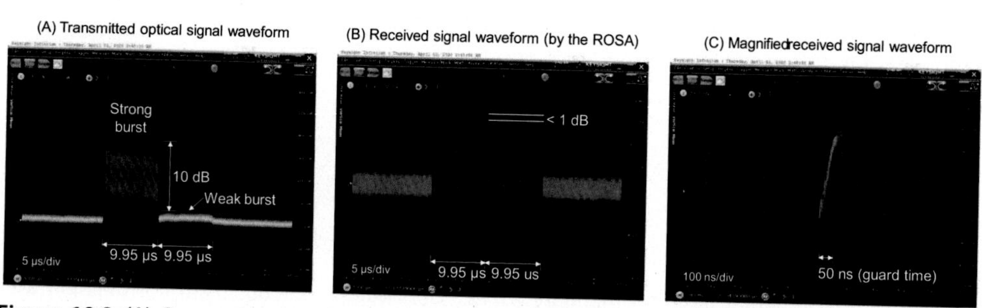

Figure 10.8 (A) Screen capture showing the transmitted optical signal waveform, (B) screen capture showing the received signal produced by the burst-mode receiver optical subassembly, and (C) screen capture showing a magnified version of the received signal waveform. *After Cheng N, Shen A, Luo Y, Zhang X, Cheng K, Steponick J, et al. Multi-rate 25/12.5/10-Gb/s burst-mode upstream transmission based on a 10G burst-mode ROSA with digital equalization achieving 20dB dynamic range and sub-100ns recovery time. In: 2020 European conference on optical communications (ECOC), paper Tu1J-7; 2020 [21].*

time, equalization based on a 32-tap FFE was also performed for the 25 Gb/s line rate, to compensate the limited bandwidth of the 10G ROSA. After equalization, the payload signal was demodulated and compared with the original payload to calculate the BER and the BER evolution with time. Fig. 10.9 shows the measured raw BER for optical transmission over 0 and 20 km SSMF as a function of the received optical power (ROP) per burst. Receiver sensitivities are measured at raw BER values of 1E-2 and 2.4E-2, respectively, which correspond to the BER thresholds of the HD- and SD-LDPC decoders considered by the ITU-T 50G-PON standard group. These key results are summarized in Table 10.3.

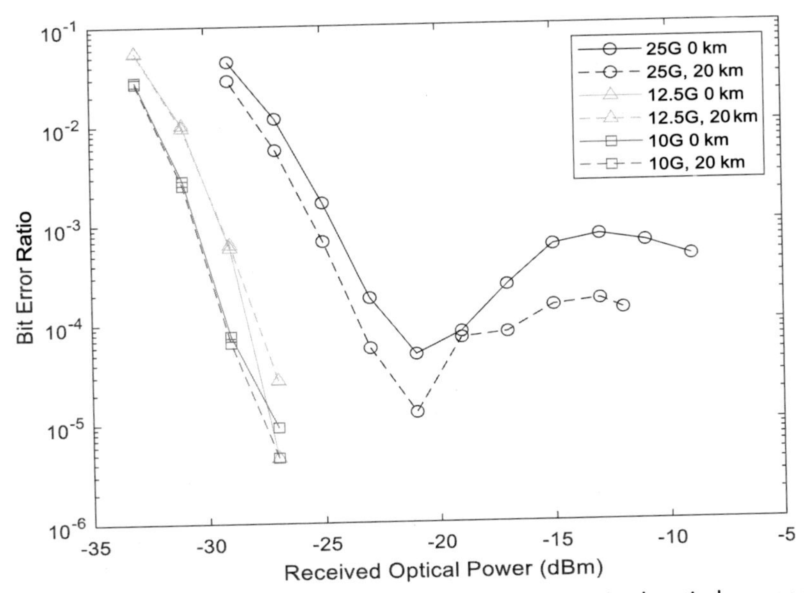

Figure 10.9 Measured raw bit error ratio as a function of the received optical power per burst at 25/12.5/10 Gb/s. *After Cheng N, Shen A, Luo Y, Zhang X, Cheng K, Steponick J, et al. Multi-rate 25/12.5/10-Gb/s burst-mode upstream transmission based on a 10G burst-mode ROSA with digital equalization achieving 20dB dynamic range and sub-100ns recovery time. In: 2020 European conference on optical communications (ECOC), paper Tu1J-7; 2020 [21].*

Table 10.3 Measured burst-mode receiver sensitivities for three data rates.

	Receiver sensitivity at BER = 1E-2 (2.4E-2)	
	$L = 0$ km	$L = 20$ km
10 Gb/s	− 32.1 (−33) dBm	− 32.1 (−33) dBm
12.5 Gb/s	− 31 (−32) dBm	− 31 (−32) dBm
25 Gb/s	− 26.9 (−28) dBm	− 27.7 (−28.8) dBm

From the results shown in Table 10.3, we have the following observations:

- With receiver-side EQ, the 10G burst-mode ROSA is capable of receiving 25, 12.5, and 10 Gb/s signals with reasonably good receiver sensitivities, both for 0- and 20-km transmissions. With an ONU transmitter power of 5 dBm, a link budget of 31.9 dB can be achieved for all these cases.
- The dispersion penalty is negligible at 10 and 12.5 Gb/s, while it becomes negative (−0.8 dB) at 25 Gb/s, which is reasonable considering the signal wavelength used.
- Increasing the data rate from 10 to 12.5 Gb/s and 25 Gb/s causes sensitivity penalties of 1.1 and 4.6 dB at BER = 1E-2 after 20-km transmission, respectively.
- Increasing the BER threshold to 2.4E-2 (via SD-LDPC) increases the link budget to 33 dB for all these cases.

It is important for the upstream burst-mode receiver to have a large dynamic range of over 15 dB to accommodate the differential optical path losses in an ODN. When the ROP is between −26.9 and −6 dBm, the BER values in all the cases are below the HD-LDPC threshold of 1E-2. This indicates that the burst-mode ROSA is suitable for supporting a large dynamic range of over 20 dB. It is also important to achieve fast burst recovery to reduce T_p, thus increasing the upstream throughput and/or reducing the upstream processing latency. Fig. 10.10 shows representative recovered 25-Gb/s NRZ symbols of a weak burst at an ROP of −21 dBm. With the use of burst-mode EQ, stable burst synchronization and recovery can be realized after a short preamble time (T_p) of 92 ns and a short guard time (T_g) of 50 ns.

This proof-of-concept experimental demonstration shows that the receiver-side DSP can enable commercially available 10-GHz-class burst-mode ROSAs to be used for multirate burst-mode upstream reception at 25 and 12.5 Gb/s with over 32 dB link loss budget. The DSP-assisted burst-mode receiver also enables fast burst

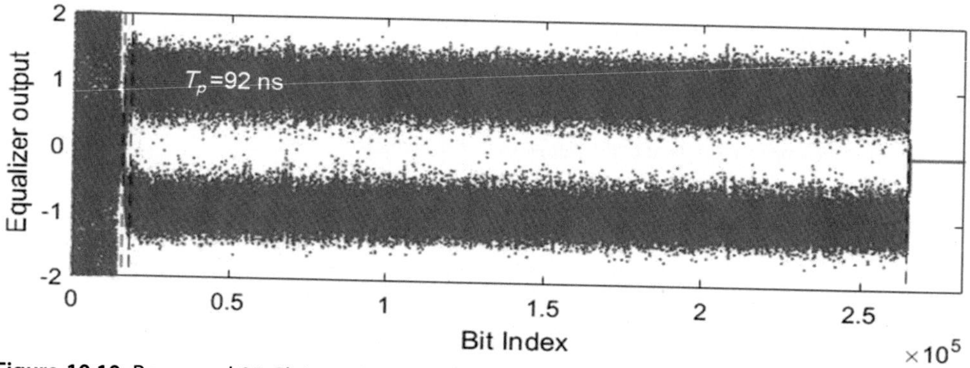

Figure 10.10 Recovered 25-Gb/s symbols in the weak burst. *After Cheng N, Shen A, Luo Y, Zhang X, Cheng K, Steponick J, et al. Multi-rate 25/12.5/10-Gb/s burst-mode upstream transmission based on a 10G burst-mode ROSA with digital equalization achieving 20dB dynamic range and sub-100ns recovery time. In: 2020 European conference on optical communications (ECOC), paper Tu1J-7; 2020 [21].*

synchronization and recovery with a preamble time (T_P) of 92 ns, as well as a large dynamic range of over 20 dB, showing the feasibility of meeting the 50G-PON upstream transmission performance targets with cost-effective implementations.

10.2.4 Use of SOA for increased transmitter power and reduced chirp

To allow the use of 50G-PON in a large variety of ODNs, the C+ OPL class with a link loss of up to 32 dB is desired to be supported. For downstream transmission at 50 Gb/s, it is very challenging to achieve such a high loss budget due to the limited output power directly from a transmitter laser, for example, a low-chirp EML. A typical EML can launch up to 6 dBm optical power into the transmission fiber. According to the results shown in Fig. 10.6, a 50-Gb/s receiver with MLSE-based channel EQ can achieve about -23 dBm sensitivity (in the average power received) at the HD-LDPC threshold of 1E-2 after 20-km transmission, thus supporting an OPL budget of $\sim$29 dB. To reach 32 dB OPL budget, an OLT transmitter using integrated EML + SOA is a viable solution [33]. In a recent study [34], it was shown that a suitably configured SOA can not only substantially increase the transmitter power but also reduce the dispersion penalty, especially when simpler FFE-based channel EQ is used, making it feasible to achieve the C+ class loss budget with additional margin.

Numerical simulations were performed by using the VPI simulation software [35]. At the OLT side, the transmitter consisted of an EML with an ER of 9 dB and a chirp parameter of 0.5, followed by a booster SOA. The booster SOA was configured to have an input power of 3 dBm from the EML and an output power of up to 13.5 dBm. At the ONU side, the receiver was based on a 25-GHz-class APD with a 3-dB bandwidth of 18.75 GHz, followed by a 31-tap FFE. Fig. 10.11 shows the generated optical signal power and frequency chirp as a function of time with and without the SOA. The SOA gain saturation and carrier dynamics cause a pattern dependence on the amplified optical waveform in the back-to-back (B2B) case. On the other hand, the SOA reduces the frequency chirp of the amplified signal in the 20-km transmission case.

Fig. 10.12 shows the simulation results for 50-Gb/s transmission with and without SOA. The receiver sensitivities for the B2B and 20-km transmission cases are summarized in Table 10.4. The key observations are:

- At the HD-LDPC BER threshold of 1.0E-2, the dispersion penalty after 20-km transmission is reduced from 2.5 dB without the SOA to 1.4 dB with the SOA.
- At the SD-LDPC BER threshold of 2.4E-2, the dispersion penalty after 20-km transmission is reduced from 2.3 dB without the SOA to 1.2 dB with the SOA.
- With the booster SOA and assuming 10-dBm power launched into the ODN, power budgets of 34.9 and 36.5 dB can be achieved at the HD- and SD-LDPC thresholds, respectively, leaving 2.9 and 4.5 dB additional margins for implementation imperfections.

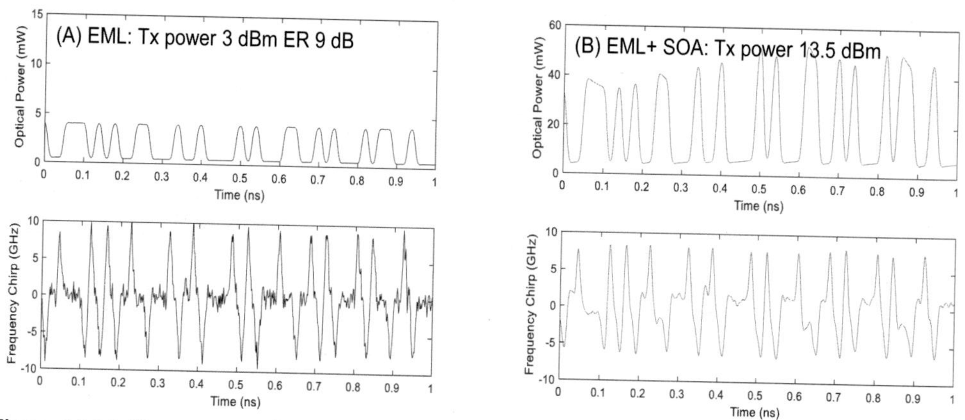

Figure 10.11 The power and frequency chirp of a 50-Gb/s signal as a function of time at the output of the electroabsorption modulated laser (EML) (A) and the booster semiconductor optical amplifier (SOA) (B). *After Cheng N, Liu X, Shen A, Effenberger F. Reducing 50G-PON downstream dispersion penalty with a suitably configured booster SOA. In: Contribution D24, ITU-T SG15/Q2 Meeting; March 2021 [34].*

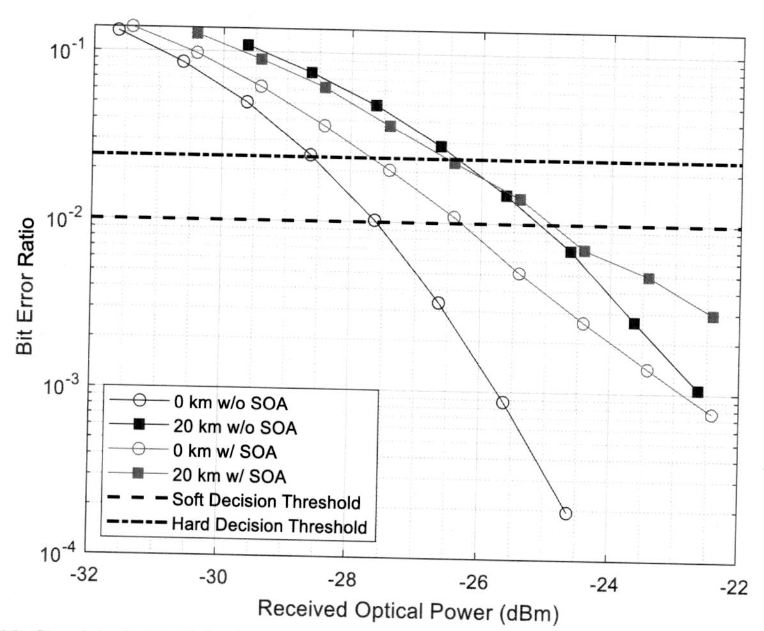

Figure 10.12 Simulated 50-Gb/s downstream transmission performances without and with the booster semiconductor optical amplifier (SOA). *After Cheng N, Liu X, Shen A, Effenberger F. Reducing 50G-PON downstream dispersion penalty with a suitably configured booster SOA. In: Contribution D24, ITU-T SG15/Q2 Meeting; March 2021 [34].*

Table 10.4 Simulated 50G-PON downstream receiver sensitivities without and with a booster SOA.

	Without SOA		With SOA	
Distance	0 km	20 km	0 km	20 km
Receiver sensitivity at BER = 1.0E-2	− 27.6 dBm	− 25.1 dBm	− 26.3 dBm	− 24.9 dBm
Receiver sensitivity at BER = 2.4E-2	− 28.6 dBm	− 26.3 dBm	− 27.7 dBm	− 26.5 dBm

BER, Bit error ratio; *PON*, passive optical network; *SOA*, semiconductor optical amplifier.
Source: After Cheng N, Liu X, Shen A, Effenberger F. Reducing 50G-PON downstream dispersion penalty with a suitably configured booster SOA. In: Contribution D24, ITU-T SG15/Q2 Meeting; March 2021 [34].

10.2.5 TDEC for assessing the transmitter quality and the dispersion penalty

PON systems requires interoperability of OLT and ONU from different vendors. This requirement is satisfied by open global standards that specify the transmitter, transmission channel, and receiver as individual elements of a complete PON system. In GPON and XG(S)-PON, the transmitter is specified by its average output power (P_{avg}) and ER. Optical modulation amplitude (OMA) is another performance indicator for optical transmitter. For NRZ OOK, OMA is defined as

$$\mathrm{OMA} = P_1 - P_0 \tag{10.3}$$

where P_1 and P_0 are the optical power levels of "ones" and "zeros," respectively. By definition, we have $\mathrm{ER} = P_1/P_0$ and $P_{avg} = (P_1 + P_0)/2$, which lead to the following relations between OMA and P_{avg}

$$\mathrm{OMA} = 2P_{avg}\frac{\mathrm{ER} - 1}{\mathrm{ER} + 1} \tag{10.4}$$

and

$$P_{avg} = \mathrm{OMA}\frac{\mathrm{ER} + 1}{2(\mathrm{ER} - 1)} \tag{10.5}$$

Fig. 10.13 shows the ratio between OMA and P_{avg} (in dB) as a function of ER. As an example, at ER = 9(6) dB, OMA is 1.91(0.78) dB larger than P_{avg}.

OMA has the advantage of being related to the received BER. Assuming an ISI-free optical signal and a thermal noise–limited receiver (e.g., a PIN receiver), OMA is related to the Q factor of the received signal as

$$Q = \frac{\mathrm{OMA}}{\sigma_1 + \sigma_0} \tag{10.6}$$

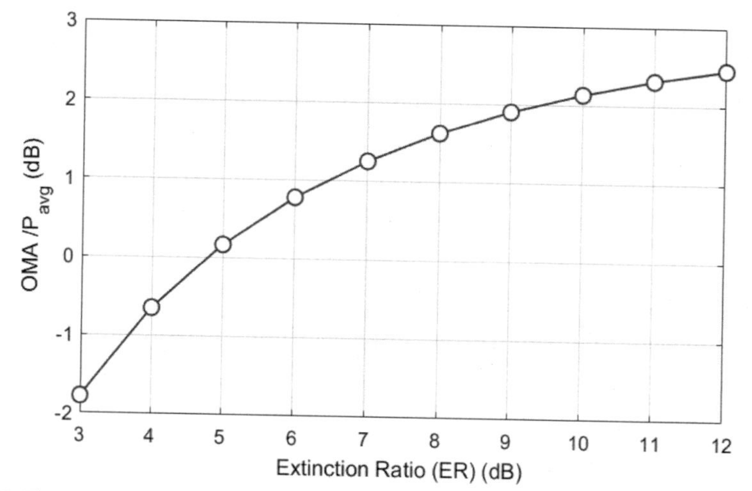

Figure 10.13 The ratio between optical modulation amplitude (OMA) and P_{avg} (in dB) as a function of extinction ratio (ER).

where σ_1 and σ_0 are the standard deviations of thermal noises on digits 1 and 0, respectively, and the Q factor is related to BER as

$$\mathrm{BER} = \frac{1}{2}\mathrm{erfc}\left(Q/\sqrt{2}\right) \tag{10.7}$$

where erfc is the complementary error function.

In real systems, however, ISI occurs because of the bandwidth limitations of the transmitter and the receiver, as well as the fiber dispersion–induced signal distortions. To allow the 50G-PON power budget specification to accommodate more implementation flexibility, the use of an "OMA-minus-penalty" method to specify the OLT transmitter parameters has been generally adopted [10]. The use of OMA gives flexibility for vendors to have different trade-offs between P_{avg} and ER, and subtracting the OPP from OMA allows for a further trade-off between the transmitter quality and the OPP. For example, if a transmitter has a larger chirp and causes a larger dispersion penalty, the vendor can compensate for the larger dispersion penalty by increasing the OMA of the transmitter. On the other hand, if a transmitter causes a smaller dispersion penalty, it can be allowed to have a reduced OMA.

It is important to determine which penalty metric to use. During the ITU-T Q2/15 Interim Meeting held in Xi'an, China, in April 2019, the use of the TDEC metric with a reference equalizer to characterize transmitters in G.HSP was proposed and approved [36]. The use of a reference receiver in measuring TDEC is based on the fact that receiver-side EQ is adopted in 50G-PON. This provides additional design flexibilities. For example, it is possible for a transmitter to use preemphasis and still pass the transmitter TDEC test.

In the following, we will present the TDEC method in detail. For high-speed transmission, advanced FEC with threshold (or reference) BER levels between 1E-4 and 1E-2 are often employed. As these reference BER levels are much higher than the required post-FEC BER levels (typically between 1E-15 and 1E-12), new methods that assess the direct impact of the transmitter and the OPP on the receiver sensitivity at these reference BER levels are both useful and feasible. The TDEC method is one such new method that has been adopted in many high-speed transmission standards such as IEEE 802.3bm, IEEE 802.3bs, and IEEE 802.3cd.

The TDEC method consists of the following five procedural steps [36]:

1. An oscilloscope is used to acquire the signal waveform. Through the eye diagram, four vertical histograms $f_{1,L}(y)$, $f_{0,L}(y)$, $f_{1,R}(y)$, and $f_{0,R}(y)$ are measured at 0.4 unit interval (UI) and 0.6 UI, and above and below P_{avg}. Here the subscript 1, 0, L, and R indicate above P_{avg}, below P_{avg}, left to the center of the eye at 0.4 UI, and right to the center of the eye at 0.6 UI, respectively.
2. A reference equalizer is used to equalize the received signal waveform. The reference equalizer is conventionally based on an FFE equalizer.
3. The amplitude eye opening is statistically examined by determining how much additive white Gaussian noise can be added to the signal to achieve a given reference BER ($\mathrm{BER_{ref}}$) when observed with a virtual receiver in the oscilloscope.
4. A similar analysis is performed on an ideal signal waveform having the same OMA as the signal under test to determine the amount of the added noise to achieve the same $\mathrm{BER_{ref}}$.
5. The ratio between the standard deviations of the two added noises is used to calculate the power penalty due to the nonideal transmitter and the OPP.

The TDEC method finds the standard deviation of the added noise for the left (or right) side of the eye diagram σ_{L} (or σ_{R}) such that

$$\frac{\int f_{1,L(R)}(y)P\left(\frac{y-P_{\mathrm{avg}}}{C_{\mathrm{EQ}}\cdot\sigma_{\mathrm{L(R)}}}\right)dy}{2\cdot\int f_{1,L(R)}(y)dy}+\frac{\int f_{0,L(R)}(y)P\left(\frac{P_{\mathrm{avg}}-y}{C_{\mathrm{EQ}}\cdot\sigma_{\mathrm{L(R)}}}\right)dy}{2\cdot\int f_{0,L(R)}(y)dy}=\mathrm{BER_{ref}} \tag{10.8}$$

where $P(x)=\mathrm{erfc}(x/\sqrt{2})/2$ is the probability of having a value of larger than x under the normal distribution and C_{EQ} is the noise enhancement factor caused by the reference equalizer. Here, $P\left(\frac{y-P_{\mathrm{avg}}}{\sigma}\right)$ and $P\left(\frac{P_{\mathrm{avg}}-y}{\sigma}\right)$ can be regarded as the error probabilities when adding a noise of standard deviation σ to "ones" and "zeros," respectively. The noise enhancement factor C_{EQ} can be expressed as

$$C_{\mathrm{EQ}}=\sqrt{\int_f N(f)\times\left|H_{\mathrm{EQ}}(f)\right|^2 df} \tag{10.9}$$

where $N(f)$ is the normalized noise spectrum, $H_{EQ}(f)$ is the frequency response of the reference equalizer with $H_{EQ}(f=0)$ normalized to 1. When no equalizer is used, we have $H_{EQ}(f)\equiv 1$ and $C_{EQ}=1$. For a typical bandwidth-limited signal, the reference equalizer is effectively a high-pass filter, for which we have $C_{EQ}>1$. For a signal with preemphasis, the reference equalizer may act like a loss pass filter, for which we have $C_{EQ}<1$.

The standard deviation of the noise that can be added to the device under test (DUT), σ_{DUT}, can then be expressed as

$$\sigma_{DUT} = \sqrt{\min(\sigma_L, \sigma_R)^2 + S^2} \tag{10.10}$$

where S is the standard deviation of the oscilloscope noise when no input signal is present. For an ideal reference signal without any noise, we can also add noise to it until the BER reaches BER_{ref} in the virtual receiver. The standard deviation of the added noise σ_{ideal} can be theoretically calculated from the OMA and the BER_{ref} by

$$\sigma_{ideal} = \frac{OMA}{2 \cdot \sqrt{2} \cdot \mathrm{erfcinv}(2 \cdot BER_{ref})} \tag{10.11}$$

where erfcinv is the inverse complementary error function. For example, with BER_{ref} being 1E-2 (or 2.4E-2), the value of $\sqrt{2} \cdot \mathrm{erfcinv}(2 \cdot BER_{ref})$ is 2.33 (or 1.98).

Finally, the TDEC (defined in dB) can be obtained as

$$TDEC = 10 \log\left(\frac{\sigma_{ideal}}{\sigma_{DUT}}\right) \tag{10.12}$$

The result is the effective power penalty for the signal under test, indicating the additional power in dB required at the receiver compared to the ideal reference transmitter [36].

The TDEC method has been well tested and adopted in several IEEE standards where thermal noise–limited PIN receivers are primarily used. For PON systems, APD receivers are usually used instead, to achieve better receiver sensitivities for meeting high power budget requirements. APD may generate may more noise on "ones" than on "zeros" owing to the beating between the signal and the amplified spontaneous emission (ASE) noise. In bandwidth-limited systems with receiver-side EQ, the noises on "ones" and "zeros" become essentially even, as experimentally observed in Fig. 10.10, and therefore the TDEC method above can be readily applied. Indeed, a reasonably good matching between the TDEC and the optical power penalty was experimentally verified when receiver-side EQ was applied [37]. The matching between the TDEC and the optical power penalty could be further improved, for example, by slightly adjusting the TDEC measurement window size, for example, from 0.2UI to 0.15UI.

With the verification of the TDEC method, the ITU-T 50G-PON standard adopts the "OMA minus TDEC" approach to define compliant transmitters. The transmitter vendors can determine the minimal transmitter powers on the basis of their own TDEC values. Different vendors can choose their own transmitter technologies and components as long as the "OMA minus TDEC" specification is satisfied. For the receiver vendors, they can also optimize their implementations and only need to meet the sensitivity requirements based on any transmitter that could pass the TDEC compliance test. This approach increases the flexibility in implementation and promotes a broader supply chain, while still guaranteeing interoperability and easy compliance qualification.

10.2.6 Jitter-related specifications

GPON and XG(S)-PON have well-defined jitter-related specifications for jitter tolerance, jitter generation, and jitter transfer [3–5]. To investigate the jitter tolerance of 50G-PON, real-time measurements based on field-programmable gate arrays (FPGAs) were performed [38]. Fig. 10.14 depicts the experimental setup. A 400-kHz sinusoidal jitter was generated by a Keysight AWG onto a 16.66 GHz clock that was used as the clock source for a real-time 25-Gb/s NRZ transmitter consisting of an FPGA (for signal generation), a 33.33-GSa/s digital-to-analog converter (DAC) and a transmitter optical subassembly (TOSA). Fig. 10.15 illustrates a sinusoidal jitter or periodical jitter (PJ) with a peak-to-peak amplitude of 10.08 ps.

The 25-Gb/s NRZ signal was then transmitted over a 20-km SSMF, before being received by a real-time 25-Gb/s NRZ receiver consisting of a ROSA, a 33.33-GSa/s ADC, and an FPGA for DSP-based signal recovery at a parallel processing speed of ~260 MHz. The real-time DSP included a 20-tap FFE. Fig. 10.16 shows the measured BER performance as a function of the ROP under four peak-to-peak jitter amplitudes (PJ_{PP}), 0, 9, 15, and 18 ps, respectively, corresponding to 0-, 0.225-, 0.375- and 0.45- UIs for the 25-Gb/s NRZ signal. We can see that the optical power penalties due to 9, 15, and 18 ps PJ_{PP} are about 0.5, 1.1, and 1.5 dB, respectively, at the HD-LDPC BER threshold of 1E-2. The penalty induced by a PJ_{PP} of 15 ps (or 0.75UI at 50 Gb/s) can be reduced to less than 1 dB for 50-Gb/s NRZ if the parallel processing speed of the

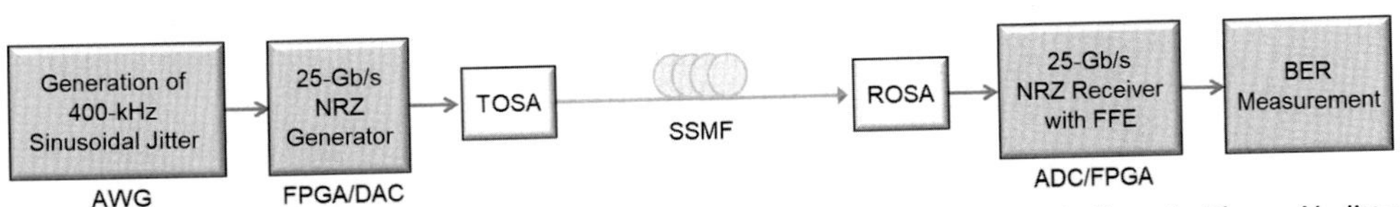

Figure 10.14 Experimental setup for the jitter tolerance study. *After Liu X, Shen A, Cheng N, Jitter tolerance mask for 50G-PON. In: Contribution D12, ITU-T SG15/Q2 Meeting; February 2021 [38].*

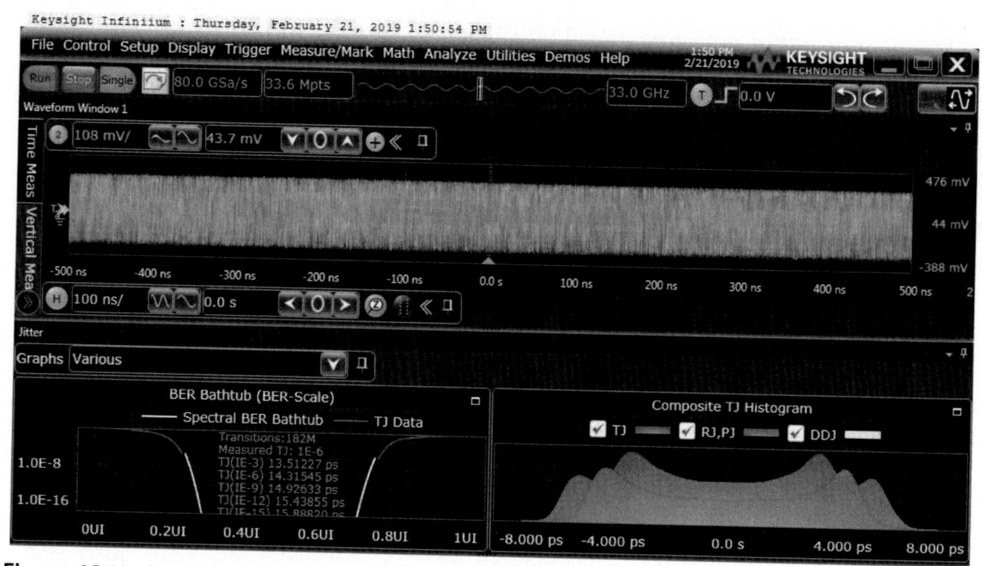

Figure 10.15 Screen capture showing the generation of a sinusoidal jitter whose peak-to-peak amplitude was 10.08 ps. *After Liu X, Shen A, Cheng N, Jitter tolerance mask for 50G-PON. In: Contribution D12, ITU-T SG15/Q2 Meeting; February 2021 [38].*

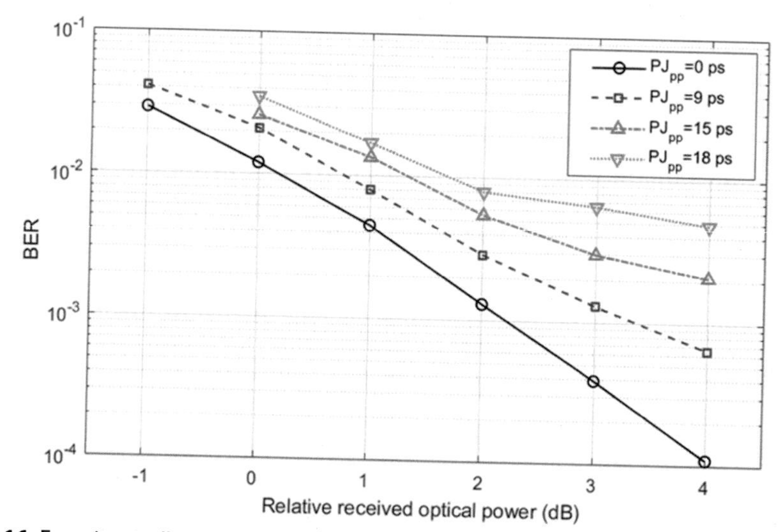

Figure 10.16 Experimentally measured bit error ratio (BER) performance as a function of received optical power under different PJ peak-to-peak amplitudes (PJ_{pp}). *After Liu X, Shen A, Cheng N, Jitter tolerance mask for 50G-PON. In: Contribution D12, ITU-T SG15/Q2 Meeting; February 2021 [38].*

DSP-based receiver is more than doubled, for example, to over 600 MHz by using an application-specific integrated circuit (ASIC) (instead of FPGA).

By dint of the recent advances in low-jitter clocks for high-speed 50-Gbaud pluggable transceivers, PJ_{PP} can be reduced to much less than 10 ps (or 0.5 UI at 50 Gb/s). We can thus reuse the jitter tolerance mask of XG(S)-PON for 50G-PON. Fig. 10.17 shows the jitter tolerance mask for 50G-PON [10]. As specified in previous ITU-T PON systems, the jitter tolerance is defined as the peak-to-peak amplitude of sinusoidal jitter applied on the input of 50G-PON signal that causes a 1-dB optical power penalty at the optical equipment. Note that it is a stress test to ensure that no additional penalty is incurred under operating conditions. The ONU (OLT) shall tolerate, as a minimum, the input jitter applied according to the mask in Fig. 10.17, with the parameters specified in that figure for the downstream (upstream) line rate. The jitter tolerance specification for the OLT is informative as it can only be measured in a setting that permits continuous operation of the upstream.

Regarding jitter generation, Table 10.5 shows a set of recently proposed 50G-PON jitter generation requirements [39]. In order to align the jitter generation mask with the jitter tolerance mask, the upper high-pass frequencies for all the 50G-PON line rates are set to 4000 kHz, while the low-pass frequencies are simply scaled proportional to the line rate. For the lower high-pass frequencies, there are two specifications for each line rate, one scaling proportionally with the line rate and the other fixed at 10 kHz, to confine the jitter generation based on the common receiver jitter tracking capability.

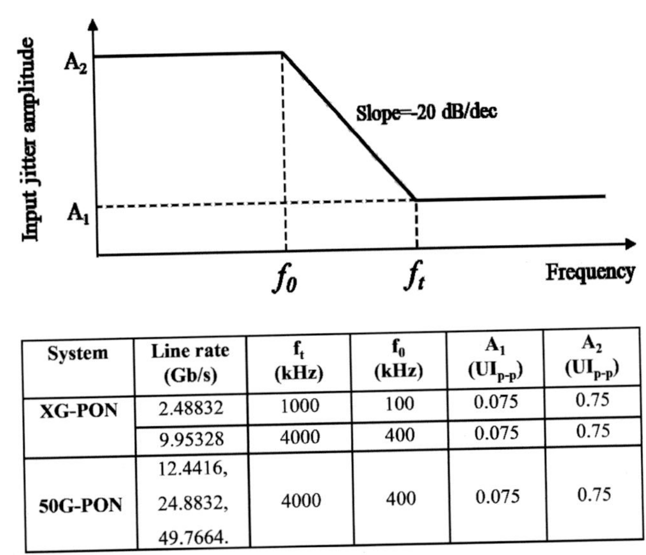

System	Line rate (Gb/s)	f_t (kHz)	f_0 (kHz)	A_1 (UI_{p-p})	A_2 (UI_{p-p})
XG-PON	2.48832	1000	100	0.075	0.75
	9.95328	4000	400	0.075	0.75
50G-PON	12.4416, 24.8832, 49.7664.	4000	400	0.075	0.75

Figure 10.17 Jitter tolerance mask specifications for 50G-passive optical network (50G-PON) [39] and 10-Gigabit-capable-passive optical network (XG-PON) [4].

Table 10.5 Jitter generation requirements for 50G-PON [39].

Line rate (Gbit/s)	Measurement band (−3 dB frequencies)[a]		Peak-to-peak amplitude (UI)[b]
	High-pass (kHz)	Low-pass (MHz) −60 dB/dec	
12.4416	10	100	0.5
	25		0.3
	4000		0.1
24.8832	10	200	0.7
	50		0.3
	4000		0.1
49.7664	10	400	1
	100		0.3
	4000		0.15

[a]The high- and low-pass measurement filter transfer functions are defined in the clause 5 of [ITU-T G.825].
[b]The measurement time and pass/fail criteria are defined in the clause 5 of [ITU-T G.825].

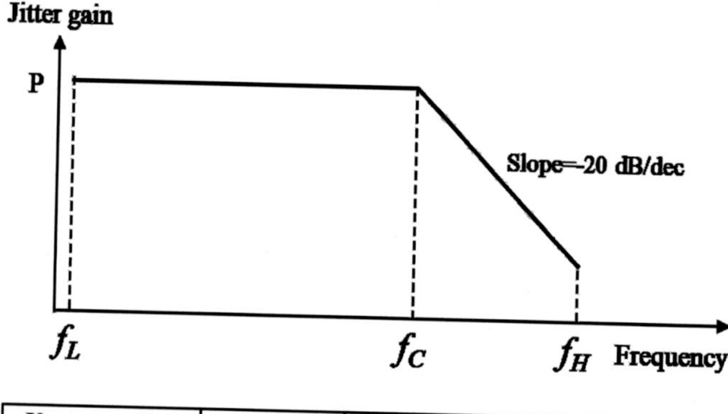

Upstream line rate (Gbit/s)	f_L (kHz)	f_C (kHz)	f_H (kHz)	P (dB)
12.4416	10	4000	100000	0.1
24.8832	10	4000	200000	0.1
49.7664	10	4000	400000	0.1

Figure 10.18 Jitter transfer mask for 50G-passive optical network optical network unit [39].

Jitter transfer applies only to the ONU. The jitter transfer function is defined as $20 \cdot \log_{10}(J_{US}/J_{DS})$, where J_{US} and J_{DS} are the jitter amplitudes in an absolute time unit (e.g., picoseconds) of the upstream signal and the (50-Gb/s) downstream signal, respectively. The jitter transfer function of an ONU shall be under the curve given in Fig. 10.18, when input sinusoidal jitter up to the tolerance mask level in Fig. 10.17 is applied.

As discussed above, all the line rates in 50G-PON share the same corner frequency of 4000 kHz (or 4 MHz), which helps align all the jitter masks for accurate jitter specification. The high-frequency edge (f_H) scales with the line rate, while the low-frequency edge (f_L) is fixed at the same value as that used in XG(S)-PON to fully cover the relevant jitter frequency range and ensure reliable system operation. Overall, the jitter-related specifications in 50G-PON accommodate DSP-based receivers and can be readily met in modern 50-Gbaud-class optical transceivers, thus ensuring both high performance and practical implementation.

10.3 50G-PON transmission convergence layer aspects

10.3.1 LDPC design considerations and performances

FEC is widely adopted in PON systems to improve receiver sensitivity and increase link budget. In GPON, the Reed–Solomon (RS) code (255,239) is used with a BER threshold of 1E-4 [3]. In XG(S)-PON, the downstream transmission adopts RS(248, 216), which is a truncated form of RS(255, 223), achieving an increased BER threshold of 1E-3 [4,5]. In the IEEE 802.3ca 50G-EPON standard, high-coding-gain LDPC is adopted, achieving a further increased BER threshold of 1E-2 [23]. The LDPC code matrix is a 12 × 69 quasicyclic matrix with a circulant size of 256. For the mother code, the codeword length, payload length, and parity length are 256 × 69 (=17,664) bits, 256 × 57 (=14,592) bits, and 256 × 12 (=3072) bits, respectively, as illustrated in Fig. 10.19. The LDPC mother code is thus represented as LDPC(17,664, 14,592). The IEEE 802.3ca standard adopts this mother code but has 512 parity bits punctured and 200 payload bits shortened, resulting in LDPC(16,952, 14,392) with a code rate of 0.849. In the ITU-T 50G-PON standard, the same LDPC mother code is used but with 384-bit puncturing and no shortening, resulting in LDPC(17,280, 14,592) with a code rate of 0.844. This LDPC design choice was made based on the following considerations [40]:

- *Inclusion of the physical synchronization block downstream (PSBd) in the first LPDC codeword*

  With the use of the high-coding-gain LDPC, the 50G-PON system is operating at a raw BER level that is too high for the 13-bit hybrid error control (HEC) to reliably protect the superframe counter (SFC) and the operation control (OC)

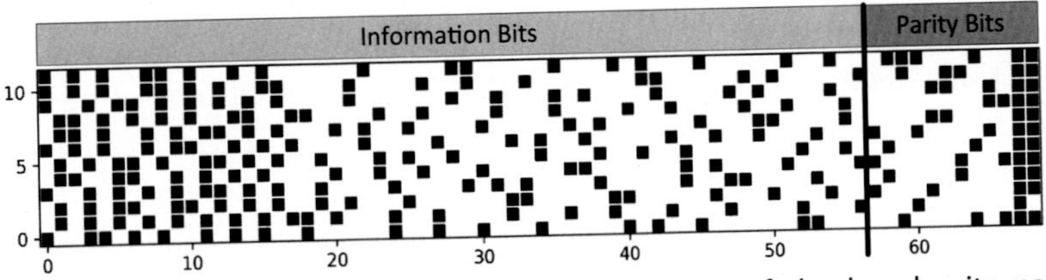

Figure 10.19 Illustration of the mother code matrix structure of the low-density parity check adopted by both IEEE 50G-Ethernet PON and international telecommunications union telecommunication 50G-passive optical network.

structure inside the PSBd. This issue is resolved by including the PSBd in the first LPDC codeword to better protect the SFC and the OC structure.

- *Integer number of codewords per 50G-PON frame*

  XG(S)-PON specifies that each downstream frame contains an integer number of FEC codewords, which makes the implementation easy and avoids the need to specify a fractional codeword. It is desired to have the same feature in 50G-PON. Given that the 125-μs downstream frame length in 50G-PON is 6,220,800 bits and PSBd is inside the first codeword, we only need to make the LDPC codeword length (in bits) to be a factor of 6,220,800. The five largest FEC codeword lengths that are both (1) factors of 6,220,800 and (2) not larger than the mother codeword length of the mother code (17,664) are 17,280, 16,200, 15,552, 15,360, and 14,400, from which we shall select an appropriate codeword length.
- *Codeword length being a multiple of internal processing bus width*

  In high-speed ASIC implementations, it is desirable for the codeword length to be a multiple of the internal processing bus width. Assuming that a 10-Gb/s Serializer/Deserializer (SerDes) has a typical output bit width of 16 or 32 bit, the internal processing bus width for 50-Gb/s SerDes is expected to be increased to 64 or 128 bits. It is thus desired for the codeword length to be divisible by 128. Thus the codeword length options are narrowed down to 17,280 and 15,360.
- *High code rate and low computational overhead*

  To achieve high code rate and low computational overhead for a given mother code, we shall select the largest possible codeword length. Thus it is appropriate to select the codeword length to be 17,280 bits, which is only 384 bits fewer than the codeword length of the mother code (17,664). The codeword length of 17,280 bits can be realized by (1) shortening the payload by 384 bits to have LPDC(17,280, 14,208) with a code rate of 0.822, (2) puncturing the parity by 384 bits to have LPDC(17,280, 14,592) with a code rate of 0.844, or something in between. To have the highest code rate, LDPC(17,280, 14,592) is preferred if its HD and SD decoding performances are satisfactory and do not exhibit error floors.
- *Satisfactory HD and SD decoding performances*

For PON systems, it is important to ensure that there is no error floor at the FEC output BER of 1E-12. For the LPDC without puncturing, it has been experimentally verified that no error floor at output BER of 1E-12 is present for both HD decoding [41] and SD decoding [42,43]. When the LPDC is punctured by 512 bits, error floor appears for SD decoding [44]. For the LDPC(17,280, 14,592) with 384-bit puncturing and no shortening, it has been verified that no error floor is present for both HD and SD deciding, as shown in Fig. 10.20 [45]. The HD and SD BER thresholds for an output BER of 1E-12 are measured to be 1.1E-2 and 2.4E-2, respectively, which are higher than those of the IEEE 50G-EPON LDPC(16,952, 14,392). Remarkably, the HD and SD decoding performances of the LDPC(17,280, 14,592) are less than 1.2 dB

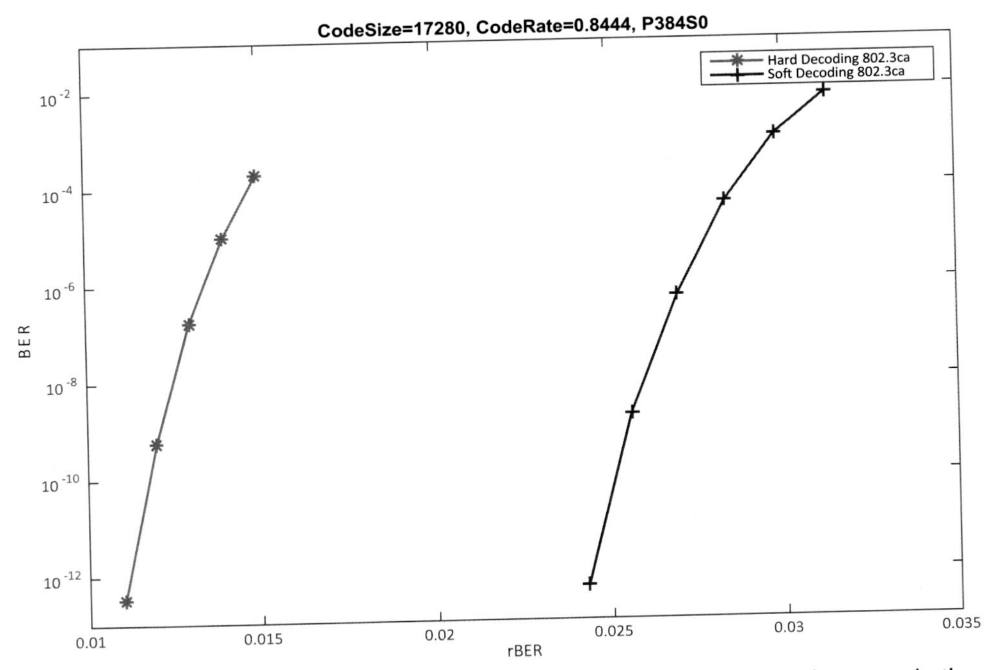

Figure 10.20 Output bit error ratio (BER) versus input BER for the international telecommunications union telecommunication 50G-passive optical network low-density parity check (17,280, 14,592) with hard-decision and soft-decision decoding. *After Han Y, Wilson B, Amitai A. HSP LDPC performance curves. In: Contribution D13, ITU-T SG15/Q2 Meeting; May 2020 [45].*

away from their respective Shannon limits, as shown in Fig. 10.21. This shows the superior HD and SD decoding performances of the LDPC(17,280, 14,592).

In accordance with the above, the LDPC(17,280, 14,592), constructed from the mother code LDPC(17,664, 14,592) with 384 bit puncturing and no shortening, has been selected by the ITU-T 50G-PON standard [9].

The scope of ITU-T G.hsp.ComTC states that the TC layer will have support for a range of downstream and upstream line rates, such as 50, 25, and 12.5 Gb/s, as well as futuristic data rates such as 100 and 75 Gb/s. It is desirable for the ComTC specification to define the operation of HSP systems in a manner that is independent of a particular transmission rate. One way is to have a parameterized specification, where the parameter values can be set according to the requirements of a particular PMD recommendation. 50G-PON uses a parameterized specification based on the following method [9]:

- Setting 12.4416 Gb/s as a *fundamental line rate*, ρ_0, and defining a *line rate factor*, ϕ (which is a positive integer), to represent a particular line rate of $\rho_0\phi$ in the PON system.

Table 10.6 shows the number of LDPC(17,664, 14,592) codewords per 125-μs PON frame versus the line rate factor ϕ. Conveniently, each PON frame contains an integer number of LDPC codewords for all the data rates of interest, making it easy to

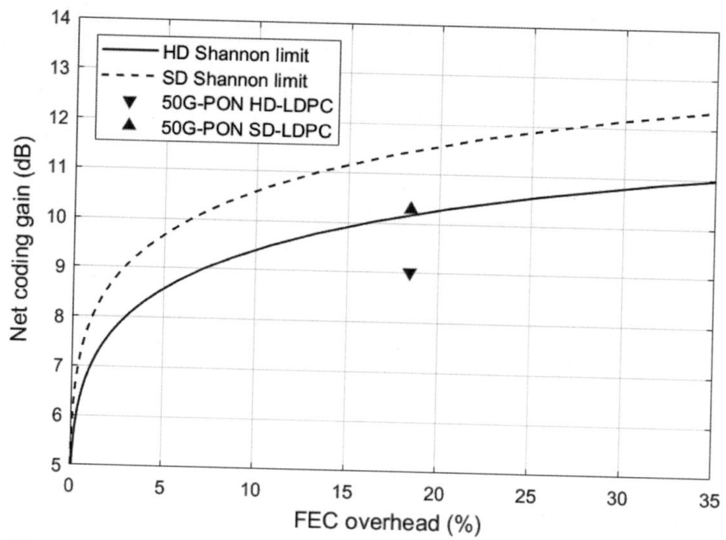

Figure 10.21 The net coding gains the international telecommunications union telecommunication 50G-passive optical network (PON) low-density parity check LDPC(17,280, 14,592) with hard-decision (HD) and soft-decision (SD) decoding as compared to the Shannon limits.

Table 10.6 The number of low-density parity check (LDPC) codewords per passive optical network (PON) frame versus the line rate and the line rate factor.

Line rate in Gb/s (*R*)	Line rate acronym	Line rate factor (ϕ)	No. of LDPC codewords per 125-μs PON frame
12.4416	12.5G	1	90
24.8832	25G	2	180
49.7664	50G	4	360
74.6496	75G	6	540
99.5328	100G	8	720

scale line rate without worrying about shortening the last codeword of each PON frame and shortening different numbers of payload bits for different line rates. Thus the choice of LDPC(17,280, 14,592) additionally offers convenient line rate scaling in G.hsp.ComTC specifications.

10.3.2 Performance improvement via block-interleaving over four LDPC codewords

As discussed in Sections 10.2.2 and 10.3.1, DSP-based receiver-side EQ and high-coding-gain LDPC are introduced to 50G-PON improve the receiver sensitivity. However, it was found recently that the error statistics in bandwidth-limited and

digitally equalized 50G-PON transmission deviates from the additive white Gaussian noise model and causes an extra optical power penalty of up to 0.6 dB [46]. A block interleaver performing bit-interleaving of every four adjacent LDPC codewords was proposed and verified numerically to be effective in reducing the correlated errors from a decision-feedback equalizer (DFE) [47]. More recently, an experimental study was conducted on the error statistics of three common types of DSP-based equalizers, namely, FFE, DFE, and MLSE for 50G-PON downstream transmission with bandwidth-limited
25-GHz-class transmitter and receiver [48]. It was shown that the four-codeword block interleaver effectively reduces the correlated errors and the variance of the distribution of the input errors per LDPC codeword, and thereby improves the overall performance in the presence of equalizer-induced correlated errors. More details on this study are presented in the following.

- *Experimental setup*

Fig. 10.22 shows the experimental setup. An electric 50-Gb/s NRZ signal containing 125-μs PON frames was generated by an AWG. Each 50G-PON downstream frame contained 360 LDPC codewords each having 17,280 bits. The transmitter-side signal processing included LDPC encoding, scrambling, and the four-codeword block-interleaving, which is abbreviated as 4 × interleaving here. With the 4 × interleaving, every four adjacent codewords, (A_1 A_2 ... $A_{17,280}$ B_1 B_2 ... $B_{17,280}$ C_1 C_2 ... $C_{17,280}$ D_1 D_2 ... $D_{17,280}$) were interleaved to (A_1 B_1 C_1 D_1 A_2 B_2 C_2 D_2 ...... $A_{17,280}$ $B_{17,280}$ $C_{17,280}$ $D_{17,280}$). The electric NRZ signal was used to drive a 25-GHz-class EML in the O-band. The output power of the EML-based transmitter can exceed 8 dBm. The optical signal then passed through a 20-km SSMF. A VOA was used to vary the optical signal power for receiver sensitivity measurements. The optical

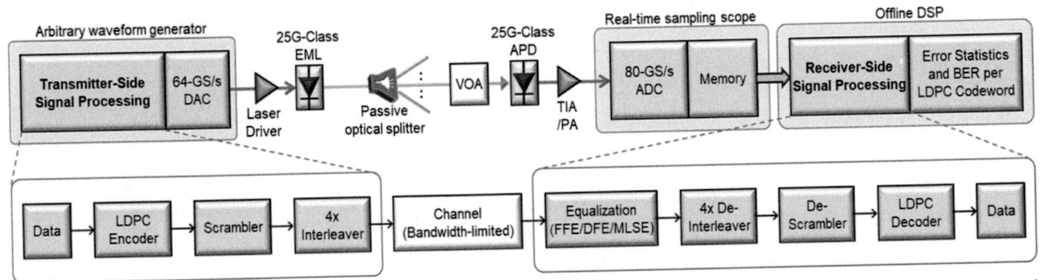

Figure 10.22 Experimental setup for measuring the error statistics in 50G-passive optical network (PON) downstream transmission with digital equalization [based on feed-forward equalization (FFE), decision-feedback equalizer (DFE), and maximum-likelihood sequence estimation (MLSE) and block-interleaving over four low-density parity check (LDPC) codewords]. *After Liu X, Shen A, Cheng N, Luo Y, Effenberger F. Performance improvements in bandwidth-limited and digitally-equalized 50G-PON downstream transmission via block-interleaving over four LDPC codewords. In: Optical fiber communication conference, paper M3G.6; 2021 [48].*

signal was then detected by a commercially available 25-GHz-class APD with a 3-dB bandwidth of about 18 GHz. After photo-detection, an 80-GSamples/s ADC embedded in a real-time sampling scope was used to sample the received signal. The sampled signal was equalized by a 20-tap FFE, a 1-tap DFE (following a 20-tap FFE), or an MLSE (following a 20-tap FFE). For each case, 14,400 LDPC codewords worth of data were measured. The receiver-side signal processing included 4 × deinterleaving (which reversed the effect of the 4 × interleaving), descrambling, and LDPC decoding. The input bit errors in each 17,280-bit LDPC codeword were recorded for error analysis. To study the impact of the 4 × interleaving, the authors also compared the case with the 4 × interleaving and deinterleaving turned off. The recovered signal bits were compared to the original signal bits to calculate pre-FEC BER for various received optical power levels.

- *Experimental results*

Fig. 10.23 shows the experimentally measured mean BER performances of the 50-Gb/s NRZ signal as a function of received optical power for the three EQ cases. Representative recovered 50-Gb/s NRZ symbols without EQ and with FFE are shown as insets. Without EQ, the two-level NRZ signal becomes a three-level signal, indicating substantially bandwidth-limited transmission. At the 50G-PON, LDPC input BER

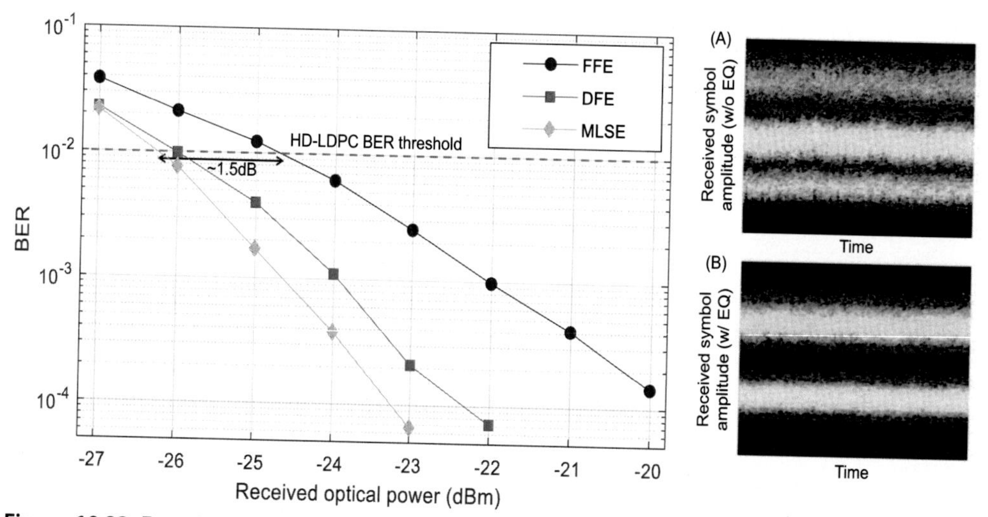

Figure 10.23 Experimentally measured mean bit error ratio (BER) performances with receiver-side equalization (EQ) based on the feed-forward equalization (FFE), decision-feedback equalizer (DFE), and maximum-likelihood sequence estimation (MLSE). Insets: Representative recovered 50-Gb/s nonreturn-to-zero symbols without EQ (A) and with FFE (B) at −20 dBm received power. *After Liu X, Shen A, Cheng N, Luo Y, Effenberger F. Performance improvements in bandwidth-limited and digitally-equalized 50G-PON downstream transmission via block-interleaving over four LDPC codewords. In: Optical fiber communication conference, paper M3G.6; 2021 [48].*

threshold for HD of 1×10^{-2}, the received optical powers for FFE, DFE, and MLSE are required to be -24.7, -26, and -26.2 dBm, respectively. DFE and MLSE outperform FFE by about 1.3 and 1.5 dB, respectively. Fig. 10.24 shows the measured probabilities of consecutive errors with and without the $4\times$ interleaving for all the three EQ cases at the same mean BER of 1×10^{-2}. For comparison, simulated probabilities of consecutive errors assuming random errors are also plotted. Evidently, the probabilities of consecutive errors are noticeably reduced by the $4\times$ interleaving, showing its effectiveness in reducing the error correlation.

The ROP was then set to -24.7 dBm, at which the mean BERs in the FFE, DFE, and MLSE cases were 1×10^{-2}, 2.28×10^{-3}, 1.23×10^{-3}, respectively. Fig. 10.25 shows the probability of the BER per-codeword for each of the three EQ cases. For comparison, an uncorrelated binary symmetric channel (BSC) curve following the Poisson distribution is generated for the same average BER for each case. At the worst-case BER-per-codeword (occurring at $\sim 7 \times 10^{-5}$ probability), the BER performance degradations in FFE, DFE, and MLSE from their corresponding Poisson distributions (1), (2), and (3), are 0.3, 0.9, and 0.7 dB, respectively. From these experimental results, we can see that DFE suffers most from the correlated errors, which is reasonable because of the decision-error contamination. When the $4\times$ interleaving is applied, the BER performance degradations in FFE, DFE, and MLSE from their corresponding Poisson distributions are reduced to 0.1, 0.4, and 0.2 dB, respectively, indicating the effectiveness of the $4\times$ interleaving in improving the receiver performance in the presence of equalizer-induced correlated errors. With the $4\times$ interleaving, MLSE outperforms DFE and FFE by 0.7 and

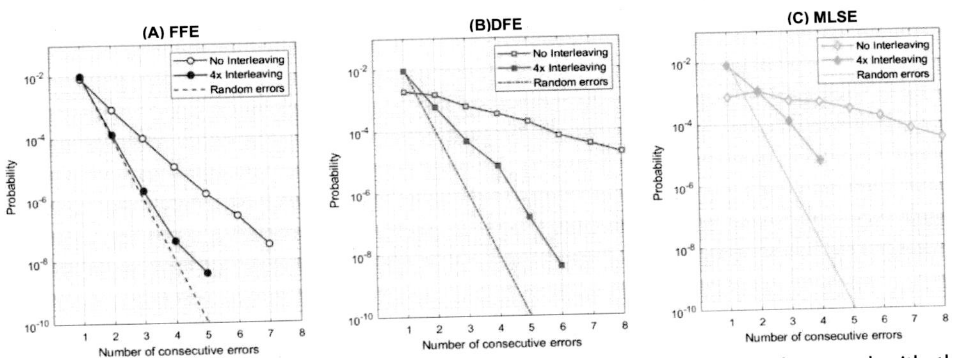

Figure 10.24 Experimentally measured probabilities of consecutive errors without and with the $4\times$ interleaving for feed-forward equalization (FFE), decision-feedback equalizer (DFE), and maximum-likelihood sequence estimation (MLSE). *After Liu X, Shen A, Cheng N, Luo Y, Effenberger F. Performance improvements in bandwidth-limited and digitally-equalized 50G-PON downstream transmission via block-interleaving over four LDPC codewords. In: Optical fiber communication conference, paper M3G.6; 2021 [48].*

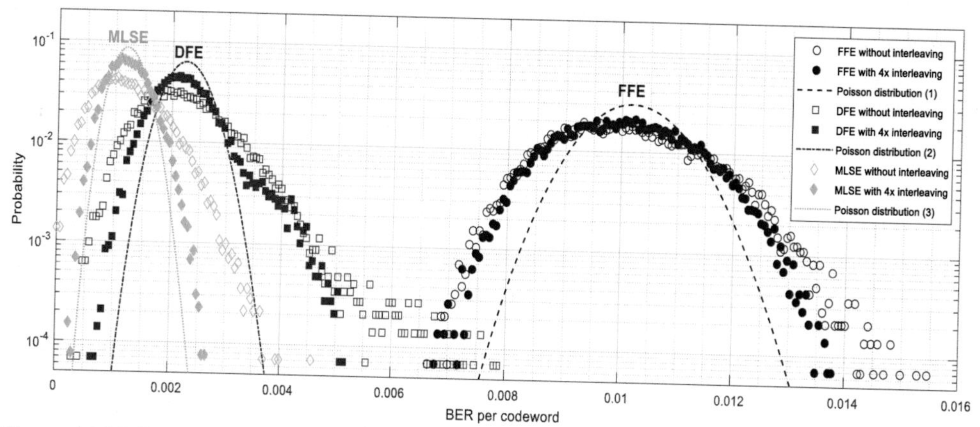

Figure 10.25 Experimentally measured probability of bit error ratio-per-codeword for feed-forward equalization (FFE), decision-feedback equalizer (DFE), and maximum-likelihood sequence estimation (MLSE) at −24.7 dBm received optical power (ROP). *After Liu X, Shen A, Cheng N, Luo Y, Effenberger F. Performance improvements in bandwidth-limited and digitally-equalized 50G-PON downstream transmission via block-interleaving over four LDPC codewords. In: Optical fiber communication conference, paper M3G.6; 2021 [48].*

2 dB, respectively, in terms of the worst-case BER-per-codeword. MLSE with the 4× interleaving provides the best performance and supports a power budget of about 34 dB with 8-dBm transmitter power, making it feasible to support the N2 power budget (31 dB) with additional margin for other OPPs. With SD-LDPC, the C+ power budget (32 dB) can also be supported. On the other hand, FFE without interleaving offers the simplest implementation with low processing latency and can be used to support the N1 (29 dB) power budget. Thus an OLT-configurable bit-interleaving approach has been adopted for 50G-PON downstream transmission to better support different system requirements [49,50]. To accommodate this OLT-configurable bit-interleaving approach, the downstream synchronization state machine is slightly amended such that while in the Hunt state, the ONU searches for a physical synchronization sequence (PSync) in all possible alignments (both bit and byte) *for all possible OLT bit-interleaving modes* within the downstream signal [50].

In summary, the performance improvements in bandwidth-limited 50G-PON downstream transmission with receiver-side EQ based on FFE, DFE, and MLSE by the 4× interleaving over LDPC codewords have been experimentally quantified [48]. It has been shown that at 8-dBm transmitter power, MLSE with the 4× interleaving can achieve the N2 and C+ power budgets, while FFE without interleaving offers a simple and low-latency implementation capable of supporting the N1 power budget. These results show the promise of using receiver-side EQ, advanced LDPC, 4× interleaving, and high-power optical transmitter to make 50G-PON a viable solution for broadband access applications in the 5G era.

10.3.3 Enhanced synchronization state machine

In 50G-PON, high-coding-gain LDPC with input BER threshold of over 1E-2 is adopted, as described in previous sections. At this high input BER level, the SFC can no longer be reliably protected by the 13-bit HEC, which was used in XG(S)-PON to protect the SFC, thus causing the synchronization (Sync) state machine used in XG-PON to be unreliable for 50G-PON. It is thus necessary to improve the downstream synchronization mechanism in 50G-PON. One approach is to use the FEC decoding status, instead of SFC, in the Sync state machine to avoid the issue of unreliable SFC reading [51]. Another approach is to protect the SFC with the same LDPC that is being used to protect the payload bits such that reliable SFC reading can be obtained after LDPC decoding and used in the 50G-PON Sync state machine [52]. With the availability of a reliable SFC, the 50G-PON synchronization state machine can reuse most of the well-established features of the XG-PON synchronization state machine. More details on this synchronization approach are as follows.

- *Downstream synchronization state machine*

Fig. 10.26 shows the downstream Sync state machine specified in the ITU XG-PON [4]. The Sync state machine is designed to be reasonably immune to both false-lock on an independent uniformly random bitstream and false-loss of synchronization under a raw (or uncorrected) BER of up to 1E-3. However, the FEC threshold in 50G-PON is increased to about

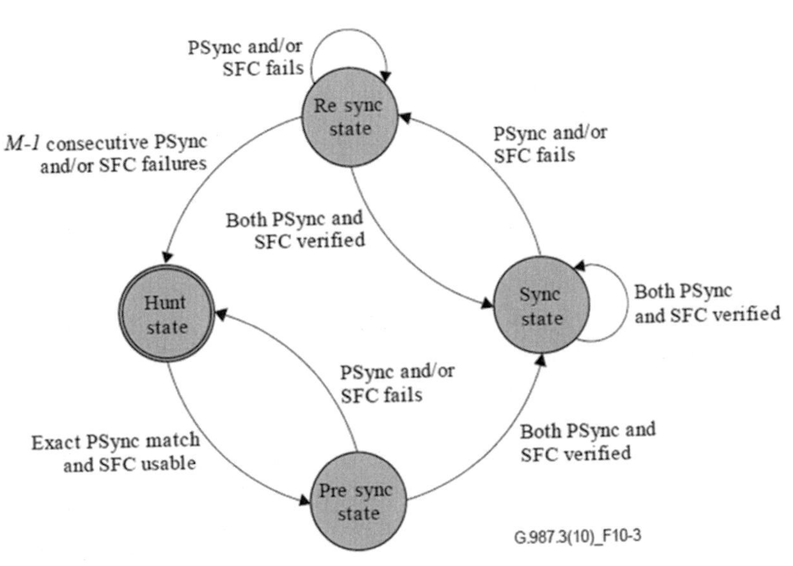

Figure 10.26 10-Gigabit-capable-passive optical network downstream Sync state machine. *After ITU-T Recommendation G.983.1. Broadband optical access systems based on Passive Optical Networks (PON); 2005 [2].*

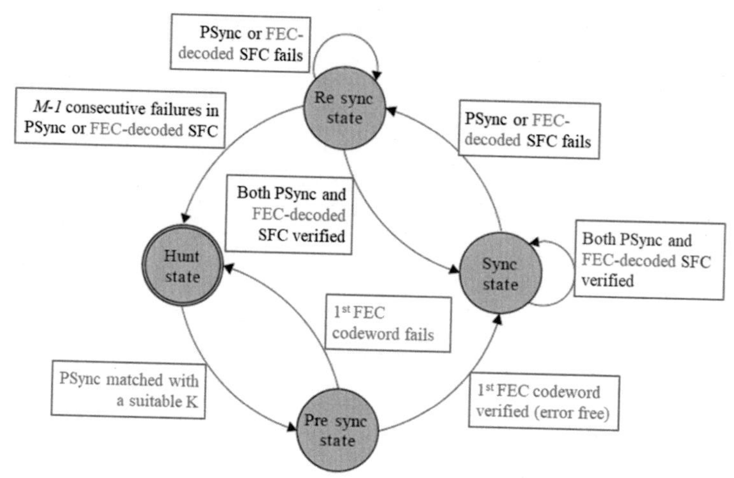

Figure 10.27 An improved Sync state machine for 50G-passive optical network. *After Shen A, Liu X, Luo Y, Effenberger F. 50G-PON downstream synchronization state machine. In: D46, ITU-T SG15/Q2 Meeting; April 2020 [52].*

3E-2 and the SFC can no longer be reliably protected by the HEC. Therefore we can use an improved Sync diagram where the SFC is protected by the powerful LDPC [52], as shown in Fig. 10.27.

In the Sync state, it is possible to replace the verification of "both PSync and SFC" used in XG(S)-PON with the verification of LDPC FEC decoding success [51]. However, SFC is a key parameter that is used in several important PON system features. In the XG-PON encapsulation method (XGEM), SFC is one of the factors to determine the initial counter block value. In time-of-day (ToD) distribution, SFC is used to represent the reference downstream PHY frame. The SFC verification is thus needed to support these system features and doing this in the Sync state would give a high confidence for using SFC in other system functions. It is desirable to keep the use of SFC verification in the Sync state. Thus it is desirable for the improved Sync state machine to reuse most of the XG(S)-PON features, as shown in Fig. 10.27. The key aspects are as follows:

- At the OLT, protect the SFC and its HEC in the first FEC codeword of each PON frame.
- From the Hunt state to the Presync state, use Psync matching only.
- From the Presync state to the Sync state, use FEC decoding status.
- In the Sync state and the Resync state, use the same verification methods as those used in XG-PON, with the only difference being the use of FEC-decoded SFC, instead of non-FEC-protected SFC, for SFC verification.

The false-lock and false-loss of sync probabilities will be quantified in the following.

- *False-lock probability*

For G-PON, the false-lock probability for each 125-μs PON frame can be expressed as

$$P_{\text{false-lock}} = P_{\text{false-presync}} \cdot P_{\text{false-sync}} \quad (10.13)$$

where $P_{\text{false-sync}}$ is the probability of false match with a N-bit PSync with up to K bits of errors allowed

$$P_{\text{false-sync}} = \sum_{k=0}^{K} \binom{N}{k} \frac{1}{2^N} \quad (10.14)$$

and $P_{\text{false-presync}}$ is the probability of false PSync match in the PreSync state where each PON frame contains L bits

$$P_{\text{false-presync}} = (L - N + 1) \cdot P_{\text{false-sync}} \quad (10.15)$$

In G-PON, we have $N=32$ and $K=0$, resulting in a $P_{\text{false-lock}}$ of 1.69E-14, which corresponds to a mean time per false-lock of 235 years.

For XG-PON, the false-lock probability for each 125-μs PON frame can be expressed as

$$P_{\text{false-lock}} = P_{\text{false-presync}} \cdot P_{\text{false\_2}} \cdot P_{\text{sfc2}} \quad (10.16)$$

where $P_{\text{false-presync}}$ is obtained by

$$P_{\text{false-presync}} = (L - (64 + 64) + 1) \cdot P_{\text{false\_0}} \cdot P_{\text{sfc1}} \quad (10.17)$$

$P_{\text{false\_0(2)}}$ is the probability of false match with a N-bit PSync with up to 0(2) bits of errors allowed, P_{sfc1} is the probability for the first SFC to be correctable based on random match

$$P_{\text{sfc1}} = \sum_{k=0}^{2} \binom{64}{k} \frac{1}{2^{13}} \quad (10.18)$$

and P_{sfc2} is the probability for the second SFC to falsely match its expected bit pattern with up to 2 bit errors

$$P_{\text{sfc2}} = \sum_{k=0}^{2} \binom{64}{k} \frac{1}{2^{64}} \quad (10.19)$$

We have $P_{\text{false-lock}}$ for XG-PON of 2.2E-46, which corresponds to a mean time per false-lock of 1.8E34 years.

For the aforementioned 50G-PON synchronization scheme, the false-lock probability can be expressed as

$$P_{\text{false-lock(50G)}} = (L - N + 1) \cdot \sum_{k=0}^{K} \binom{N}{k} \frac{1}{2^N} \cdot \sum_{e=0}^{E} \binom{n}{e} \frac{1}{2^p} \tag{10.20}$$

where n is the LDPC codeword length, p is the number of parity bits, and E is the nominal number of errors that can be corrected in each FEC codeword. As per the current specifications being discussed in the ITU 50G-PON standards, we have $n = 17{,}280$, $p = 2688$, and $E = 345$ (for a FEC BER threshold of 2E-2). Then, the false-lock probability is calculated to be around 2E-80, which is much lower than that in either GPON or XG-PON.

- *False-loss of sync probability*

  For G-PON, the false-loss of sync probability can be expressed as

$$P_{\text{false-loss-sync}} = (P_{\text{miss}})^{M2} \tag{10.21}$$

where $M2$ is the number of consecutive incorrect PSync allowed in the Sync state and P_{miss} is the probability of missing one Psync pattern

$$P_{\text{miss}} = 1 - \sum_{k=0}^{K} \binom{N}{k} (1 - P_e)^{N-k} P_e^{\,k} \tag{10.22}$$

where P_e is the raw BER threshold set by the FEC used. In G-PON, we have $N = 32$, $K = 0$, $P_e = 1\text{E-}4$, and $M2 = 5$, resulting in a $P_{\text{false-loss-sync}}$ of 3.33E-13, which corresponds to a mean time per false-loss of sync of 11.9 years.

For XG-PON, the probability of missing the P_{sync} pattern and the SFC verification is

$$P_{\text{miss}} = 1 - \quad P_{\text{sync}} \cdot P_{\text{sfc}} \tag{10.23}$$

where P_{sync} is the probability of finding the P_{sync}

$$P_{\text{sync}} = \sum_{k=0}^{K} \binom{N}{k} (1 - P_e)^{N-k} P_e^{\,k} \tag{10.24}$$

and P_{sync} is the probability of verifying the SFC

$$P_{\text{sfc}} = \sum_{k=0}^{2} \binom{N}{k} (1 - P_e)^{N-k} P_e^{\,k} \tag{10.25}$$

In XG-PON, we have $N = 64$, $K = 2$, $M2 = 3$, and $P_e = 1\text{E-}3$, resulting in a $P_{\text{false-loss-sync}}$ of 5.04E-13, which corresponds to a mean time per false-loss of sync of 7.86 years.

For 50G-PON, P_e can be as high as 2E-2, resulting in a $P_{\text{false-loss-sync}}$ of greater than 2E-3, which corresponds to a mean time per miss-sync of less than 0.1 seconds, which is unacceptable. With the use of the proposed scheme where LDPC-decoded (or corrected) SFC is used, P_e effectively becomes the maximum post-FEC BER, which is specified to be 1E-12. With the much reduced BER in the LDPC-decoded SFC, $P_{\text{false-loss-sync}}$ becomes less than 1E-13 for $K=7$, leading to a mean time per false-loss of sync of over 50 years, which is better than that in either G-PON and XG-PON.

- *Calculation results*

Fig. 10.28 shows the calculated false-lock and false-loss of sync probabilities as a function of K for the proposed 50G-PON downstream synchronization scheme. For comparison, G-PON and XG-PON results are also plotted. When the maximum number of allowed bit errors in the Psync matching process (K) is larger or equal to 7, the false-loss of sync probability of the proposed scheme becomes less than 1E-13, better than that in either G-PON or XG-PON.

Fig. 10.29 shows the calculated mean occurring times of false-lock and false-loss of sync events as a function of K. When K is larger or equal to 7, the mean time for synchronization failure is greater than 50 years, longer than that of either G-PON or XG-PON, as well as the common telecommunication equipment lifespan of 20 years. Table 10.7 summarizes the results with $K=7$ for 50G-PON.

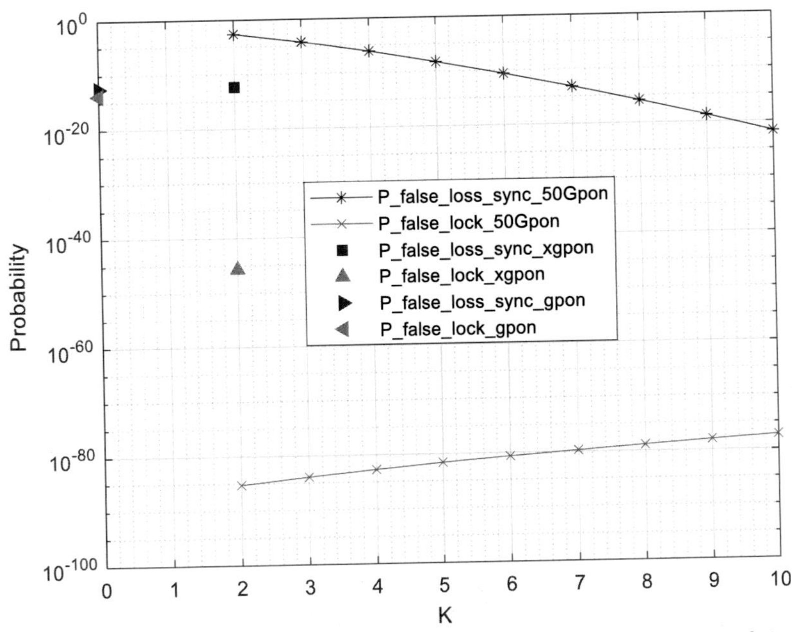

Figure 10.28 Calculated false-lock and false-loss of sync probabilities as a function of *K*.

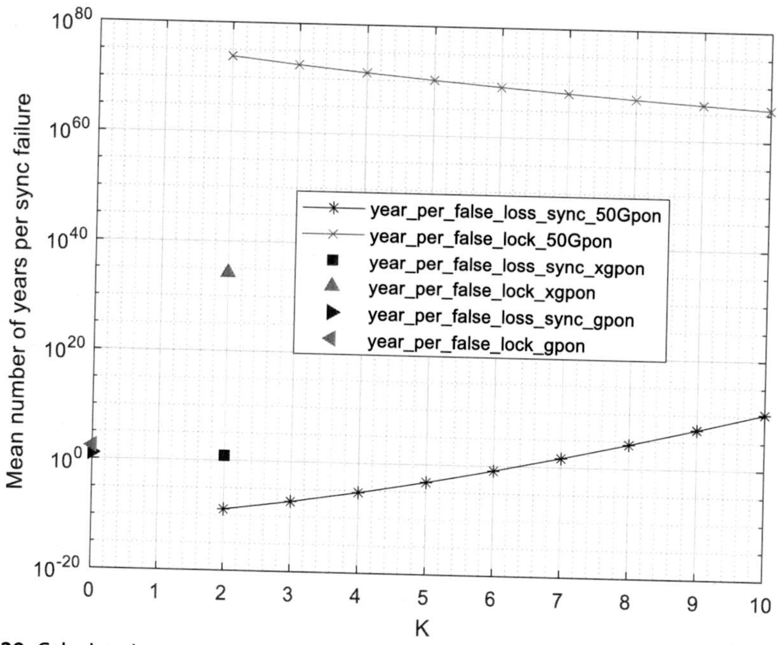

Figure 10.29 Calculated mean occurring times of false-lock and false-loss of sync events as a function of *K*.

Table 10.7 False-lock and false-loss of sync probabilities in GPON, 10-Gigabit-capable symmetric-passive optical network(PON) and 50G-PON.

	False-lock probability (mean time to incorrect lock)	False-loss-of-sync probability (mean time to incorrect unlock)
GPON	1.7E-14 (235 years)	3.3E-13 (12 years)
XG(S)-PON	2.2E-46 (1.8E34 years)	5E-13 (7.9 years)
50G-PON	~2E-80 (>1E68 years)	<1E-13 (>50 years)

In summary, the downstream synchronization scheme for 50G-PON can be enhanced to achieve lower false-lock and false-loss of sync probabilities than G-PON and XG-PON even at a much-increased FEC threshold of up to ~2E-2. This helps make 50G-PON a reliable broadband access technology in the 5G era.

10.3.4 Low-latency 5G X-haul–related aspects

As described in Chapter 6, Point-to-Multipoint Transmission, TDM-PON can be used to cost-effectively support 5G front-, mid-, and back-haul services. To meet the stringent latency requirements of 5G X-haul, three enabling technologies have been introduced to 50G-PON as follows.

- *Cooperative dynamic bandwidth allocation (DBA)*

In traditional TDM-PON systems, a negotiation between the ONU and the OLT is needed for dynamic bandwidth allocation to meet the real-time traffic needs, and it usually takes much longer than the PON frame period of 125 μs, which is unacceptably long for low-latency X-haul services. With cooperative DBA, the radio access network (RAN) scheduling can be shared with the PON scheduling in advance, so that the upstream traffic from wireless user equipment (UE) can be seamlessly carried over by the TDM-PON system without having to wait for the negotiation between the ONU and the OLT, as illustrated in Fig. 10.30. In effect, cooperative DBA eliminates the latency associated with the DBA negotiation in PON systems. Along with the ITU-T 50G-PON standardization effort, the OLT capabilities for supporting cooperative DBA are defined as an ITU-T Recommendation Series G Supplement [12]. For the specific use case on low-latency mobile front-haul over PON, the cooperative transport interface (CTI) defined by the Open RAN Alliance (O-RAN) can be used for the communication between the PON OLT and the mobile distributed unit [13].

- *Accelerated DBA scheduling with multiple bursts per ONU per 125-μs frame*

To further reduce the DBA latency, accelerated DBA scheduling can be applied to allow multiple bursts (or time slots) for an ONU to transmit during each 125-μs PON

Table 10.8 Some standardized slice/service types.

SST	Value	Slice application
eMBB	1	The handling of 5G enhanced mobile broadband (eMBB).
URLLC	2	The handling of ultra-reliable low-latency communications (URLLC).
MIoT	3	The handling of massive internet of things (MIoT).
V2X	4	The handling of vehicle-to-everything (V2X) services.

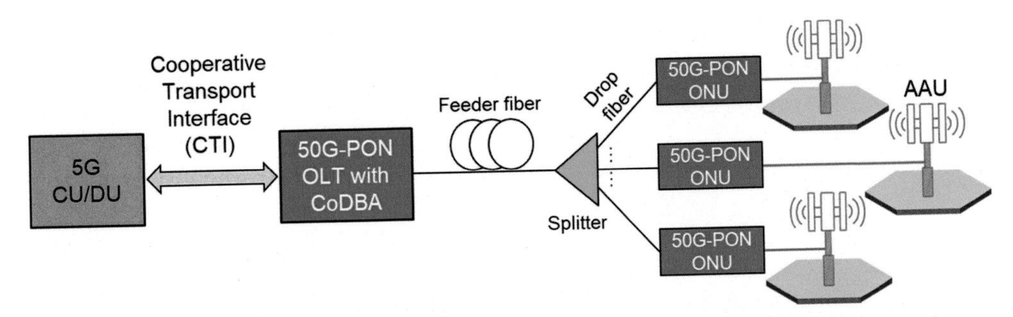

Figure 10.30 Illustration of the use of 50G-passive optical network (50G-PON) to support low-latency 5G front-haul communications via cooperative dynamic bandwidth allocation/assignment (CoDBA).

frame. For example, when each ONU is allowed to transmit eight bursts per PON frame, the maximum waiting time is reduced by eight times from 125 to ~16 μs, as illustrated in Fig. 10.31. Assuming that the OLT is serving 16 ONUs, each burst can last up to ~1 μs. To limit the burst-mode overhead time to below 10% of the burst duration, the OLT needs to complete the tracking of each upstream burst within ~100 ns. This calls for fast burst-mode tracking that includes synchronization and channel equalization for data recovery. As described in Section 10.2.3, DSP-assisted burst-mode receiver is capable of fast burst-mode data recovery with a preamble time of <100 ns. Thus the accelerated DBA scheduling to further reduce the DBA latency can be readily supported by the DSP technology being adopted in 50G-PON.

- *Elimination of the ranging window by a dedicated activation wavelength (DAW)*

In a typical PON system, the OLT needs to periodically open a ranging window to discover any new ONUs that need to join the system and find their distances from the OLT. The duration of this ranging window is determined by the maximum differential range of the PON (e.g., 20 km) and is typically set to 250 μs. During the ranging window, working ONUs must temporarily suspend their transmissions, which adversely affects the ongoing low-latency communications of working ONUs. To address this issue, a DAW is used for ranging. As 50G-PON can coexist with either GPON or XG(S)-PON, the DAW can be conveniently chosen to be either the GPON upstream wavelength centered at 1310 nm or the XG(S)-PON upstream wavelength centered at 1270 nm. Fig. 10.32 shows an implementation example where 50G-PON is used for high-speed low-latency X-haul communications and the DAW is set as the GPON upstream wavelength at 1310 nm. The 50G-PON medium access

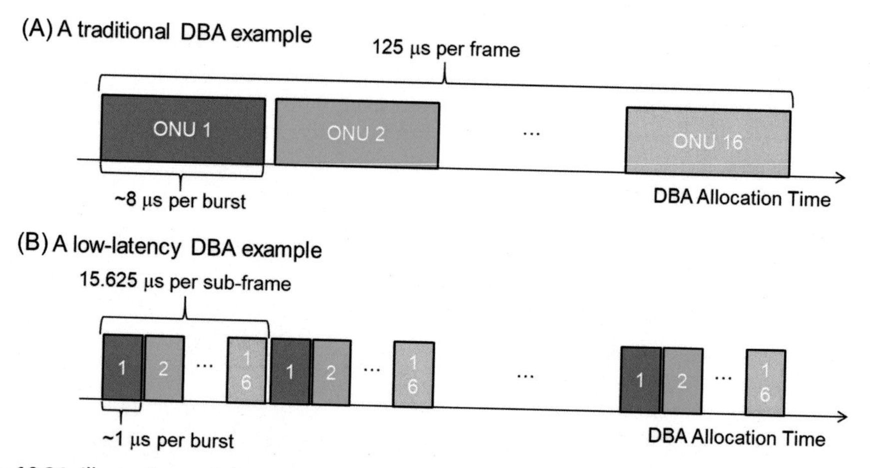

Figure 10.31 Illustrations of the traditional dynamic bandwidth allocation/assignment (DBA) allowing only one burst per optical network unit (ONU) per frame (A) and the low-latency DBA allowing multiple bursts per ONU per frame.

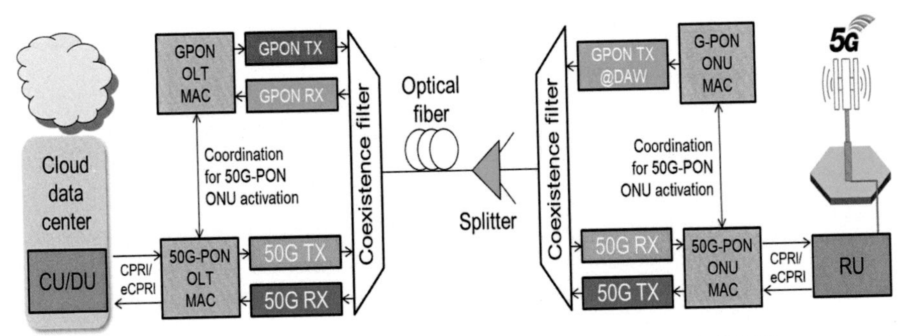

Figure 10.32 Illustration of the use of a dedicated activation wavelength (DAW) to support 50G-passive optical network (PON) to achieve uninterrupted low-latency communication for 5G front-haul. *CU*, Centralized unit; *DU*, distributed unit; *RU*, remote unit.

control (MAC) coordinates with the GPON MAC to discover new ONUs, obtain their serial numbers, and conduct the requisite ranging to achieve uninterrupted low-latency communication. The use of the DAW to eliminate the ranging window for low-latency PON has been included in the ITU-T G.9804.2 Appendix IX [9].

With the above three enabling technologies, the overall processing latency in the PON segment of the network (including the OLT and ONU processing latencies) can be limited to be much less than 100 μs, sufficient for low-latency X-haul applications.

Another important requirement for carrying 5G wireless services over PON is the precise distribution of the ToD synchronization information from the central office to all remote sites via PON. The IEEE 1588 precision time protocol (PTP) [53] can be used. Since the IEEE 1588 PTP requires precise assessment of the time delay difference between the downstream and upstream directions, the PON system needs to conduct the following procedures:

- Precise assessments of the internal delays in the OLT and the ONUs
- Precise assessments of the asymmetric propagation delays over the fiber due to different downstream and upstream wavelengths
- The OLT collaborating with each ONU to accurately compensate for the time delay difference between the downstream and upstream directions

The above precise ToD distribution method has been included in the modern PON standards including the ITU-T 50G-PON standard [8,9].

The ability to perform network slicing is another key technical requirement of 5G. There are several standardized slice/service types (SSTs) [54], some of which are shown in Table 10.8.

The above SSTs are expected to be common across different segments of an end-to-end network, and thus it is desirable for the PON segment to also support these SSTs to facilitate end-to-end network slicing. This objective has been included in the ITU-T HSP requirements [8].

10.4 Future perspectives on higher speed PON

Since the establishment of the ITU-T HSP project in 2018, the global PON community has worked closely to define the HSP standard series, the G.9804.x series. The HSP requirement recommendation G.9804.1 had been published in 2020, followed by the publication of the HSP common TC recommendation G.9804.2 and 50G-PON PMD recommendation G.9804.3 in 2021. In parallel, active research and development efforts have been made to commercialize 50G-PON systems. In October 2020, Swisscom conducted a live trial of a prototype 50G-PON line card that supported 50-Gb/s downstream and 25-Gb/s upstream [55]. The 50G-PON system is expected to initially address the business customer market and facilitate access to mobile communication masts, particularly for 5G [55]. With the widespread deployment of 50-Gbaud optics and electronics in data center optical connections, the 50G-PON standardized by the ITU-T is well positioned to become a viable optical access solution beyond the XG(S)-PON in the mid-2020s.

Going forward, HSP systems are expected to continue evolving with the following potential enhancements:

- Further increased line rates, for example, by using higher level modulation formats [56–58].
- Further increased system capacities by using higher speed time-and-wavelength-division multiplexing.
- Flexible modulation and FEC coding to address a large variety of link loss budget requirements [56–58].
- Better support of CoDBA for low-latency applications such as 5G front-haul.
- Better energy saving, for example, by enhancing the sleep mode operation.
- Better security, for example, by using enhanced data encryption and key security
- Support of network slicing.

As ultra-broadband access to homes and businesses is becoming increasingly important because of essential services such as telecommuting and distance education, higher speed PON technologies such as 50G-PON are expected to complement 5G wireless technologies to together support the essential broadband communication needs of our global society in the years to come.

References

[1] The Full Service Access Network (FSAN) Group, <https://www.fsan.org/>.
[2] ITU-T Recommendation G.983.1. Broadband optical access systems based on Passive Optical Networks (PON); 2005.
[3] ITU-T Recommendation G.984. Gigabit-capable passive optical networks (GPON); 2008.
[4] ITU-T Recommendation G.987. 10-Gigabit-capable passive optical networks (XG-PON); 2016.
[5] ITU-T Recommendation G.9807.1. 10-Gigabit-capable symmetric passive optical network (XGSPON); 2017.

[6] Ovum Report. Overview—wireline access equipment market; 2020.
[7] ITU-T G.sup.HSP. Supplement to ITU-T G-series recommendations: PON transmission technologies above 10 Gb/s per wavelength; 2018.
[8] ITU-T Recommendation G.9804.1. Higher speed passive optical networks – requirements; 2019.
[9] ITU-T Recommendation G.9804.2. Higher speed passive optical networks: common transmission convergence layer specification; 2021.
[10] ITU-T Recommendation G.9804.3. Higher speed passive optical networks: physical media dependent layer specification; 2021.
[11] ITU-T Work Item G.hsp.TWDMpmd. Higher speed passive optical networks: TWDM PMD; 2017–2020.
[12] ITU-T Work Item G.sup.CoDBA. OLT capabilities for Cooperative DBA; 2017–2020.
[13] O-RAN Specification O-RAN.WG4.CTI-TCP.0-v01.00. Cooperative transport interface transport control plane specification; 2020.
[14] ITU-T G.Sup.66. 5G wireless fronthaul requirements in a passive optical network context; 2020.
[15] Yoshimoto N, Kani J, Kim S, Iiyama N, Terada J. DSP-based optical access approaches for enhancing NG-PON2 systems. IEEE Commun Mag 2013;51(3):58–64.
[16] Liu X. Technology options for high-speed PON at 25Gb/s per wavelength and beyond. Invited talk at the FSAN Workshop on Future Access Networks, Atlanta, Georgia, October 7, 2015.
[17] Liu X, Effenberger F. Emerging optical access network technologies for 5G wireless [invited]. IEEE/OSA J Opt Commun Netw 2016;8:B70.
[18] Houtsma V, van Veen D. A study of options for high-speed TDM-PON beyond 10G. J Light Technol 2017;35(4):1059–66.
[19] Tao M, Zheng J, Dong X, Zhang K, Zhou L, Zeng H, et al. Improved dispersion tolerance for 50G-PON downstream transmission via receiver-side equalization. In: Optical fiber communication conference, paper M2B.3; 2019.
[20] Rosales R., et al. First demonstration of an E2 class downstream link for 50Gb/s PON at 1342nm. In: 2020 European conference on optical communications (ECOC), paper Tu2J-5; 2020.
[21] Cheng N, Shen A, Luo Y, Zhang X, Cheng K, Steponick J, et al. Multi-rate 25/12.5/10-Gb/s burst-mode upstream transmission based on a 10G burst-mode ROSA with digital equalization achieving 20dB dynamic range and sub-100ns recovery time. In: 2020 European conference on optical communications (ECOC), paper Tu1J-7; 2020.
[22] Li B, Zhang K, Zhang D, He J, Dong X, Liu Q, et al. DSP enabled next generation 50G TDM-PON. J Opt Commun Netw 2020;12:D1–8.
[23] IEEE. P802.3c.50Gb/s Ethernet passive optical networks task force. Available from: http://www.ieee802.org/3/ca/index.shtml.
[24] Knittle C. IEEE 50Gb/s EPON (50G-EPON), In: Optical fiber communications conference and exhibition (OFC), paper Th1B.2; 2020.
[25] Zhang D, Liu D, Wu X, Nesset D. Progress of ITU-T higher speed passive optical network (50G-PON) standardization. IEEE/OSA J Opt Commun Netw 2020;12(10):D99–108.
[26] Bahl L, Cocke J, Jelinek F, Raviv J. Optimal decoding of linear codes for minimizing symbol error rate (Corresp.). IEEE Trans Inf Theory 1974;20(2):284–7.
[27] Zeng H, Shen A, Cheng N, Chand N, Liu X, Effenberger F. High performance 50G-PON burst-mode upstream transmission with fast burst synchronization and recovery. In: ITU-T SG15/Q2, Geneva; July 2019.
[28] Zeng H, Shen A, Cheng N, Chand N, Liu X, Effenberger F. High-performance 50G-PON burst-mode upstream transmission at 25Gb/s with DSP-assisted fast burst synchronization and recovery. In: Asia communications and photonics conference, paper T1G.3, Chengdu, China; November 2019.
[29] Shen A, Cheng N, Luo Y, Liu X. 50G-PON burst-mode upstream transmission at 25Gb/s using a commercially available 10G APD/TIA/AGC ROSA. In: Contribution C1737, ITU-T SG15/Q2, Geneva; February 2020.
[30] Funada T, Tanaka N. G.hsp50Gpmd: preamble time of 50G-PON upstream burst-mode receiver. In: Contribution C1539, ITU-T SG15/Q2, Geneva; February 2020.

[31] Ohtomo Y, Kamitsuna H, Katsurai H, Nishimura K, Nogawa M, Nakamura M, et al. High speed circuit technology for 10-Gb/s optical burst-mode transmission. In: Optical fiber communications conference and exhibition (OFC), paper OWX1; 2010.

[32] Qiu XZ, Yin X, Verbrugghe J, Moeneclaey B, Vyncke A, Van Praet C, et al. Fast synchronization 3R burst-mode receivers for passive optical networks [invited tutorial]. J Light Technol 2014;32:644–59.

[33] Li C, Chen J, Li Z, Song Y, Li Y, Zhang Q. Demonstration of symmetrical 50-Gb/s TDMPON in O-band supporting over 33-dB link budget with OLT-side amplification. Opt Exp 2019;2:18343–50.

[34] Cheng N, Liu X, Shen A, Effenberger F. Reducing 50G-PON downstream dispersion penalty with a suitably configured booster SOA. In: Contribution D24, ITU-T SG15/Q2 Meeting; March 2021.

[35] VPIphotonics: Simulation Software and Design Services, <https://www.vpiphotonics.com/>.

[36] Le Cheminant G, Zhang K, Houtsma V, Harstead E, Liu X. TDEC (transmitter dispersion and eye-closure) method for equalizer-enabled 50G-PON. In: Contribution D37, Q2/15 Interim Meeting, Xi'an, China; April 2019.

[37] Nesset D, Cano I, Leyba D, Le Cheminant G. Experimental investigation of modified TDEC approach to PON Tx characterization in G.HSP. In: Contribution D128, Q2/15 interim meeting, Dusseldorf, Germany; October 2019.

[38] Liu X, Shen A, Cheng N. Jitter tolerance mask for 50G-PON. In: Contribution D12, ITU-T SG15/Q2 Meeting; February 2021.

[39] Oksman V, Coomans W, Liu X, Shen A, Strobel R. Proposal for jitter specification. In: Contribution C-2590, ITU-T SG15/Q2; April 2021.

[40] Liu X, Luo Y, Effenberger F. 50G-PON FEC codeword length options. In: Contribution C-1733, ITU-T SG15/Q2, Geneva; February 2020.

[41] Laubach M, Yang S, Han Y, Hirth R, Kramer G. LDPC adjustments from Motion #6, Chicago. In: IEEE P802.3ca meeting, contribution laubach_3ca_1_0518; May 2018.

[42] Yang M, Li L, Liu X, Djordjevic IB. FPGA-based real-time soft-decision LDPC performance verification for 50G-PON. In: Optical fiber communications conference and exhibition (OFC), paper W3H.2; 2019.

[43] Yang M, Li L, Liu X, Djordjevic IB. Real-time verification of soft-decision LDPC coding for burst mode upstream reception in 50G-PON. J Light Technol 2020;38(7):1693–701.

[44] Schaefer NFJ, Schedelbeck G, Strobel R, Hoof WV, Lefevre Y. G.HSP: LDPC code proposal. In: ITU-T Q2 conference call; November 19, 2019.

[45] Han Y, Wilson B, Amitai A. HSP LDPC performance curves. In: Contribution D13, ITU-T SG15/Q2 Meeting; May 2020.

[46] Mahadevan A, et al. 50G PON FEC evaluation with error models for advanced equalization. In: Optical fiber communications conference and exhibition (OFC), paper Th1B.6; 2020.

[47] Schaefer NFJ, Schedelbeck G, Strobel R. Correlated errors in G.HSP. In: Contribution D10, ITU-T SG15/Q2 meeting; May 12, 2020.

[48] Liu X, Shen A, Cheng N, Luo Y, Effenberger F. Performance improvements in bandwidth-limited and digitally-equalized 50G-PON downstream transmission via block-interleaving over four LDPC codewords. In: Optical fiber communication conference, paper M3G.6; 2021.

[49] Van Hoof W, Lefevre Y, Fredricx F. On the need for configurable bit interleaving for downstream 50GG.hsp. In: D16, ITU-T SG15/Q2 meeting; October 2020.

[50] Shen A, Liu X, Effenberge F, Luo Y. Downstream synchronization state machine supporting OLT-configurable bit-interleaving in 50G-PON. In: D20, ITU-T SG15/Q2 meeting; October 2020.

[51] Schaefer FJ, Schedelbeck G, Strobel R, Van Hoof W, Lefevre Y. G.HSP: ONU downstream synchronization state machine. In: Contribution to the ITU-T Q2 conference call on March 24, 2020.

[52] Shen A, Liu X, Luo Y, Effenberger F. 50G-PON downstream synchronization state machine. In: D46, ITU-T SG15/Q2 Meeting; April 2020.

[53] IEEE Standard 1588-2019. IEEE standard for a precision clock synchronization protocol for networked measurement and control systems; 2019.

[54] ETSI TS 123 501 V16.6.0 (2020-10). 5G; System architecture for the 5G system (5GS) (3GPP TS 23.501 version 16.6.0 Release 16).
[55] See for example, <https://www.lightwaveonline.com/fttx/pon-systems/article/14185094/swiss-com-trials-50g-pon>.
[56] Zeng H, Shen A, Liu X, Effenberger F. Supporting a wide range of ODN loss budgets in 50Gb/s PON by link-adaptive modulation and coding. Presentation at the April 2018 FSAN Meeting.
[57] van Veen D, Houtsma V. Strategies for economical next-generation 50G and 100G passive optical networks [Invited]. IEEE/OSA J Opt Commun Netw 2020;12(1):A95–103.
[58] Borkowski R, Straub M, Ou Y, Lefevre Y, Jeliü Ž, Lanneer W, et al. World's first field trial of 100 Gbit/s flexible PON (FLCS-PON). In: European Conference On Optical Communications 2020, PDP2.2; 2020.

CHAPTER 11

The fifth generation fixed network (F5G)

11.1 Overview of the fifth generation fixed network

11.1.1 Technical scope of the fifth generation fixed network

Over the last 40 years, mobile communication network has evolved from 1G, 2G, 3G, and 4G to 5G. The widespread deployments and applications of 5G are underway in the current decade. In the 5G era, fixed network, which includes optical access network and optical transport network (OTN) segments, is playing an increasingly important role in supporting broadband access to 5G base stations, homes, offices, shopping centers, business buildings, factories, smart cities, and much more. According to an OVUM report, by the first half of 2019, 570 million fiber-to-the-home (FTTH) users have been registered worldwide [1]. It is also estimated that 700 million households will have implemented optical access by 2023. Reaching deeper to the final access points such as rooms, office desks, and factory machines, optical fiber will realize its full potential to support a fully connected, intelligent world with high bandwidth, high reliability, low latency, and low energy consumption.

With the fiber-to-everywhere vision, in 2020, the European Telecommunications Standards Institute (ETSI) established an Industry Specification Group (ISG) dedicated to the definition and specification of the fifth generation fixed network (F5G) [2]. Similar to mobile network, fixed network had entered its fifth generation around the year 2020 and is characterized by the use of 10-Gb/second-capable passive optical network (10G-PON) for the optical access network segment and 5G-oriented OTN for the optical metro and core network segments. Further teaming up with advanced Wi-Fi technologies, the fiber-to-everywhere approach aims to provide people with connectivity and machines with flexibility and mobility.

Similar to 5G, F5G has three main application categories:

- Enhanced fixed broadband (eFBB), which supports applications that require large bandwidth.
- Guaranteed reliable experience (GRE), which supports applications that require high reliability.
- Full-fiber connection (FFC), which supports applications that require massive fiber connections.

Fig. 11.1 illustrates the technical characteristics of F5G. Overall, F5G aims to provide over 10 times higher bandwidth in eFBB, 10 times better reliability and latency in GRE, and 10 times denser fiber connections in FFC [3].

Optical Communications in the 5G Era
DOI: https://doi.org/10.1016/B978-0-12-821627-9.00002-4

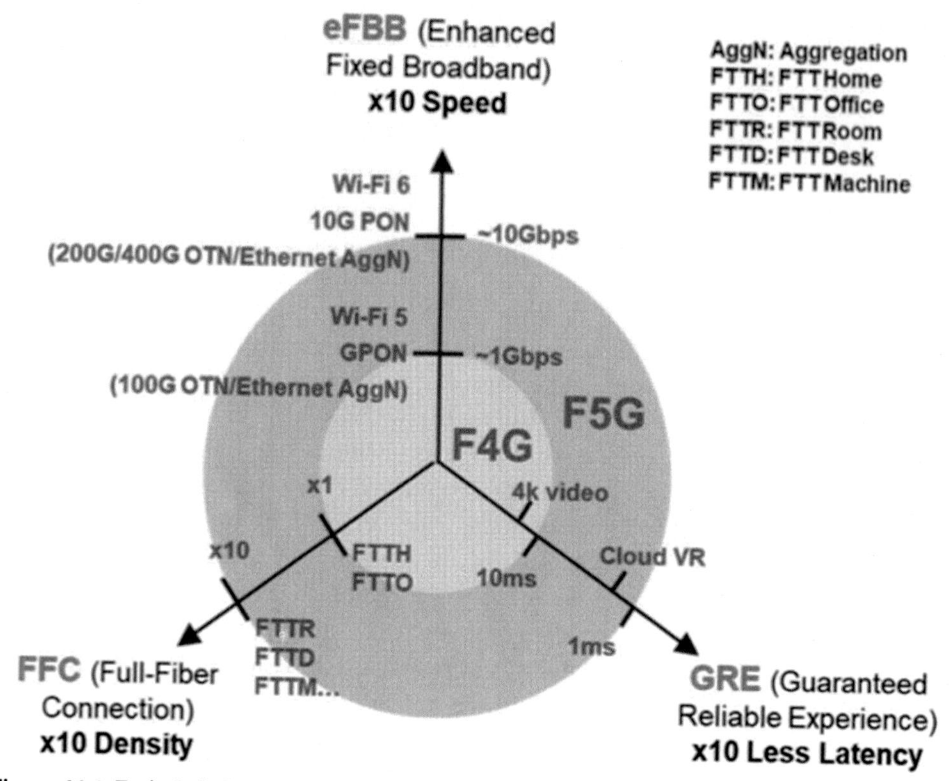

Figure 11.1 Technical characteristics of fifth generation fixed network. *(After ETSI F5G white paper. Available from: https://www.etsi.org/images/files/ETSIWhitePapers/etsi_wp_41_F5G_ed1.pdf, 2020).*

For eFBB, F5G increases the bandwidth over the previous generation fixed network by applying new technologies in multiple network segments:

- In the Customer Premise Network (CPN) segment, Wi-Fi 6 is replacing Wi-Fi 5 to improve performance, stability, and coverage. Combined with fiber-to-the-room (FTTR), both residential and business users can experience seamless Gb/second broadband access.
- In the Access Network (AN) segment, 10G-PON is replacing Gigabit-capable passive optical networks (GPON) to become the dominant FTTH technology, while high-speed OTN is reaching deeper to access sites to establish end-to-end (E2E) optical connections for high-bandwidth applications.
- In the Aggregation Network (AggN) segment and the core network (CN) segment, both OTN and internet protocol (IP) network are providing more bandwidth by using high-speed serial interfaces at 200, 400 Gb/second, and beyond, and dense wavelength-division multiplexing over a wide optical amplification window, as described in Chapter 7, Cloud Data Center Optics, Chapter 8, High-Capacity

Long-Haul Optical Fiber Transmission, and Chapter 9, Superchannel Transmission and Flexible-Grid Wavelength Routing.

For GRE, F5G improves the network reliability and the quality of experience (QoE) via a multifaceted approach:

- The Wi-Fi technology is improved to achieve reduced the air interface latency, enhanced reliability, and interaction experience.
- Optical access and transport networks are improved to guarantee high-quality data transportation, minimum packet loss, low latency, and high-precision clock synchronization.
- Network controller is enhanced, for example, by software-defined network (SDN), network function virtualization (NFV), big data analytics, and artificial intelligence (AI), to achieve automatic management and operation of the E2E network.
- E2E network slicing capability is introduced to guarantee the key performance indicators (KPIs) of different services with the optimized total network cost and/or power consumption.

For FFC, F5G uses the fiber-to-everywhere infrastructure to support ubiquitous connections, including FTTH, FTTR, fiber-to-the-office (FTTO), fiber-to-the-desk (FTTD), and fiber-to-the-machine (FTTM). The footprint of fiber connections is expected to expand by 10 times. At the same time, the fiber connection density may increase by 10 times. Thus the total number of fiber connections may increase by 100 times in the F5G era. The application scenarios of F5G can be dramatically expanded, for example, to include various emerging vertical industry applications, enabling the era of full-fiber connectivity.

11.1.2 Network architecture of fifth generation fixed network

Fifth generation fixed network has a well-defined network architecture, which consists of three planes [4]:

- **Management, control, and analysis (MCA) Plane**, which is responsible for the management, control and operation analysis of the E2E network and contains three logical components.
- **Service plane**, which provides service connections to customers and broadband services.
- **Underlay plane**, which provides the physical connections to transport the network traffic via multiple network segments such as the CPN, AN, AggN, and CN segments.

Fig. 11.2 illustrates the overall network architecture of F5G. In the MCA plane, there are three key elements [4]:

- Digital twin, which provides a real-time representation of the E2E network for autonomous operation and intelligent analysis.
- Autonomous management and control, which enables autonomous network configuration, service deployment, and network operation.

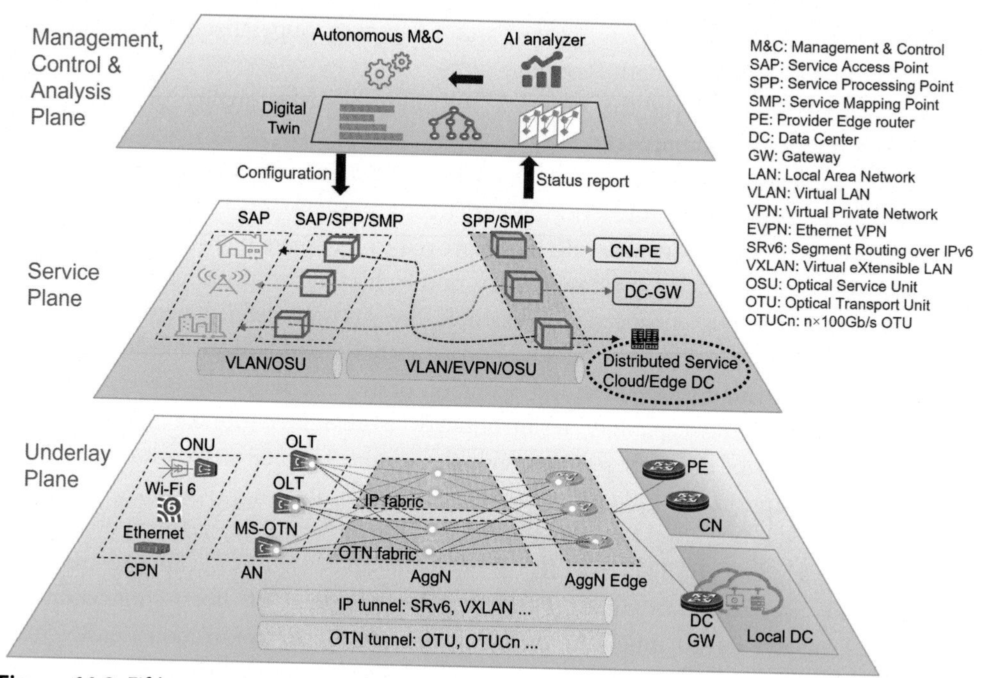

Figure 11.2 Fifth generation fixed network overall network architecture. *(After F5G architecture, https://www.etsi.org/committee/f5g).*

- AI analyzer, which uses AI to perform network analysis and reasoning, enabling advanced functions such as network optimization, QoE evaluation, and fault identification and prediction.

In the service plane, new service connections can be dynamically created and configured by coordinating with the MCA plane and the underlay plane. The service plane contains the following key elements:

- Service access point (SAP), which provides service access to the end user
- Service processing point (SPP), which performs L1/L2/L3 service processing
- Service mapping point (SMP), which maps service traffic to proper underlay channels.

CPN typically contains SAPs only. AN typically contains SAPs, SPPs, and SMPs because AN not only performs the access function but also processes the service traffic and maps it to proper underlay channels. AggN and CN typically contain SPPs and SMPs at the network segment boundaries for service tunnel stitching. Service with different service-level agreements (SLAs) can be supported by communicating the SLAs from the service plane to the underlay plane and querying the necessary resources with the coordination of the MCA plane.

In the underlay plane, there are physical layer equipment and devices to provide the physical connections and dynamic programmable path selection under the control of the MCA plane. The underlay plane consists of four network segments:

- CPN, which is mainly based on Wi-Fi and optical network unit (ONU)
- AN, which is mainly based on PON and multiservice OTN
- AggN, which is mainly based on OTN and IP network
- CN, which is mainly based on OTN and IP network.

There are technology boundaries between network segments, but all the network segments are controlled by the MCA plane to achieve E2E network operation and optimization. The ETSI ISG-F5G defines an E2E fixed network and service management framework for multidomain, multitechnology and multilayer networking with hierarchical service management domains [4].

Fig. 11.3 shows the architecture of F5G E2E management and control system, which consists of an E2E orchestrator and multiple domain controllers on the top of the F5G network topology [5]. In the domain controller layer, four domain controllers are introduced for the four network segments in the F5G network topology [5]:

- CPN controller, which controls the CPN that includes ONUs, Wi-Fi devices, etc. The CPN controller functional block can either reside locally in the control software inside the ONU or reside remotely in the management and control system of the AN. In the remote case, the management and control of the ONU is delegated to the network operator. A northbound interface (NBI) of the CPN controller,

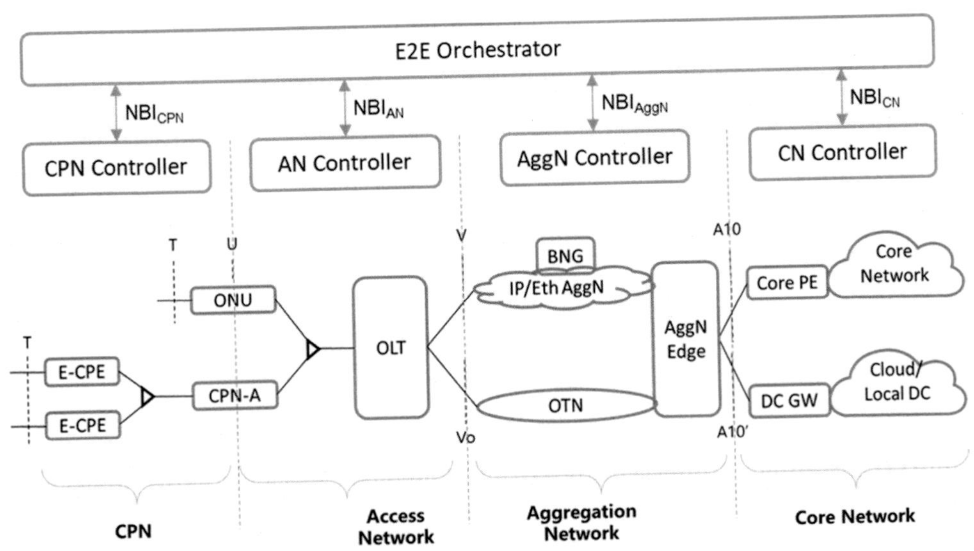

Figure 11.3 Illustration of fifth generation fixed network end-to-end management and control architecture. *(After Lin Y. E2E management architecture. In: Contribution F5G(21)005035, ETSI ISG-F5G Meeting F2F#05; 2021 March).*

NBI_{CPN}, is established to communicate with the orchestrator for E2E service provisioning.

- AN controller, which controls the AN that includes optical line terminal (OLT), optical distribution network (ODN), ONUs, etc. The AN controller function includes management of the underlay PON network and the management and control of the SAP, SPP, and SMP in the service layer of the AN. An NBI of the AN controller, NBI_{AN}, is established to communicate with the orchestrator for E2E service provisioning.
- AggN controller, which controls different types of AggN. In F5G, both OTN and IP/Ethernet network can be used for the AggN segment. The AggN controller may provide the resource and service orchestration function if both OTN and IP/Ethernet network coexist in the same AggN. An NBI of the AggN controller, NBI_{AggN}, is established to communicate with the orchestrator for E2E service provisioning.
- CN controller, which controls the CN that may include OTN, IP network, as well as data center interconnections. An NBI of the CN controller, NBI_{CN}, is established to communicate with the orchestrator for E2E service provisioning.

The E2E orchestrator cooperates with the domain controllers through NBI_{CPN}, NBI_{AN}, NBI_{AggN}, and NBI_{CN}, and performs the resource orchestration and service provisioning functions as follows:

- The E2E resource orchestration function chiefly focuses on the orchestration of the F5G underlay plane. This function includes gathering the (abstracted) topology, resource, and status information, triggering the creation of tunnel in each network segment, resource optimization across multiple network domains, identification and location of network failures, analysis of status changes, and prediction of failures.
- The E2E service provisioning function mainly focuses on the management and control of the F5G service plane. The E2E orchestrator is capable of configuring the SAPs, SPPs, and SMPs for the needed service access, process and mapping functions to enable the creation, activation, modification, and deletion of services in an autonomous manner. Furthermore, the E2E orchestrator is capable of monitoring the performance of the services and taking necessary actions when service degradations occur, according to the SLAs of the corresponding services.

As interoperability is vital for ensuring a diverse supply chain, standardized network interfaces are desired. Within the scope of F5G, network interfaces such as NBI_{CPN}, NBI_{AN}, NBI_{AggN} and NBI_{CN} will be standardized. In addition, reference service models will be provided to enable faster and easier integration and interoperation of all the network segments of F5G.

Network slicing is seen as a foundational 5G capability [6,7]. With the coordination among the E2E orchestrator and the domain controllers in F5G, E2E network slicing for any given service can be supported, enabling differentiated experience assurance. In the CPN segment, the network slicing capability has been implemented in modern Wi-Fi technologies through carrier- and space-based slicing [8]. In the AN segment, PON ports support slicing in accordance with the service type. The network slicing capability in a high-speed PON is being enhanced [9], as mentioned in Chapter 10, 50-Gb/second Passive Optical Network. In the AggN segment, IP network naturally supports *soft* network slicing, while OTN is being enhanced to support *hard* network slicing with guaranteed bandwidth and deterministically low latency [10–15]. Particularly, the next-generation OTN (NG-OTN) uses optical service units (OSUs) with fine bandwidth granularity of ~2 Mb/second to more efficiently support a diverse set of services, including high-quality private line services [16]. We will discuss OTN and its enhanced features in detail in the following section.

11.2 Advances in optical transport network

11.2.1 Evolution of optical transport network

Optical transport network comprises a set of optical network elements connected by optical fiber links, capable of providing functionalities such as transport, multiplexing, switching, management, supervision, and survivability of optical channels that carry client signals [17–21]. OTN wraps each client signal transparently into a container for transport across optical networks, preserving the client's native structure, timing information, and management information.

OTN supports wavelength-division multiplexing (WDM) of optical channels with various line rates. Fig. 11.4 shows the construction of an OTN system [19], which consists of the following procedures:

- Optical channel payload unit (OPU) is formed by combining client signal with OPU-related overhead (OH_{OPU}), where the client signal can be of any existing protocol, for example, synchronous optical networking (SONET) synchronous digital hierarchy (SDH), generic framing procedure (GFP), gigabit Ethernet, and IP.
- Optical channel data unit (ODU) is formed by combining OPU and ODU-related overhead (OH_{ODU}).
- Optical transport unit (OTU) is formed by combining ODU, OTU-related overhead (OH_{OTU}), framing alignment signal, and forward error correction (FEC) overhead for parity bits (OH_{FEC}).
- Optical channel (OCh) is formed by modulating multiple OTUs onto an optical channel carrier.
- Multiple OChs are then wavelength-division multiplexed in optical multiplexing sections and transmitted over optical transport sections.

The Telecommunication Standardization Sector of International Telecommunications Union (ITU-T) has standardized the formation of an OTU frame in ITU-T

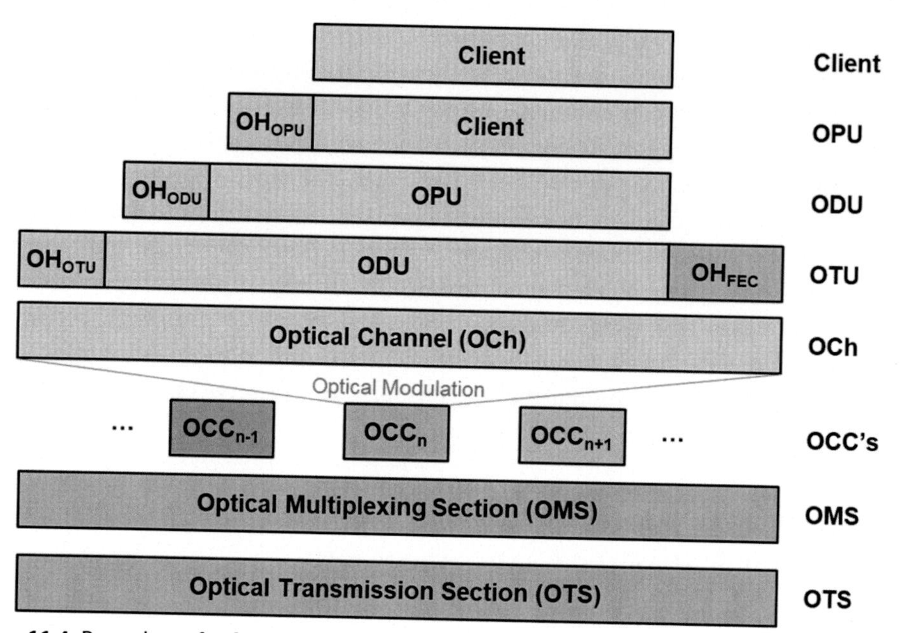

Figure 11.4 Procedures for forming an optical transport unit frame in an optical transport network system.

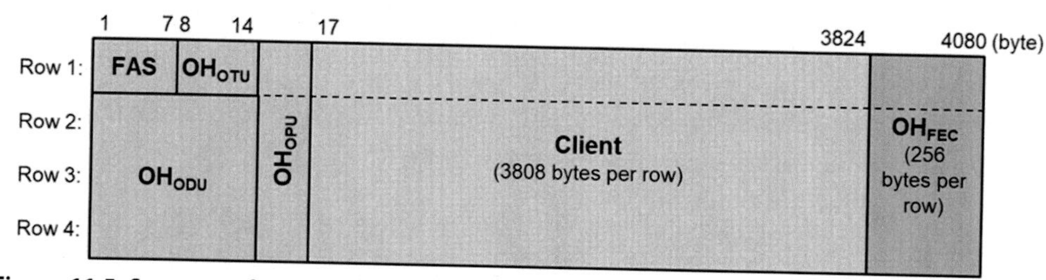

Figure 11.5 Structure of an optical transport unit frame.

Recommendation G.709 [19], as shown in Fig. 11.5. Each OUT frame uses 16-byte-interleaved Reed–Solomon (RS) FEC (255,239) with the following structure:

- Each row: 16 × 255 bytes or 4080 bytes
- Payload/row: 16 × 239 bytes or 3824 bytes
- Client/row: 16 × 238 bytes or 3808 bytes
- OH/row: 16 × 16 bytes or 256 bytes
- Number of rows: 4
- Total bits/frame: 4 × 16 × 255 × 8 = 130,560 bits.

Table 11.1 shows the main ODUk (where k is a subindex to specify the ODU) types ranging from ODU0 (~1.24 Gb/second) to ODU4 (~105 Gb/second), as well as flexible ODU (ODU_{flex}) for constant bit rate (CBR) mapping and for GFP.

Table 11.1 The main ODUk types, bit rates, and applications.

ODU Type	Bit Rate (Gb/s)	Typical transport examples
ODU0	1.24416	– One Gigabit Ethernet (GbE) signal.
ODU1	2.49877512605042	– Two ODU0 signals. – One SONET OC-48 or SDH STM-16 signal.
ODU2	10.0372739240506	– 8 ODU0 signals. – 4 ODU1 signals. – One SONET OC-192 or SDH STM-64 signal.
ODU2e	10.3995253164557	– One 10 Gigabit Ethernet (10GE) signal. – One 10 Gigabit Fiber Channel (10GFC) signal.
ODU3	40.3192189830509	– 32 ODU0 signals. – 16 ODU1 signals. – 4 ODU2 signals. – One SONET OC-768 or SDH STM-256 signal. – One 40 Gigabit Ethernet (40GE) signal.
ODU3e2	41.7859685595012	– 4 ODU2e signals.
ODU4	104.794445814978	– 80 ODU0 signals. – 40 ODU1 signals. – 10 ODU2 signals. – 2 ODU3 signals. – One 100 Gigabit Ethernet (100GE) signal.
ODUflex (CBR)	239/238 × (client bit rate)	– A CBR signal such as a common public radio interface (CPRI) signal.
ODUflex (GFP)	Flexible and configurable	– A stream of Ethernet/IP packets with any configured rate using the GFP.

Table 11.2 The main OTUk types, bit rates and applications.

Signal	Bit rate (Gb/s)	Typical transport examples
OTU1	2.66	– One SONET OC-48 or SDH STM-16 signal.
OTU2	10.70	– One SONET OC-192 or SDH STM-64 signal. – One 10GE signal.
OTU2e	11.09	– One 10GE signal.
OTU2f	11.32	– One 10GFC signal.
OTU3	43.01	– One OC-768 or SDH STM-256 signal. – One 40GE signal.
OTU4	112	– One 100GE signal.
OTUCn	$n \times 239/226 \times 99.5328$	– n × logically interleaved 100-Gb/s data streams – A plurality of 100GE and 400GE signals

Table 11.2 shows the main OTUk types ranging from OTU1 (~2.66 Gb/second) to ODU4 (112 Gb/second). Representative applications are also enumerated in the tables.

For optical channels at per-channel data rate of 100 Gb/second and beyond, advanced FEC codes with a higher overhead than that of RS (255,239) are typically used. To effectively carry client signals with rates greater than 100 Gb/second, ITU-T G.709 introduced the definition of OTUCn, where C stands for 100 and *n* represents the number of logically interleaved 100-Gb/second data streams [19]. Fig. 11.6 shows the formation of an OTUCn frame. Evidently, the OTUCn frame format is highly scalable through the interleaving of n OTUC signals, each of which has the format of a standard OTUk signal without the FEC overhead columns. The combined OTUCn frame has n instances of OH_{OPU}, OH_{ODU}, and OH_{OTU}.

A fundamental difference between the OTUk frame format and the OTUCn frame format is that the OTUCn frame has no dedicated area for FEC overhead columns. This means that the OTUCn and ODUCn frame formats are identical, except for the OTUC-specific overhead fields. This allows different OTUC interface types where each interface can choose its own FEC with the required coding gain, which is desirable for high-speed channels to meet various transmission distance requirements [19]. For example, commonly used FEC codes for 100-Gb/second and beyond transmission include RS(544,514) [19], stair-case FEC [22], concatenated FEC (CFEC) [23,24], CFEC + [25], open FEC [26], and high-coding-gain LDPC [27,28].

Accurate time synchronization is an important feature in 5G-oriented optical networks [29]. Frequency and time synchronization over OTN has been standardized by the ITU-T Study Group 15 (SG15) in multiple recommendations such as ITU-T G.709, G.781, G.805, G.810, G.8260, G.8261, G.8262, G.8263, G.8264, G.8265, G.8271, G.8272, G.8273, and G.8275 [30–45]. More specifically, the ITU-T G.8275 standard is a precise time synchronization telecommunications standard based on IEEE 1588v2 precision time protocol [46]. The ITU-T G.8275 standard defines network equipment models, source selection algorithms, and packets to facilitate interconnection and interworking. The ITU-T G.8273 standard defines the time synchronization performance of each single device in terms of time error, noise generation, noise tolerance, noise transfer, transient response, and holdover.

Another major benefit provided by OTN is enhanced operation, administration, and maintenance (OAM) for wavelength channels. OTN-based core and metro networks offer advantages over traditional WDM transponder-based networks by

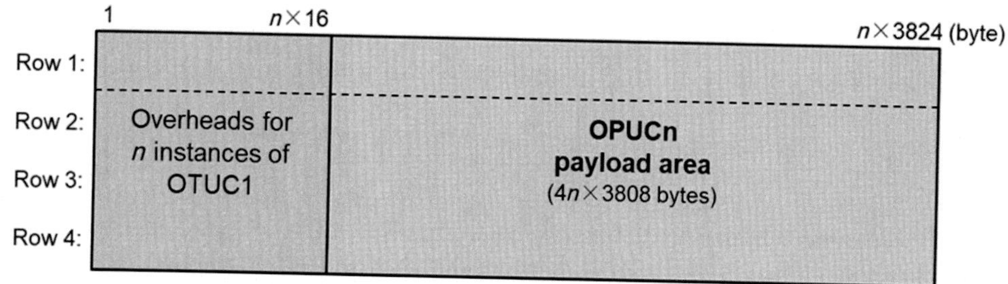

Figure 11.6 Structure of an optical transport unit for n × 100 Gb/s (OTUCn) frame.

providing enhanced OAM features such as embedded communications channel, performance monitoring, fault detection, and multiplexing of lower rate client signals into higher speed payloads. The IP-over-OTN architecture also offers reduced hops, better management and monitoring, and increased protection of services.

In the scope of F5G, OTN plays a crucial role in traffic aggregation and transport [4]. To transport the ever-increasing network traffic, OTN is evolving to support per-wavelength channel data rates of 200, 400 Gb/second, and beyond, achieving an aggregated per-fiber transmission capacity of over 48 Tb/second when the super C band is used, as presented in Chapter 8, High-Capacity Long-Haul Optical Fiber Transmission. To reduce the transport latency and energy consumption, optical cross connects with multidegree reconfigurable optical add/drop multiplexers are used for wavelength routing. To address the need for aggregating a diverse set of client services within each wavelength channel, OSU is under specification by ITU-T to reduce the bandwidth granularity from ~1.24 Gb/second (ODU0) down to ~2 Mb/second [15], which will be discussed in more depth in the following section.

11.2.2 Optical service unit-based optical transport network

To better support a diverse set of client services, OTN has evolved to multiservice optical transport network (Ms-OTN), as illustrated in Fig. 11.7. In Ms-OTN, OTN, SDH, and Ethernet/IP packet services can be carried over the same wavelength, with their corresponding bandwidths allocated flexibly. The containers for these services can be based on ODUk, virtual container (VC), virtual local area network (VLAN), and multiprotocol label switching (MPLS). Ms-OTN supports multiple services with a unified cross-connection and transmission, as well as centralized maintenance via L0/L1/L2 orchestration.

As OTN further expands from core network to metro and access networks, more and more client services with different KPIs, such as private line services of enterprise users, need to be supported. To meet this demand, the OTN container is being transformed from ODU to OSU, which is attached to a tributary port number (TPN) for dynamic identification of the carried service unit [15]. In OSU-based OTN, unified fine-granularity service grooming and switching are supported, as illustrated in

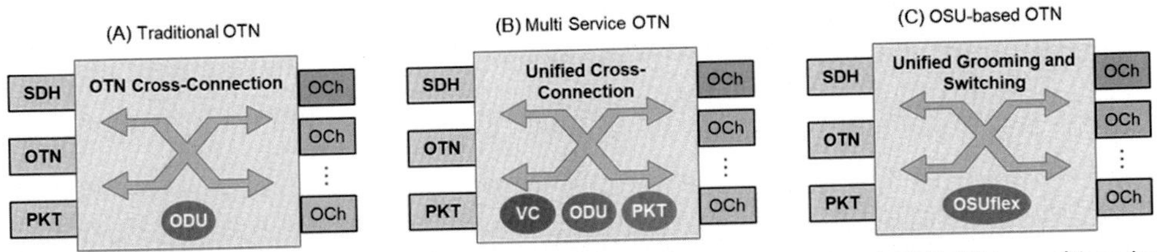

Figure 11.7 Optical transport network (OTN) evolution from the traditional OTN (A) to multiservice OTN (B) and optical service unit-based OTN (C) (*PKT*, packet services).

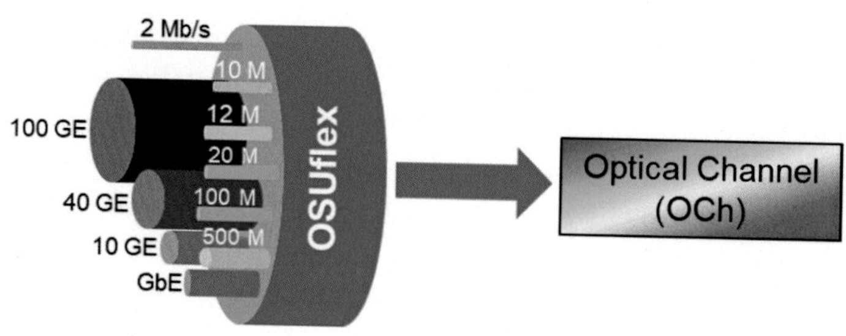

Figure 11.8 Aggregation of various services with fine granularity by using optical service unit flex.

Fig. 11.7. Compared to the traditional OTN, OSU-based OTN offers the following benefits.

- *Finer bandwidth granularity*

  The smallest container of traditional OTN is ODU0 (~1.24 Gb/second), which makes it inefficient to carry a client service whose bit rate is less than 1 Gb/second. In OSU-based OTN, a flexible OSU container (OSUflex) is used to provide flexible bandwidth allocation with a fine bandwidth granularity of ~2 Mb/second, thereby dramatically increasing the OTN bandwidth efficiency for carrying typical client services with bit rates between 2 and 500 Mb/second. Note that in OSU-OTN, hitless bandwidth adjustment from 2 to 100 Gb/second can be supported with no service interruption [47]. Fig. 11.8 illustrates the aggregation of various services with fine granularity by using OSUflex.

- *More client services supported*

  In a traditional OTN, because of the minimum ODU size of ~1.24 Gb/second, a 100-Gb/second wavelength channel based on OTUC1 can only support up to 80 services. In OSU-based OTN, on the other hand, the TPN can specify over 1000 OSU containers in a single wavelength. Thus a 100-Gb/second wavelength channel can support over 1000 client services as long as the aggregated bandwidth of these services is less than 100 Gb/second. Note that the bandwidth resources are allocated to services on-demand and can be reconfigured when new services are added or existing services are completed. Fig. 11.9 illustrates the support of various types of services using OSUflex in cooperation with OTUk and OTUCn.

- *Reduced latency for service grooming and switching*

  In traditional OTN, multiple layers of mapping and encapsulation are needed for a given client service, often leading to a large service encapsulation latency. In OSU-based OTN, OSUflex encapsulation is implemented in a unified manner where each client service is directly mapped and encapsulated into a higher-order ODUk container, thus greatly reducing the service encapsulation latency, regardless of the service bandwidth. In addition, OSU-based OTN supports sequential forwarding to implement the first-in first-out mechanism during centralized cross-connection, thereby

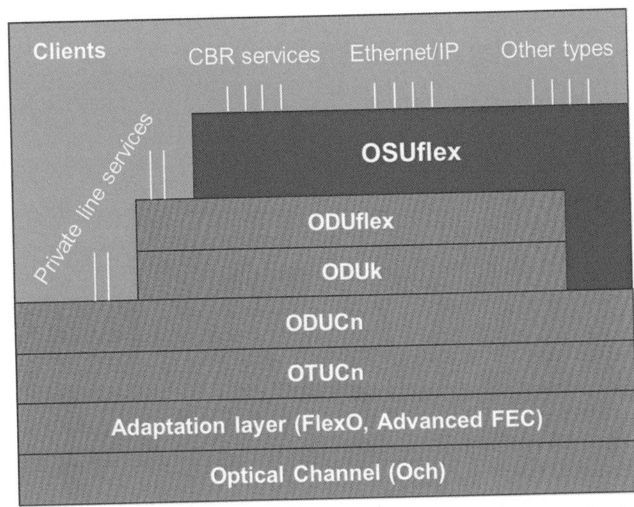

Figure 11.9 Support of various types of services using optical service unit flex (OSUflex) in cooperation with OTUk and OTUCn.

greatly reducing the switching latency. In a proof-of-concept demonstration [15], the E2E latency for a 2-Mb/second service was reduced from 4398 μs using the traditional OTN technology to 1289 μs using the OSU-based OTN technology, representing a remarkable latency reduction of over 70%.

- *Better support of E2E network slicing*

  In an OSU-based OTN, the OSUflex container supports *hard* network slicing with flexibly adjustable bandwidth, which provides guaranteed service bandwidth that is independent of other services in the same OTN. Network slicing at a fine granularity (of ~2 Mb/second) also reduces bandwidth resource fragmentation and improves bandwidth utilization efficiency, especially for E2E networking slicing for typical client services whose bit rates are less than 500 Mb/second. Once the E2E network slicing path is established, the E2E latency experienced by a service is fully deterministic.

The implementation of the OSU-based OTN is shown in Fig. 11.9. The OSUflex container can efficiently carry services whose bit rates range from ~2 Mb/second to over 100 Gb/second, including CBR services, packet services, and other types of services. For low-speed services, the OSUflex containers can be multiplexed into ODUflex/ODUk and then into ODUCn for transmission. For high-speed services, the OSUflex container can be multiplexed directly into ODUCn for transmission. Each OSU frame has an overhead section that carries the TPN corresponding to the service, which is unique at the server layer to ensure that the receiving side can correctly identify the service. The timestamp information and the byte quantity information are also carried in the OSU overhead section. The OSU frame can be based on a 192-byte format [15] that strikes a good balance between low overhead and low processing latency.

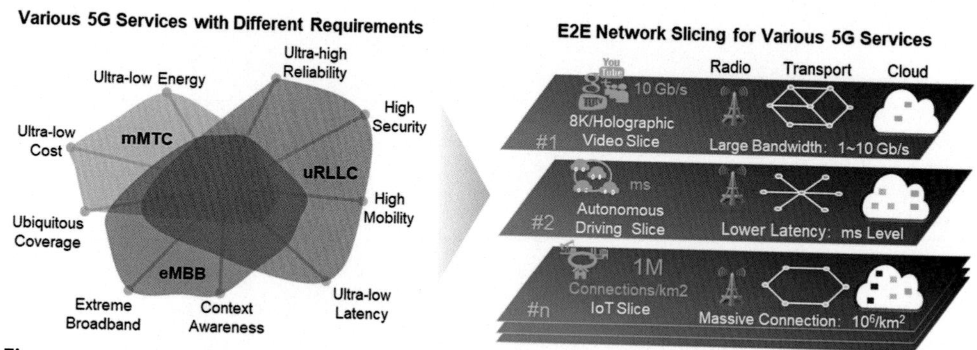

Figure 11.10 Illustration of the use of end-to-end network slicing for service-specific optimization in 5G networks. *eMBB*, enhanced mobile broadband; *mMTC*, massive machine type communications; *uRLLC*, ultra-reliable and low-latency communications; *IoT*, internet of things. *(After Liu X, Deng N. Emerging optical communication technologies for 5G. In: Willner A, editor. Chapter 17 of Optical fiber telecommunications VII; 2019).*

In January 2020, China Mobile conducted the industry's first OSU-based OTN trial in Qingdao, China, focusing on the support of fine service granularity, flexible bandwidth adjustment, and low processing latency. The trial results revealed that the new OSU-OTN supports hitless bandwidth adjustment from 2 to 100 Gb/second, with fine granularity service grooming at a step of 2 Mb/second. In addition, the OSU-OTN technology reduced the E2E latency by over 70% for a 2-Mb/second service and by over 35% for a 100-Mb/second service [15]. The forwarding latency of a single node was substantially reduced to only 10 μs. The OSU-OTN technology is regarded as a promising NG-OTN technology to better carry end-user services and achieve the KPIs of a diverse set of client services with reduced overall capital expenditure (CapEx) and operating expense (OpEx).

A key feature of 5G mobile network is the ability to perform E2E network slicing to achieve service-specific optimization in terms of both quality of service and resource utilization, as shown in Fig. 11.10. With the use of SDN, intelligent optical transport and access networks can be built with network virtualization and slicing capabilities for 5G [29]. Fig. 11.11 illustrates such a 5G-oriented optical network with E2E network slicing capability based on OSU-OTN. Because of the *hard* network slicing capability of OTN, guaranteed bandwidth and deterministically low latency can be achieved to effectively meet the KPI requirements of any given 5G service.

At ITU-T SG15 plenary meeting held in January 2020, ITU-T agreed to initiate a new project named G.osu to clearly specify the OSU service layer including the switching and bandwidth adjustment functions [48]. Concurrently, the China Communications Standards Association initiated the development of OSU-OTN—related standards [11,14]. In the scope of ETSI ISG-F5G, the use of OSU-OTN to better support emerging applications such as high-quality private line services,

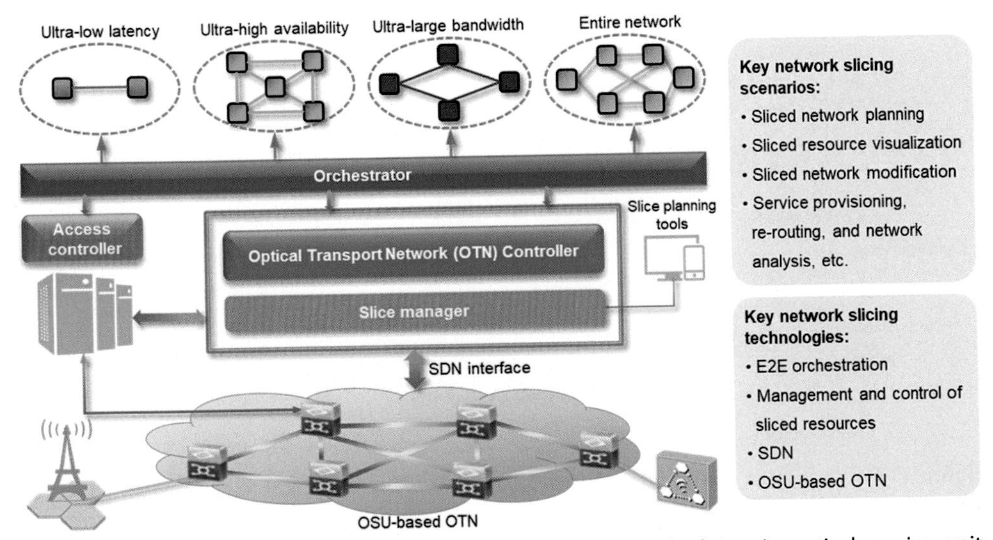

Figure 11.11 Software-defined network-based intelligent network slicing in optical service unit-optical transport network for 5G mobile services with various slicing requirements. *(After Liu X, Deng N. Emerging optical communication technologies for 5G. In: Willner A, editor. Chapter 17 of Optical fiber telecommunications VII; 2019).*

scenario-based broadband services, and E2E network slicing is also under active study [16]. We will discuss the use cases of F5G and highlight their demands on the next-generation fixed network in the following section.

11.3 Fifth generation fixed network use cases

11.3.1 Overview of fifth generation fixed network use cases

With the technological advances in all of its network segments, CPN, AN, AggN including OSU-OTN, and CN, F5G is well positioned to support many emerging use cases in the 5G era. In February 2021, ETSI ISG-F5G published the first release of F5G use cases [16]. There are 14 use cases identified in the first release under the three main use case categories of F5G, namely, eFBB, GRE, and FFC, as illustrated in Fig. 11.12. Table 11.3 enumerates all these use cases in three application types, new/enhanced services to users, expanded fiber infrastructure and services, and management and optimization.

It can be seen from Fig. 11.12 and Table 11.3 that F5G offer a diverse set of use cases to directly deliver value to end users, network operators, and vertical industries. In the following sections, we will briefly describe six exemplary use cases in terms of their backgrounds, objectives, and the technologies that are needed to enable them.

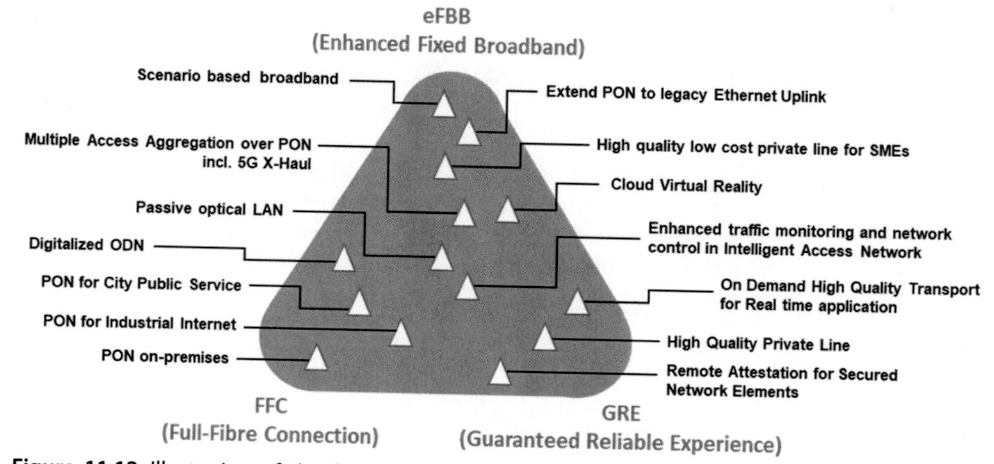

Figure 11.12 Illustration of the first 14 fifth generation fixed network (F5G) use cases within the three dimensions of F5G, namely, enhanced fixed broadband, guaranteed reliable experience, and full-fiber connection. *(After ETSI F5G Use Cases Release #1. Available from: https://www.etsi.org/deliver/etsi_gr/F5G/001_099/002/01.01.01_60/gr_F5G002v010101p.pdf, 2021).*

Table 11.3 List of the fifth generation fixed network use cases according to their application types.

New/Enhanced services to users
Use case #1: Cloud virtual reality (VR) Use case #2: High-quality private line Use case #3: High-quality low-cost private line for small and medium enterprises (SMEs)
Expanded fiber infrastructure and services
Use case #4: PON on-premises Use case #5: Passive optical local area network (LAN) Use case #6: PON for industrial internet Use case #7: PON for city public service Use case #8: Multiple access aggregation over PON Use case #9: Extend PON to legacy Ethernet uplink
Management and optimization
Use case #10: Scenario-based broadband Use case #11: Enhanced traffic monitoring and network control in intelligent access network Use case #12: On-demand high-quality transport for real-time applications Use case #13: Remote attestation for secured network elements Use case #14: Digitalized ODN

11.3.2 Cloud virtual reality

Cloud VR offloads computing and rendering functions in VR services from local VR equipment to a shared cloud infrastructure, as shown in Fig. 11.13. Powerful cloud-computing capabilities improve the user experience and reduce the cost and energy consumption of local VR equipment, thus promoting the widespread adoption of cloud VR.

The main advantages of cloud VR are:

- Cloud computing provides better logical computing and image processing capabilities, thus improving the user experience.
- User equipment is simplified to only have basic functions, thus reducing the cost of ownership.
- Content management and provisioning are implemented in the cloud, facilitating the protection of VR content copyrights.
- Leveraging the widespread cloud infrastructure and the reduced cost of cloud VR equipment, the ecosystem of cloud VR can be quickly developed.

The main technologies required are:

- Cloud computing resources for hosting the VR services
- An E2E optical network for transporting the cloud VR services with guaranteed performance and efficient resource utilization.

To carry a cloud VR service with guaranteed performance and efficient resource utilization, the cloud VR service can be isolated from other services via E2E hard slicing over PON and OTN. Accurate matching between the cloud VR service bandwidth and its corresponding slice bandwidth is needed to maximize the transport efficiency. The four levels of VR experiences, fair, comfortable, ideal, and ultimate, require channel data rates of over 40, 65, 270, and 770 Mb/second, respectively [49]. Consequently, the slice bandwidth needs to be flexible and of fine granularity. As described in Section 11.2.2, OSU-based OTN supports network slicing with flexible bandwidth and fine bandwidth granularity of 2 Mb/second and is thus well suited for carrying cloud VR services.

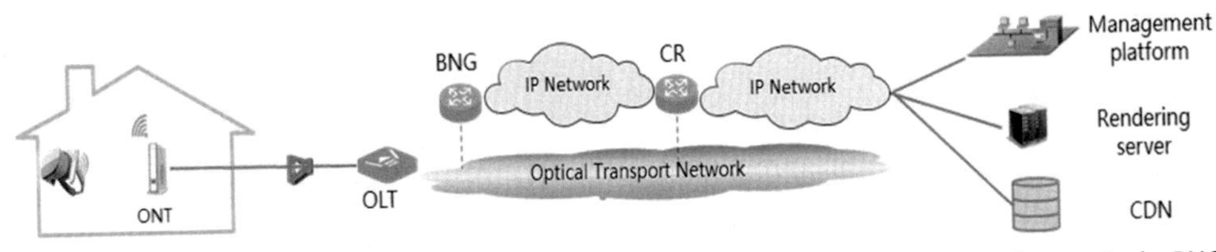

Figure 11.13 Cloud virtual reality network architecture (*ONT*, optical network terminal; *BNG*, broadband network gateway; *CR*, core router; *CDN*, content delivery network). *(After ETSI F5G Use Cases Release #1. Available from: https://www.etsi.org/deliver/etsi_gr/F5G/001_099/002/01.01.01_60/gr_F5G002v010101p.pdf, 2021).*

11.3.3 High-quality private line for small and medium enterprises

Private line services that provide internet access, cloud access, and connectivity with remote offices are essential for a large number of SMEs. High-quality private line services ensures that all the service requirements are met for various enterprise communication needs.

The main advantages of high-quality private line for SMEs are:

- Assurance in bandwidth, latency, coverage, reliability, and security for a better user experience.
- Ability to support a large number of end users and terminals.
- Better utilization of cloud services to help enterprises reduce the CapEx.
- E2E service provisioning and maintenance to help enterprises reduce the OpEx.

E2E network slicing is important for high-quality private line services to meet the KPIs of various enterprise communication needs with different priorities. Fig. 11.14 shows a network configuration where communication traffics are routed through different AggN segments based on their requirements on bandwidth and latency. The priority of a given enterprise traffic is first identified and labeled in the CPN segment. In the AggN segment, high-priority traffic is carried by OTN (e.g., OSU-OTN) with guaranteed bandwidth and low latency, while other traffic is carried by IP network. Thus all the enterprise communication needs can be supported with satisfactory performance and optimal network resource utilization (e.g., in terms of cost efficiency and energy efficiency).

11.3.4 Passive optical network on-premises

10-Gb/second PON (XG-PON) and 10-Gb/second symmetric (XGS-PON) had been standardized by ITU-T in 2016 and 2017, respectively [50,51]. Wide deployment of XG(S)-PON has started in 2020. With XG(S)-PON in the AN segment,

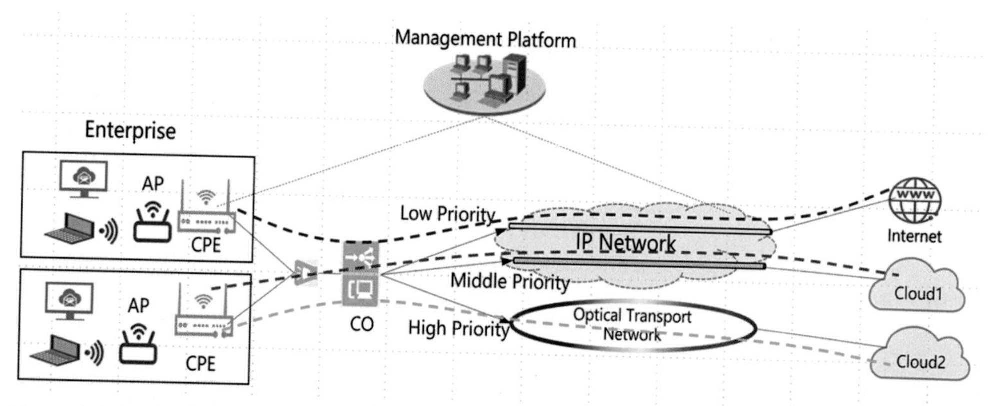

Figure 11.14 Small and medium enterprise private line service identification and distribution of application traffic (*AP*, access point; *CO*, central office). *(After ETSI F5G Use Cases Release #1. Available from: https://www.etsi.org/deliver/etsi_gr/F5G/001_099/002/01.01.01_60/gr_F5G002v010101p.pdf, 2021).*

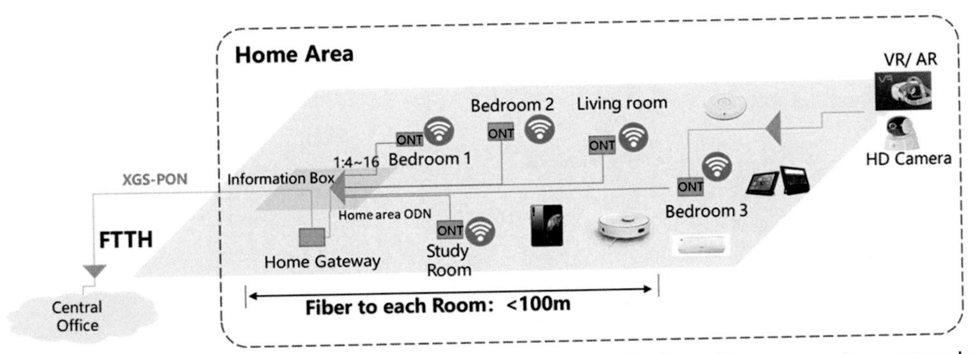

Figure 11.15 Illustration of the use of fiber-to-the-room for high-quality on-premises network. *(After ETSI F5G Use Cases Release #1. Available from: https://www.etsi.org/deliver/etsi_gr/F5G/001_099/002/01.01.01_60/gr_F5G002v010101p.pdf, 2021).*

each user can obtain a peak data rate of over 1 Gb/second. Usually, end user devices are directly connected to CPN, which is primarily based on Wi-Fi and copper connections whose bandwidths are often limited to much below 1 Gb/second. To fully utilize the large bandwidth provided by XG(S)-PON, the on-premises network needs to be upgraded, and FTTR is a promising upgrade option for the on-premises network, as shown in Fig. 11.15.

The main advantages of the FTTR approach are:

- High bandwidth, enabled by high-speed optical transceivers
- Low energy consumption, enabled by the low loss of optical fiber
- Easy installation because of the light weight and small diameter of optical fiber cable
- Immunity to electro-magnetic interference (EMI).

To enable a viable deployment of FTTR, the cost of the optical transceivers used in FTTR needs to be minimized and the PON system needs to be extended to connect with more optical network terminals. Furthermore, the fiber cables that connect the rooms are desired to be sufficiently flexible and invisible. Moreover, tight integration between optical transceiver and Wi-Fi module is desirable. More technical innovations are expected in the forthcoming years to make FTTR a viable solution.

11.3.5 Passive optical local area network

Local area network is widely used in business areas and campuses to provide communication services such as voice, Internet access, Wi-Fi connectivity, and videoconferencing. Traditional LAN systems use electrical switches for routing and Ethernet cables for connection. These electrical switches consume power and need to be hosted in equipment rooms. With the increase of LAN traffic, more power consumption and equipment room space are needed. It is becoming challenging for this approach to satisfy the ever-

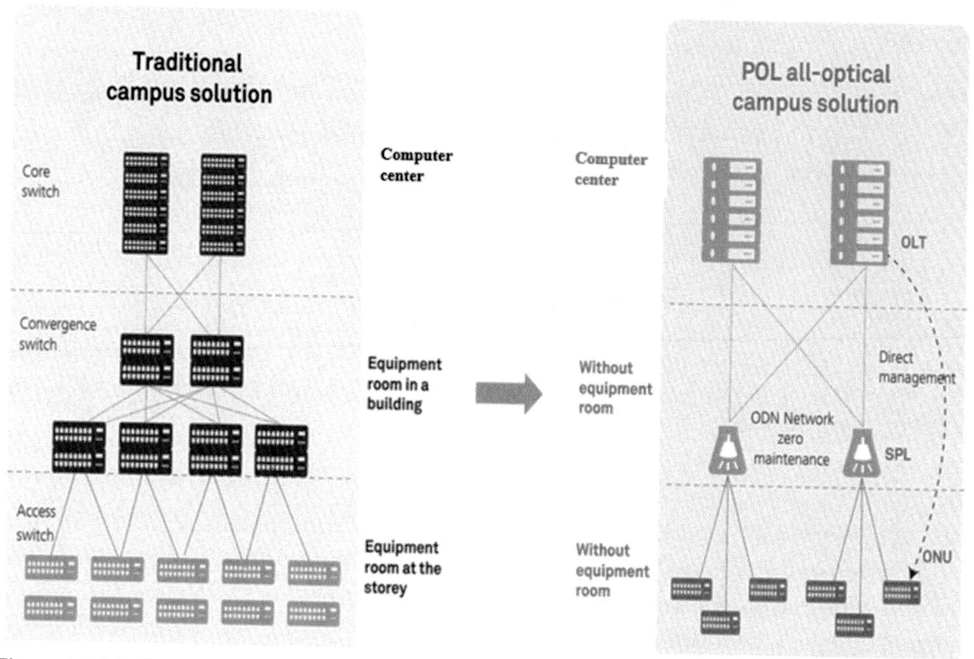

Figure 11.16 Comparison between a traditional local area network (LAN) and a passive optical LAN for a campus network (*SPL*, passive optical splitter). *(After ETSI F5G Use Cases Release #1. Available from: https://www.etsi.org/deliver/etsi_gr/F5G/001_099/002/01.01.01_60/gr_F5G002v010101p.pdf, 2021).*

increasing capacity demands of LAN. To address with this challenge, passive optical LAN (POL) based on PON technology can be used, as shown in Fig. 11.16 [16].

In POL, the OLT and ONU function as the convergence and access switches in the traditional LAN. Similar to electrical switches, the OLT and ONU forward data based on Ethernet or IP addresses. The OLT is connected upstream to the core switch. Ethernet interfaces are available on both user and network sides in the POL system. Remarkably, the equipment rooms used to host the convergence and access switches in the traditional LAN are no longer needed, as the ODN of PON is both passive and compact.

The main advantages of POL over the traditional LAN are:

- No equipment rooms required for switching in the access and convergence layers
- Reduced power consumption, enabled by the passive ODN
- High bandwidth, enabled by high-speed fiber optical connections
- Large coverage area, enabled by the low loss of optical fiber.

POL has already been deployed in certain scenarios. With the large coverage area of PON and the increased availability of local data centers, POL is expected to be more extensively deployed. With the continued increase in LAN traffic volume, higher speed PON systems, such as 50G-PON [52–54], can be used.

11.3.6 Industrial passive optical network

We are currently in the so-called Industry 4.0 era, where the automation of traditional manufacturing and industrial practices by using modern information and communications technology is ongoing [55,56]. The enabling technologies for Industry 4.0 include large-scale machine-to-machine communication, industrial Internet of Things (IIoT), industrial robots, industrial cloud, computer vision, and augmented reality and VR, all of which necessitate efficient, intelligent, and reliable communications.

Advanced PON technology can benefit industrial manufacturing by effectively fulfilling the following tasks [16]:

- Support of high-bandwidth and low-latency real-time communications in the industry process field data network.
- Connection with numerous machines and devices via a diverse set of interfaces.
- Construction of reliable intra-factory networks that are immune to EMI.

Fig. 11.17 illustrates the use of PON for simultaneously supporting the communication needs of an industry process field data network, an office network, and a

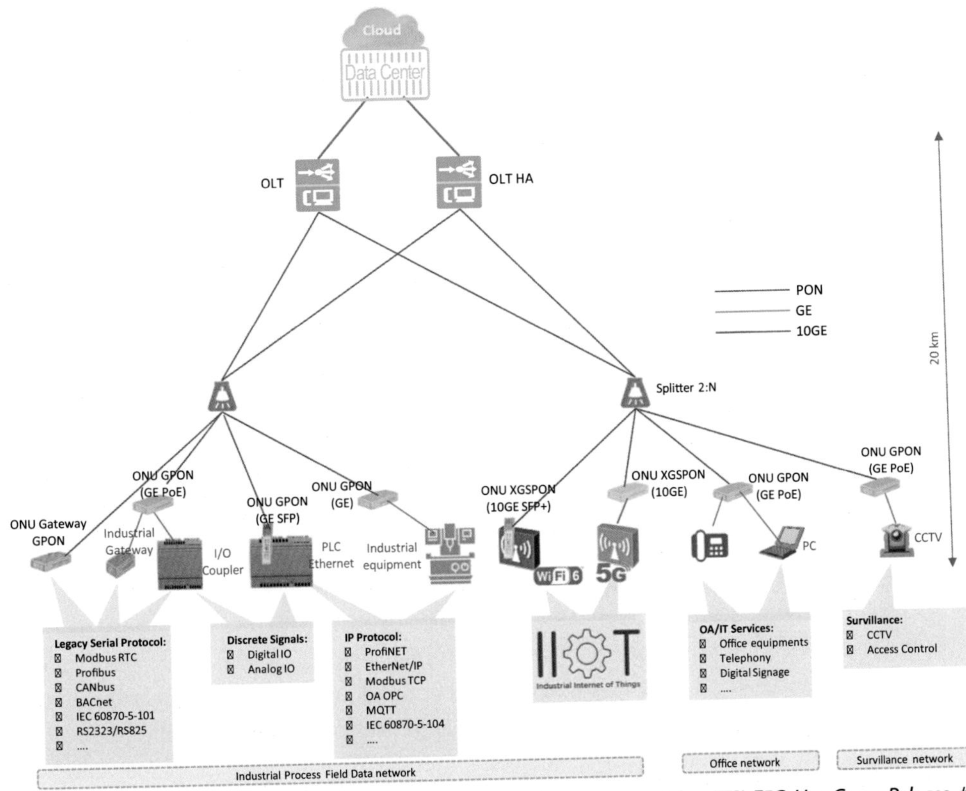

Figure 11.17 Passive optical network-based industrial networks. *(After ETSI F5G Use Cases Release #1. Available from: https://www.etsi.org/deliver/etsi_gr/F5G/001_099/002/01.01.01_60/gr_F5G002v010101p.pdf, 2021).*

surveillance network in an industrial setting. A dual-OLT configuration is used to ensure high availability for mission-critical communications. The main advantages of using PON for industrial manufacturing are:

- Long reach (e.g., 20 km) by using low-loss ODN
- Low cost by leveraging the well-established FTTH ecosystem
- Low latency by eliminating intermediate electrical switches
- Low power consumption by using passive ODN
- High availability by using dual-OLT for failure protection
- High bandwidth by dint of high-speed fiber optics
- High immunity to EMI by dint of optical fiber in the ODN
- High scalability by establishing a uniform intra-plant factory network capable of carrying any traffic generated in the factory
- Universal connectivity to simultaneously support the communication needs of industry processes, offices, and surveillances.

To fully unleash the potential of industrial PON, the optical modules and components used in the industrial setting need to meet the corresponding environmental requirements. The collaboration between PON and Wi-Fi needs to be further explored to realize low-cost and low-power-consumption IIoT. E2E network slicing can be used to provide guaranteed KPIs for different communication data streams with the optimal network resource utilization. With further advances in this field and the expansion of the cloud infrastructure, industrial PON is expected to find valuable applications in the era of Industry 4.0 and F5G.

11.3.7 Scenario-based broadband

On-line education, on-line gaming, telecommuting, and teleconferencing are common services of residential broadband, while dedicated high-quality private line services are often used for business broadband. These broadband services require the underlay network to satisfy different KPIs. Fig. 11.18 illustrates multiple scenario-based broadband services carried by an E2E managed optical access and aggregation network. Different services are differentially treated such that the KPIs of all the services are met. An E2E network management plane is used to manage all the network operations based on a given policy. For example, the policy may be to optimize network resource utilization or minimize the overall power consumption.

In essence, the requirements of the application scenarios of the end users need to be considered and differentiated in order to optimally satisfy all the users who share a same underlay network. This scenario-based broadband approach effectively configures an E2E network to optimally support multiple application scenarios. As the application scenarios of the end users change, the E2E network management plane needs to reconfigure the underlay network to reoptimize for the upcoming application

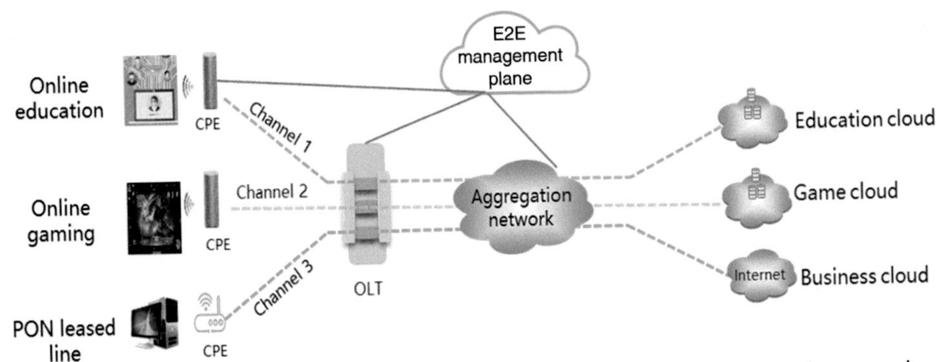

Figure 11.18 Multiple scenario-based broadband services carried by an end-to-end managed optical access and aggregation network. *(After ETSI F5G Use Cases Release #1. Available from: https://www.etsi.org/deliver/etsi_gr/F5G/001_099/002/01.01.01_60/gr_F5G002v010101p.pdf, 2021).*

scenarios. Thus the scenario-based broadband use case calls for E2E network slicing and network automation, which can leverage big data analytics and AI. Indeed, the concept of autonomous network or self-driving network is becoming increasingly important to help network operators manage more and more sophisticated networks, save OpEx, and provide the end users with constantly optimized user experience.

11.4 Future perspectives on fifth generation fixed network

With the collective effort by its over 60 member companies and participants, ETSI ISG-F5G has published the first release of F5G documents consisting of:

- The first Edition of F5G white paper entitled "The Fifth Generation Fixed Network (F5G): Bringing Fiber to Everywhere and Everything," published in September 2020 [3].
- F5G Generation Definition Release #1, published in December 2020 [57].
- F5G Use Cases Release #1, published in February 2021 [16].

The second release of F5G documents is expected to include [2]:

- F5G Architecture
- F5G Technical Landscape
- F5G Use Cases Release #2
- Industry PON
- Quality of Experience
- F5G E2E management and control.

The second release of F5G use cases is expected to include rural broadband, media transport, virtual music, extended reality based virtual presence, cloudification of medical imaging, private line connectivity to multiple clouds, edge/cloud-based visual inspection for automatic quality assessment in production, edge/cloud-based control of automated guided vehicles, and intelligent optical cable monitoring and management etc. More use cases are also forthcoming to enable F5G to bring values directly to end users and businesses.

In a recent Omdia report entitled "Global Fiber Development Index Analysis 2020" [1], it is highlighted that fiber-based broadband access technology provides an optimized and futureproof quality service, which is essential for the development of future digital services and applications across all verticals including (but not limited to) entertainment, education, home working, corporate services, smart city, and health. It is also observed that high-speed broadband drives not only consumer satisfaction but also national GDP, with a growth in GDP of 0.25%–1.5% for every 10% increase in household penetration and a further 0.3% increase for every doubling of speed [1].

The unfortunate situation of COVID-19 further reinforced the importance of a reliable and high-performance fixed network. The increased communication traffic due to more people working and studying from home places unprecedented demands on the communication network in terms of bandwidth, latency, QoE, reliability, etc. It can be expected that 5G and F5G communication technologies will complement each other and jointly support a fully connected, intelligent world for the benefit of our global society for years to come.

References

[1] Omdia report. Global fiber development index analysis. https://omdia.tech.informa.com/OM014270/Global-Fiber-Development-Index-Analysis-2020, 2020.
[2] ETSI ISG-F5G, https://www.etsi.org/committee/f5g.
[3] ETSI F5G white paper. Available from: https://www.etsi.org/images/files/ETSIWhitePapers/etsi_wp_41_FSG_ed1.pdf, 2020.
[4] F5G architecture, https://www.etsi.org/committee/f5g.
[5] Lin Y. E2E management architecture. In: Contribution F5G(21)005035, ETSI ISG-F5G Meeting F2F#05; 2021 March.
[6] 3GPP TR 28.801. Study on management and orchestration of network slicing for next generation network. http://www.3gpp.org/DynaReport/28801.htm
[7] See for example, GSMA white paper. An introduction to network slicing. https://www.gsma.com/futurenetworks/resources/an-introduction-to-network-slicing-2/, 2017.
[8] IEEE Standard 802.11ax-2021. https://standards.ieee.org/standard/802_11ax-2021.html
[9] ITU-T Recommendation G.9804.1amd. Higher speed passive optical networks – requirements; 2021.
[10] Vilalta R, López-de-Lerma AM, Muñoz R, Martínez R, Casellas R. Optical networks virtualization and slicing in the 5G era. In: Optical fiber communication conference (OFC), paper M2A.4, San Diego, USA; 2018.
[11] China Telecommunication Industry Standard YD/T 1990–2019: General technical requirements for optical transport network (OTN); 2019.
[12] Liu X, Deng N. Emerging optical communication technologies for 5G. In: Willner A, editor. Chapter 17 of Optical fiber telecommunications VII; 2019.
[13] Jing R, Man X, Mok W, Zhao W, Schubert A, Wu Q, et al. Consideration on a new work item for optical service unit (OSU). In: Contribution 1540, ITU-T Q11/SW15 Meeting; 2020 January.
[14] China Telecommunication Industry Standard YD/T YD/T 2484–2020: Technical requirement on packet enhanced optical transport network (OTN) equipment; 2020.
[15] Bai L. Optical service unit (OSU)-based next generation optical transport network (NG OTN) technology and verification. In: MATEC web of conferences 336, paper 04014; 2021.

[16] ETSI F5G Use Cases Release #1. Available from: https://www.etsi.org/deliver/etsi_gr/F5G/001_099/002/01.01.01_60/gr_F5G002v010101p.pdf, 2021.
[17] Bonaventura G, Jones G, Trowbridge S. Optical transport network evolution: hot standardization topics in ITU-T including standards coordination aspects. IEEE Commun Mag 2008;46(10):124–31.
[18] Carroll M, Roese J, Ohara T. The operator's view of OTN evolution. IEEE Commun Mag 2010;48(9):46–52.
[19] ITU-T G.709 Recommendation. Optical transport network interfaces; 2016.
[20] Gorshe SS. OTN interface standards for rates beyond 100 Gb/s. J Light Technol 2018;36:19–26.
[21] Trowbridge S. Hot topics in optical transport networks. In: ITU-T study group 15 presentation at optical fiber communication conference (OFC). Available from: https://www.itu.int/en/ITU-T/studygroups/2017-2020/15/Documents/OFC2019-3-otn.pdf, 2019.
[22] Smith BP, Farhood A, Hunt A, Kschischang FR, Lodge J. Staircase codes: FEC for 100 Gb/s OTN. J Light Technol 2012;30(1):110–17.
[23] Optical Internetworking Forum (OIF). Implementation agreement 400ZR [Online]. https://www.oiforum.com/wp-content/uploads/OIF-400ZR-01.0_reduced2.pdf, 2020.
[24] Nagarajan R, Lyubomirsky I. Low-complexity DSP for inter-data center optical fiber communications. In: 2020 European conference on optical communications (ECOC), tutorial paper SC04; 2020.
[25] Hirbawi Y, Zhu Q, Caia J-M. Continuation & results of FEC proposals evaluation for ITU G.709.3 200–400G 450km black link. In: Contribution to the ITU G.709.3, CD11-M10; 2019 May.
[26] Roese J, et al. Proposal to specify OFEC for FlexO-LR 450 km application. In: Contribution to the ITU study group 15, Q11, SG15-C1345R1; 2019 July.
[27] Loussouarn Y, Pincemin E. Probabilistic-shaping DP-16QAM CFP-DCO transceiver for 200G upgrade of legacy metro/regional WDM infrastructure. In: Optical fiber communication conference (OFC), paper M2D.2; 2020.
[28] Sun H, et al. 800G DSP ASIC design using probabilistic shaping and digital sub-carrier multiplexing. J Light Technol 2020;38(17):4744–56.
[29] Liu X. Evolution of fiber-optic transmission and networking toward the 5G era. iScience 2019;22:489–506.
[30] Recommendation ITU-T G.805. Generic functional architecture of transport networks; 2000.
[31] ITU-T Recommendation G.810. Definitions and terminology for synchronization networks; 1996.
[32] ITU-T Recommendation G.8260. Definitions and terminology for synchronization in packet networks; 2020.
[33] ITU-T Recommendation G.8261/Y.1361. Timing and synchronization aspects in packet networks; 2019.
[34] ITU-T Recommendation G.8261.1/Y.1361.1. Packet delay variation network limits applicable to packet-based methods (Frequency synchronization); 2012.
[35] ITU-T Recommendation G.8262/Y.1362. Timing characteristics of a synchronous equipment slave clock; 2018.
[36] ITU-T Recommendation G.8263/Y.1363. Timing characteristics of packet-based equipment clocks; 2017.
[37] ITU-T Recommendation G.8264/Y.1364. Distribution of timing information through packet networks; 2017.
[38] ITU-T Recommendation G.8265/Y.1365. Architecture and requirements for packet-based frequency delivery; 2010.
[39] ITU-T Recommendation G.8265.1/Y.1365.1. Precision time protocol telecom profile for frequency synchronization; 2014.
[40] ITU-T Recommendation G.8271/Y.1366. Time and phase synchronization aspects of telecommunication networks; 2020.
[41] ITU-T Recommendation G.8271.1/Y.1366.1. Network limits for time synchronization in packet networks with full timing support from the network; 2020.
[42] ITU-T Recommendation G.8272/Y.1367. Timing characteristics of primary reference time clocks; 2018.

[43] ITU-T Recommendation G.8273.2/Y.1368.2; Timing characteristics of telecom boundary clocks and telecom time slave clocks; 2017.

[44] ITU-T Recommendation G.8275.1/Y.1369.1. Precision time protocol telecom profile for phase/time synchronization with full timing support from the network; 2020.

[45] ITU-T Recommendation G.8275.2/Y.1369.2. Precision time protocol telecom profile for phase/time synchronization with partial timing support from the network; 2020.

[46] IEEE Standard 1588. Standard for a precision clock synchronization protocol for networked measurement and control systems; 2019.

[47] See for example, https://networkmatter.com/2020/03/11/huaweis-liquid-otn-promises-more-flexible-and-granular-optical-transport/.

[48] See for example, https://ee-paper.com/the-g-osu-standard-project-led-by-china-telecom-has-been-successfully-approved/.

[49] F5G Technical Landscape, https://www.etsi.org/committee/f5g.

[50] ITU-T Recommendation G.987. 10-Gigabit-capable passive optical networks (XG-PON); 2016.

[51] ITU-T Recommendation G.9807.1. 10-Gigabit-capable symmetric passive optical network (XGSPON); 2017.

[52] ITU-T Recommendation G.9804.1. Higher speed passive optical networks – requirements; 2019.

[53] ITU-T Recommendation G.9804.2. Higher speed passive optical networks: common transmission convergence layer specification; 2021.

[54] ITU-T Recommendation G.9804.3. Higher speed passive optical networks: physical media dependent layer specification; 2021.

[55] See for example, https://en.wikipedia.org/wiki/Fourth_Industrial_Revolution.

[56] See for example, https://www.techradar.com/news/what-is-industry-40-everything-you-need-to-know.

[57] ETSI F5G Generation Definition Release #1. Available from: https://www.etsi.org/deliver/etsi_gr/F5G/001_099/001/01.01.01_60/gr_F5G001v010101p.pdf, 2020.

CHAPTER 12

Future trends and concluding remarks

12.1 Continued evolution of 5G

Years 2020 and 2021 have witnessed a fast deployment of 5G networks. As of April 2021, there have been 140 5G networks deployed globally, serving over 330 million 5G users [1]. During the early stage of 5G deployment, the promised values of 5G have been delivered to our global society. Particularly, in fighting the COVID-19 pandemic, profound impact has been made by 5G networks and the supporting optical communication networks, as discussed in Chapter 1, Introduction. In a 2021 survey on 5G forecast conducted by the IEEE Future Networks Initiative [2], the following three trends were accentuated:

- Realization of 5G value through global pandemic response
- Expedited 5G rollout to support applications driven by pandemic needs
- Pervasive connectivity fabrics to support transformations driven by COVID-19

Indeed, 5G is continuing to evolve to provide more value to our global society via 5G services to consumers (5G2C) and 5G services to businesses (5G2B), as illustrated in Fig. 12.1. In the area of 5G2C, remote medical diagnosis, online education, remote working, and online shopping are becoming more common. In the area of 5G2B, intelligent and remote operations in factories, farms, mining sites, and shipping ports are minimizing human labor, improving workplace safety, and increasing productivity. To better address these emerging applications, 5G is evolving toward 5.5G through enhancements in three additional application categories or scenarios [3–5].

As described in Chapter 1, Introduction, the original 5G application scenarios are:

- Enhanced mobile broadband (eMBB), which supports applications such as ultra-high definition (UHD) video, 3D video, work and play in the cloud, augmented reality (AR), and virtual reality (VR).
- Ultra-reliable and low-latency communications (uRLLC), which supports applications such as self-driving cars and drones, industry automation, remote medical procedures, and other mission-critical applications.
- Massive machine type communications (mMTC), which supports applications such as smart home, smart building, smart city, and massive internet of things (IoT).

In the vision of 5.5G, three additional application scenarios are added:

- Uplink-centric broadband communication (UCBC), which supports large bandwidth and wide-coverage uplink communication for applications such as video uploading in machine vision and massive broadband IoT, which are essential to industrial applications and for improved mobile user experience by providing a higher uplink speed and wider coverage.

Optical Communications in the 5G Era
DOI: https://doi.org/10.1016/B978-0-12-821627-9.00001-2

Figure 12.1 Illustration of the multifaceted 5G values being delivered to our society, promoting the evolution toward 5.5G with enhanced capabilities.

- Real-time broadband communication (RTBC), which supports reliable, low-latency, large-bandwidth communication to deliver an immersive, true-to-life experience. Potential applications include UHD AR, VR, mixed reality, and extended reality.
- Harmonized communication and sensing (HCS), which supports communication and sensing functions simultaneously for applications such as precise positioning and vehicle-to-everything (V2X) communications, which are important to autonomous driving of cars and drones.

In effect, the three new scenarios introduced for 5.5G, UCBC, RTBC, and HCS, can be regarded as the enhancements of the three original 5G scenarios, with UCBC on the foundation of eMBB and mMTC, RTBC on the foundation of uRLLC and eMBB, and HCS on the foundation of mMTC and uRLLC. Thus the 5.5G application space is broadened from the original 5G triangle to a hexagon, as illustrated in Fig. 12.2.

To support the technology evolution from 5G to 5.5G, more sub-100 GHz spectrum is needed for full-band uplink and downlink decoupling, as well as full-band carrier aggregation on demand [3]. In addition, intelligent network control and management, enabled by artificial intelligence and big data analytics, is needed to address the complexities associated with more application scenarios, services, devices, and frequency bands. Unified standards and industry collaboration will continue to be essential for the success of the global mobile communications industry in delivering the great promises of 5G and 5.5G.

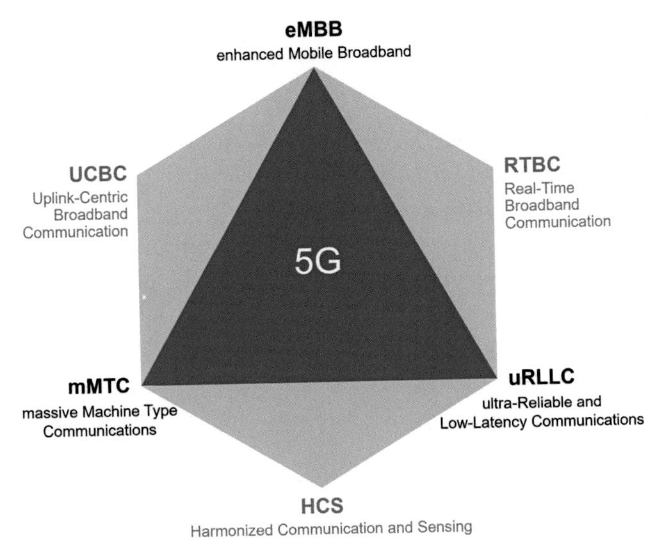

Figure 12.2 Illustration of the six application scenarios of 5.5G.

Over the last 40 years, mobile communication network has evolved from 1G, 2G, 3G, and 4G to 5G, which spans over the decade of the 2020s. The sixth generation mobile network (6G) is expected to be launched around 2030 [5,6]. There have been many studies aiming to envision what 6G will be [7–13]. Among the envisioned benefits of 6G are:

- Continued improvements in downlink and uplink speeds, spectrum efficiency, latency, traffic density, connection density, mobility, and positioning, to better support existing and new application scenarios
- Better global coverage by integrating satellite communication networks with terrestrial wireless networks to help reduce the digital divide
- Better network energy efficiency to help achieve carbon neutrality sooner
- More intelligence throughout the network to better serve our society

The natural way to better prepare us for 6G is to fully realize the values and promises of 5G and 5.5G, and along the way we will undoubtedly discover new valuable use cases and come up with innovative solutions to address them.

12.2 Continued evolution of the fifth generation fixed network

The beginning of the decade of the 2020s has also witnessed a fast deployment of fifth generation fixed network (F5G) technologies such as:

- For the access network segment, it is estimated that 700 million households will have registered for fiber-to-the-home (FTTH) services by 2023 [14], and 10G-PON is becoming the primary technology for FTTH.

- For the metro network and core network segments, coherent transmission with 200- and 400-Gb/s channels is becoming the mainstream solution.
- For data center optical connections, energy-efficient 400-Gb/s pluggable optical transceivers are being extensively adopted for various reach requirements.
- For high-capacity long-haul transmission, broadened optical amplification bands such as the super C band have been utilized.

As described in Chapter 11, F5G supports three application scenarios [15–18]:

- Enhanced fixed broadband (eFBB), which supports applications that require large bandwidth.
- Guaranteed reliable experience (GRE), which supports applications that require high reliability.
- Full fiber connection (FFC), which supports applications that require massive fiber connections.

Similar to the evolution from 5G to 5.5G occurring in the first half of this decade, F5G is also evolving with enhanced technical capabilities and extended application scenarios. The enhanced technical capabilities are witnessed in the emerging technologies described in the previous chapters, such as high-throughput mobile front-haul via wavelength-division multiplexing (WDM), coherent point-to-multipoint for efficient network traffic aggregation, 100-Tb/-class electrical cross-connect (EXC) in data centers, 1-Pb/s-class optical cross-connect (OXC) in backbone networks, capacity-approaching long-haul transmission via advanced coding and probabilistic constellation shaping (PCS), low-latency 50-Gb/s passive optical network (50G-PON), service-enabling optical transport network (OTN) capable of *hard* network slicing with fine granularity, and end-to-end (E2E) network slicing for meeting the quality of service of any given service. The extended application scenarios can be described as [19]:

- Energy-efficient broadband communication (EEBC), which supports broadband communication for remote access points, local area networks, metro-access networks, and data center networks via fiber-based energy-efficient solutions
- RTBC, which supports reliable, low-latency, large-bandwidth communication for industrial applications
- Harmonized communication and sensing (HCS), which supports communication and sensing functions simultaneously for applications such as precise indoor positioning, distributed fiber sensing, monitoring of communication and industrial infrastructures, and environmental monitoring for earthquake and tsunami, etc.

In effect, the three new scenarios, EEBC, RTBC, and HCS, can be regarded as the enhancements of the three original F5G scenarios, with EEBC on the foundation of eFBB and FFC, RTBC on the foundation of GRE and eFBB, and HCS on the foundation of FFC and GRE. Thus the F5G application space is broadened from the original triangle to a hexagon, as illustrated in Fig. 12.3 [19].

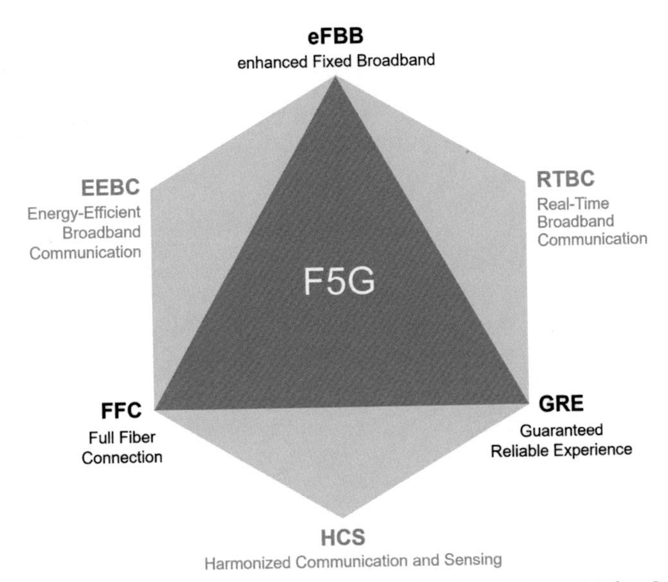

Figure 12.3 Illustration of the six application scenarios of F5G and beyond (after [19]).

For EEBC, F5G can take advantage of the low loss of optical fiber, e.g., to form passive optical local area networks for enterprises and campuses [17]. For RTBC, latency-constrained optical network can be used to support high-bandwidth and low-latency communications in the cloud-based industrial applications such as AI-based video analysis and real-time control of actuators/robots [19]. For HCS, the optical fiber communication infrastructure itself can be utilized to provide positioning and environmental monitoring. For precise indoor positioning, advanced Wi-Fi technologies can be leveraged to enable the increasingly ubiquitous optical fiber infrastructure to provide valuable positioning services. For environmental monitoring, the optical fiber communication infrastructure can be put into good use. As an example, the polarization change in optical communication traffic was recently used to detect earthquakes and water swells via a 10,000-km-long fiber-optic submarine cable, potentially enabling the global submarine fiber-optic cables to effectively serve as continuous real-time earthquake and tsunami observatories [20]. In addition, distributed acoustic sensing (DAS) can be used to provide high-accuracy detection and localization of vibrations in areas with deployed telecommunications optical fiber cables [21,22]. Moreover, when bi-directional transmission fibers are available, vibration detection and localization can be realized by analyzing the optical phase variations of the bi-directionally transmitted signals themselves [23–25]. With the vision of fiber-to-everywhere being gradually realized, the more and more ubiquitous optical fiber network infrastructure will provide both communication and sensing in a harmonized manner [26].

As in the case of the global mobile network industry, unified standards and industry collaboration will continue to be essential for the success of the global fixed network industry in delivering the great promises of F5G. In addition, the close collaboration between the mobile network community and the fixed network community will continue to be mutually beneficial in the 5G era.

12.3 Concluding remarks

The journey leading to the era of 5G and F5G has witnessed ground-breaking innovations made possible by global collaborations in the information and communications technology (ICT) community. For the journey ahead of us, we are confronted with two grand technical challenges, the communication capacity limit imposed by the Shannon theorem [27] and the slowing down of the Moore's law [28].

To address the impact of the Shannon capacity limit, our telecommunication community is exploring innovative system designs, network designs, application-specific designs, photonic integrated circuits, and better integration of photonic and electrical circuits to continue reducing the cost and energy consumption per bit to meet the-ever-increasing communication demands in a sustainable manner.

To address the impact of the noticeable slowing down of the Moore's law, our ICT is exploring innovative algorithms, software, system designs, specialized circuits, three-dimensional (3D) fabrication processes, and potentially new material platforms, in a holistic fashion, to meet the ever-increasing computing demands. However, as the Moore's second law suggested, the overall cost for producers to fulfill the Moore's law, including the costs for the R&D, manufacturing, and testing, increases exponentially over time [29]. Thus global cooperation and collaboration to drive innovations together will become increasingly necessary and beneficial.

In addition, broadening the application space of the current telecommunication technologies to emerging fields, such as 3D sensing for consumer devices, optical displays, light detection and ranging for autonomous driving, distributed fiber-optic sensing, visible light communication, intersatellite communication, optical computing, and quantum communication and computing, will further share the innovations for multiple fields and maximize their values to our society.

As many countries in our world are aiming to achieve carbon neutrality before 2060 [30], the overall energy consumption of the ICT industry needs to be carefully controlled and managed. As in the case of 5G and F5G, unified global standards that define the best technologies and practices and share them globally will continue to be essential.

With global cooperation and collaboration, we can look forward to an exciting era of 5G and F5G, where new advances in our ICT community will support a fully connected, intelligent world for the common benefit of our global society for years to come.

References

[1] See for example, https://www.huaweicentral.com/huawei-global-analyst-conference-2021-key-note-speech-of-eric-xu-huaweis-rotating-chairman/.
[2] IEEE Future Networks Initiative Article. 7 Experts cast their visions for 5G and beyond in 2021. Available from: https://futurenetworks.ieee.org/tech-focus/2021-5g-forecast, 2021 January 19.
[3] See for example, https://www.huawei.com/en/news/2020/11/mbbf-shanghai-huawei-david-wang-5dot5g.
[4] See for example, https://www.lightreading.com/asia/huawei-sets-out-its-stall-for-55g/d/d-id/765415.
[5] See for example, https://www.huaweicentral.com/huawei-ready-to-bring-5-5g-network-will-provide-gbps-downlink-rates-at-a-low-latency-of-5-ms-compared-to-5g/.
[6] See for example, https://news.cgtn.com/news/2021-04-14/50-times-faster-than-5G-Huawei-s-6G-launch-set-for-2030-ZrTkNN5fQA/index.html.
[7] David K, Berndt H. 6G Vision and requirements: is there any need for beyond 5G? IEEE Veh Technol Mag 2018;13(3):72−80.
[8] Yang P, Xiao Y, Xiao M, Li S. 6G Wireless communications: vision and potential techniques. IEEE Netw 2019;33(4):70−5.
[9] Zhang Z, et al. 6G Wireless networks: vision, requirements, architecture, and key technologies. IEEE Veh Technol Mag 2019;14(3):28−41.
[10] Saad W, Bennis M, Chen M. A vision of 6G wireless systems: applications, trends, technologies, and open research problems. IEEE Netw 2020;34(3):134−42.
[11] You X, Wang CX, Huang J, et al. Towards 6G wireless communication networks: vision, enabling technologies, and new paradigm shifts. Sci China Inf Sci 2021;64:110301.
[12] Bhat JR, Alqahtani SA. 6G Ecosystem: current status and future perspective. IEEE Access 2021;9:43134−67.
[13] Tong W, Zhu P. 6G: the Next Horizon: from Connected People and Things to Connected Intelligence. 1st ed. Cambridge University Press; 2021.
[14] Omdia report. Global fiber development index analysis. https://omdia.tech.informa.com/OM014270/Global-Fiber-Development-Index-Analysis-2020, 2020.
[15] ETSI ISG-F5G, https://www.etsi.org/committee/f5g.
[16] ETSI F5G white paper. Available from: https://www.etsi.org/images/files/ETSIWhitePapers/etsi_wp_41_FSG_ed1.pdf, 2020.
[17] F5G architecture, https://www.etsi.org/committee/f5g.
[18] ETSI F5G Use Cases Release #1. Available from: https://www.etsi.org/deliver/etsi_gr/F5G/001_099/002/01.01.01_60/gr_F5G002v010101p.pdf, 2020.
[19] Liu X. Enabling optical network technologies for 5G and beyond. J Lightwave Technol 2021; In press.
[20] Zhan Z, Cantono M, Kamalov V, Mecozzi A, Müller R, Yin S, et al. Optical polarization−based seismic and water wave sensing on transoceanic cables. Science 2021;371(6532):931−6.
[21] Wellbrock G. Fiber sensing in existing telecom fiber networks, Optical Fiber Communication Conference (OFC), Tutorial paper Tu6F.1; 2021.
[22] Ip E, et al., Distributed fiber sensor network using telecom cables as sensing media: technology advancements and applications, Optical Fiber Communication Conference (OFC), Invited paper Tu6F.2; 2021.
[23] Marra G, et al. Ultrastable laser interferometry for earthquake detection with terrestrial and submarine cables. Science 2018;361:486−90.
[24] Yan Y, Khan FN, Zhou B, Lau APT, Lu C, Guo C. Forward transmission based ultra-long distributed vibration sensing with wide frequency response. J Lightwave Technol 2021;39(7):2241−9.
[25] Ip E, et al., Field trial of vibration detection and localization using coherent telecom transponders over 380-km link, Optical Fiber Communication Conference (OFC), post-deadline paper F3B.2; 2021.
[26] Liu X. Advances in optical communication systems and networks in the 5G era. Optoelectronics and Communications Conference (OECC), Plenary Talk; 2021.

[27] Shannon CE. A mathematical theory of communication. Bell Syst Technical J 1948;27(3):379–423.
[28] Moore GE. Cramming more components onto integrated circuits. Electron Mag 1965;38(8).
[29] See for example, https://en.wikipedia.org/wiki/Moore%27s_law.
[30] See for example, https://www.cnn.com/2021/04/22/politics/white-house-climate-summit/index.html.

Index

Note: Page numbers followed by "*f*" and "*t*" refer to figures and tables, respectively.

O

S

T

U